THE ECOLOGICAL HISTORY OF EUROPEAN FORESTS

THE ECOLOGICAL HISTORY OF EUROPEAN FORESTS

Edited by

Keith J. Kirby

English Nature, Peterborough, UK
and

Charles Watkins

Department of Geography, University of Nottingham, UK

CAB INTERNATIONAL

CAB INTERNATIONAL
Wallingford
Oxon OX10 8DE
UK

Tel: +44 (0)1491 832111
Fax: +44 (0)1491 833508
E-mail: cabi@cabi.org

CAB INTERNATIONAL
198 Madison Avenue
New York, NY 10016–4314
USA

Tel: +1 212 726 6490
Fax: +1 212 686 7993
E-mail: cabi-nao@cabi.org

A catalogue record for this book is available from the British Library, London, UK

Library of Congress Cataloging-in-Publication Data
The Ecological History of European Forests / edited by K. Kirby and C.
    Watkins
        p. cm.
    Selected papers given at a international conference held at the
University of Nottingham in Sept. 1996.
    Includes bibliographical references (p.   ) and index.
    ISBN 0-85199-256-0 (alk. paper)
    1. Forests and forestry—Europe—History—Congresses. 2. Forest
ecology—Europe—History—Congresses.   I. Kirby, K. J.
    II. Watkins, C.
    SD177.E26   1998
    578.73′094—dc21
                                                                97–51763
                                                                CIP

ISBN 0 85199 256 0

Typeset in Garamond by Columns Design Ltd, Reading
Printed and bound in the UK at the University Press, Cambridge.

Contents

Preface

Most of the chapters in the book are based on presentations given at the International Conference on Advances in Forest and Woodland History which was held at the University of Nottingham in September 1996. We would like to thank the Forest Ecology Group of the British Ecological Society and the International Union of Forest Research Organisations (IUFRO) for supporting and publicizing this meeting. We would also like to thank the Department of Geography, University of Nottingham for providing secretarial and organisational assistance, especially Pat Bridges and Ben Cowell; the staff of Hugh Stewart Hall at the University of Nottingham where the conference was based; Nottingham City Council, Graham Whalley and the staff of Wollaton Hall for providing the venue for a reception; George Peterken for presenting the post-conference tour to the Wye Valley; English Nature for assisting with the preparation of the manuscript; Phil Kinsman for assistance with the bibliography, and Tim Hardwick of CAB INTERNATIONAL and his staff for their assistance with the publication of this book. Keith Kirby would like to thank David Foster and a Bullard Fellowship for the time at Harvard Forest where he did initial editing. A second book drawing on papers from the conference, *European Woods and Forests: Studies in Cultural History* edited by Charles Watkins (1998), is also published by CAB INTERNATIONAL.

Keith Kirby
English Nature, Peterborough

Charles Watkins
University of Nottingham

March 1998

Introduction – Historical Ecology and European Woodland

C. Watkins[1] and K.J. Kirby[2]

[1]*Department of Geography, University of Nottingham, University Park, Nottingham NG7 2RD, UK;* [2]*English Nature, Northminster House, Peterborough PE1 1UA, UK*

Introduction

Historical ecology has only developed as a distinct sub-discipline over the last 30 years or so. The key to the approach is the combination of very different forms of evidence to develop an understanding of the history and development of a particular piece of vegetation or habitat. This evidence may include surveys of present-day flora and fauna; historical maps and documents; oral history; land management practices; literary descriptions; pollen and soil analysis; photographs, drawings and paintings; field observation of the form of trees, wood banks, ditches and hedges, and so forth. In the UK, historical ecology has its roots in local history, natural history, historical geography and field ecology. It stems from a strong amateur tradition of interest in local landscape history which has been fostered by a wide range of local history societies and county wildlife trusts. This amateur interest has been greatly strengthened by the work of professional ecologists, historians and other writers who have excited the interest of many people in the history of plants and animals and the landscape in which they live. It is only very recently, however, that general introductory texts for historical ecology, such as Emily Russell's (1997) *People and land through time. Linking ecology and history*, have been published.

The rise of historical ecology has been strongly linked to a burgeoning interest in the history of woodland. The publication in 1976 of Oliver Rackham's *Trees and woodland in the British landscape* was to many a formative event in their understanding of the history of vegetation and the importance of linking a thorough knowledge of historical documents with a practical understanding of plant ecology. Several people had been researching, writing about and discussing historical ecology in the 1960s and 1970s. These include the historical geographer John Sheail (1980) and the ecologists Colin Tubbs (1968) and George Peterken (1981). The research and publications of

© CAB INTERNATIONAL 1998. *The Ecological History of European Forests* (eds K.J Kirby and C. Watkins)

this group, most of whom had a strong interest in woodland ecology, led to the development of an historical imagination amongst ecologists. This connection between history and ecology becomes especially potent when applied to the history of woodland, as, unlike say grassland or moorland, key components of the vegetation, the trees, themselves provide readily recognizable historical evidence. The scale of approach varies from individual trees through woods and groups of woods to estates, landscape regions and nations.

From the beginning it was recognized that historical ecology had great potential in informing how woodland could be managed in the interests of nature conservation. The validity of historical ecology as an approach was given an enormous boost in Britain by the establishment of an inventory of ancient woods in the 1980s by the Nature Conservancy Council, under the direction of George Peterken. This national inventory of all British ancient woods 2 ha or more in area was based on a comprehensive, county by county analysis of early nineteenth century topographic maps used in conjunction with existing field survey data (Spencer and Kirby, 1992). The inventory became very important in influencing how individual woods should be managed and, perhaps more importantly, was a vehicle which helped to transform the norms of woodland management. The British Forestry Commission and many private landowners began to view and value their ancient woods in a much more positive light.

In Europe and the United States there has also been a steady increase in the application of the historical ecological approach to woodland history and management. A key factor in the spread of the approach has been its inherently interdisciplinary nature. The approach has been used by forest historians, ecologists, geographers, historians and research teams which combine skills from these different disciplines. Several important international conferences have brought together specialists in woodland history and furthered the development of woodland historical ecology. These include two held as part of the European Science Foundation's *Forest Ecosystem Research Network*: one, at Trento, on human influence on forest ecosystems development in Europe (Salbitano, 1988) and one, at Freiburg, on the ecological effects of afforestation (Watkins, 1993). Others have concentrated on particular themes in woodland history, such as that organized by Jean-Paul Metaille (1992) on proto-industries and woodland. The increase in the number of meetings and publications associated with woodland historical ecology indicates the continued growth and vitality of the sub-discipline.

Methods for Studying the Historical Ecology of Woodland

Woodland ecology, perhaps more than any other branch of the subject, can suffer if we do not appreciate the long-term and large-scale dynamics of woodland vegetation and associated animal communities, and their interaction with people. However, by combining approaches that span a range of temporal and spatial scales, we can begin to understand why and how the woods we see today came to be as they are. Pollen analysis provides the longest time frame, but until relatively recently gave only a broad-scale picture of regional forest changes over centuries. Recent fine-scale resolution techniques applied to small hollow sediments (Chapters 15, 18, this volume) have allowed the shorter term trends within individual stands to be distinguished. Individual stands within the same wood may have strikingly different histories. Dendrochronology methods may provide independent evidence of past growth patterns that can be used to draw conclusions about climatic conditions (Chapter 12, this volume). Archaeological evidence points to changes in the woodland cover as areas were cleared and then abandoned (Chapter 8, this volume): perhaps fewer woods than we thought are truly primary.

In the historical era woodland becomes a regular feature of written records. There is an enormous wealth of such material, ranging from detailed descriptions of the species present in a particular wood and local forms of management to national censuses and woodland management policies. The range of maps available is also enormous: estate plans delimiting areas to be coppiced, maps of farms showing individual trees, and forest working plans. The large-scale topographic maps of the eighteenth and early nineteenth centuries have proved particularly valuable in making regional inventories of old woodland (Chapters 1, 5, 6, 9, 10, 13, 14, 15, 21, 22, 24, this volume).

This documentary evidence is supplemented by photographs, oral records and direct observation (Chapters 3, 4, 11, 16, 19, 20, 21, this volume). The present-day woods retain the imprint of past management in their banks, charcoal hearths (Chapter 15) and the shapes and ages of the trees (Chapter 4). Because woods have perhaps been more reliably recorded than most other features in the countryside, broad-scale patterns at a landscape scale can be produced, as well as individual site studies (Chapters 9, 21, 22, 25, 26, this volume). From this wealth of historical studies over the past 20 years, a number of common threads are emerging that we have tried to encapsulate in this book.

Broad Trends in European Woodland History

For many years landscape historians have tended to study the clearance of woodland, rather than woodland itself (Darby, 1956). To some extent woodland clearance has been used as a measure of increasing population, as an indication of increased agricultural production, even as a surrogate for civilization. Woodland has been seen as the 'other' land use, unknown and potentially dangerous, from which humans have created a productive and settled landscape. As such it has been treated as undifferentiated and relatively unimportant. Recent work is emphasizing the value of trees and woodland to humans and the extent of woodland management over very long time periods. Great variations in woodland type and use over time, and across Europe, are being identified and mapped, and rather than clarifying, this work is tending to bring into question any accepted clear distinction between the natural and the cultural (Chapters 1, 20, this volume).

Woodland, moreover, is increasingly being seen not as a separate category from agricultural land, rather the two forms of land use have in the past tended to merge into one another. There has, however, been a pronounced trend from the eighteenth century onwards across Europe for a move away from intimate mixtures of trees, pasture and cultivated land to a landscape where blocks of woodland and agricultural land become distinct. Complicated modes of management where individual trees and woods were used for a wide variety of purposes have tended to become simplified. Research is showing the importance of different forms of land ownership and property rights in fuelling and governing the pace of this complex land use change (Chapters 2, 3, 4, 17; Watkins, 1998a).

Many local and national studies are confirming that, in the broadest sense, the area of woodland reached its lowest level in the late seventeenth and early eighteenth centuries, although there are many local variations. From the eighteenth century onwards many areas have shown an increase in woodland. This can be linked at the large scale to changes in agricultural practices, improved agricultural yields, industrialization and the demand for different types of wood products, including the move away from firewood. Different traditions of scientific forestry developed in the eighteenth century; plantation forestry has been particularly influential and important as a means of afforesting large areas of poor grazing land of different types. The location and type of afforestation has been influenced both by land ownership and soil conditions (Chapters 17, 21, this volume). In addition to afforestation, semi-natural reversion from agricultural land to woodland takes place at different scales and for many different reasons. There are many examples of land abandonment and woodland succession (Chapters 11, 16, this volume). Often a key factor in whether forest was increasing or decreasing was the pressure from livestock.

The Interaction of Grazing Animals and Woodland

What did the primeval woodland in Europe look like? Was it dense forest or more like open parkland? In fact the question should be rephrased, since it implies a single universal state, as, how much was dense forest and how much open parkland? For centuries, if not millennia, the influence of large herbivores – in the distant past bison, the extinct auroch and the various species of deer, more recently domesticated animals – has had a major impact on the structure and composition of the woodland (Chapter 1). From Africa we can see how what we might consider as wildwood was turned gradually into more or less tame parkland (Chapter 2). An inherent tension between high levels of herbivores and tree cover can sometimes be resolved and systems develop that are stable over centuries as in the Forest at Hatfield, wooded meadows in Scandinavia, or the small farms near Vicenza (Chapters 1, 3, 4, this volume). These systems are often the refuge for species lost from woods treated as coppice or high forest (Chapter 7). However, shifts in economics may also lead to the loss of the grazing as wood-pasture is turned into coppice (Chapter 19) or woods become open heath (Chapter 5). The species composition of grazed and ungrazed coppices may differ long after the grazing has stopped (Chapter 6); particular species become restricted to inaccessible gorges (Chapter 15). As foresters in many parts of Europe struggle with the impact of increasing deer populations there are lessons to be learnt from the past: woods can survive, but not necessarily in the form they have now, nor will we be able to return them to the forms that they had in the past. Moreover, what works in one area may not work in another.

Local, Regional and National Variation in Woodland

The diversity of woodland management at local and regional scales and over time must be recognized. Wood-pastures might be in the form of wood meadows, wood coppice or parkland, while within any one of these categories there is further variation (Chapters 1, 3). For both Italy and the UK, examples are presented of how the woods and trees on one estate or farm were treated in different ways (Chapters 4, 13) and once apparently widespread practices, such as the alder cultivation in northern Italy, may in time be forgotten (Chapter 16). Yet these variations leave marks on the woodland and its flora and fauna.

Thus the survival of a large tract of forest rich in endemic species in southern Spain is linked first to its position on the Arab–Christian border in the early Middle Ages and then to the rise of the cork industry (Chapter 10). Italian chestnut orchards became coppice as fuel replaced food as their main product; later the chestnut coppice was left to grow out as high forest or was replaced by pine as the fuel market declined (Chapter 11). In Switzerland similar management trends that lead to darker, more shaded high forest conditions are associated with reductions in species diversity (Chapter 17). An impoverished ground flora in Hungarian steppe forest is linked to periods of past cultivation of forest glades (Chapter 20). Conversely, ancient woodland species survive outside woods where disturbance is not too high and the microclimate is moist enough (Chapter 23). Since it is now the value of woodland for its wildlife which is important, we need to stress the integration of the historical with ecological factors in how we describe and classify woods (Chapter 21).

Using Woodland History to Help Determine Priorities for Woodland Conservation

Recognition that woodland has changed its location and composition over the centuries should not lead us to the position where we accept any sort of change in them simply as the next stage in the evolution of the cultural landscape in which we live. For nature conservation purposes we value those sites where the human influence is lighter and less direct because they tend to contain more specialist woodland species (Chapters 7, 10, 14, 22, 23, this volume); there is then the wealth of cultural and historical meanings attached to trees and woods from the past (Chapters 1, 3, 4, 15, this volume). Simple classifications of ancient woodland have proved immensely valuable in promoting the conservation of woods through helping to capture public and professional awareness of their value (Chapters 9, 22, 26). Broad-scale patterns can be analysed to determine degrees of variation at the landscape scale (Chapters 11, 17, 24, 25).

At the same time, however, as the concept of ancient woodland becomes embedded in official usage, so new work makes its boundary more difficult to define. Research on wood-pasture systems makes the distinction between woodland and non-woodland more uncertain than ever (Chapters 1, 4, 3, 7); combined archaeological and ecological studies show how apparently 'good' ancient woods may have very disturbed histories and have developed on open ground, albeit a long while ago (Chapters 8, 15, 20, 21, 23); there is a sliding scale of naturalness which may need to be applied separately to different components of the system. We may need to hold on to simple categorization and classification of woodland for pragmatic reasons, but we must be flexible at the site level in using this to guide or control their treatment.

Conclusions

Historical knowledge is essential to the understanding of the current distribution, composition and structure of woodland across Europe. It must play a key role in guiding conservation and the development of new wooded landscapes. Comparison of the different studies described in this book shows that many methods of historical ecology are applicable throughout Europe, although care has to be taken with the interpretation of sources and with the regional nature of indicator species.

The management of woodland and trees is much more than just the conventional forestry training. It may involve pasturage, the treatment of individual trees, keeping open space here, favouring old trees there. We must all take responsibility for publicizing this concept if the huge diversity of historical management techniques and the impact that they have had on the associated wildlife and culture of our landscapes is to be maintained. At the same time historical studies show that change has always happened; the woods of the twentieth century will evolve, but if this is managed carefully the best of the past can be maintained.

A wide range of methods are potentially available for different sites, interests and periods. What is most rewarding, however, is to put together the findings from different approaches. More interdisciplinary research is required, particularly if patterns are to be built up across the countryside, and not just confined to a few intensively studied sites. Geographers and historians must work with botanists, zoologists and practical foresters. Best practice in research methods and data collation, analysis and storage needs to be shared, but also experience in promoting the use of historical approaches in landscape conservation. The policies adopted in the UK with its paucity of woods may not be directly applicable to France or Germany, but there may be some useful lessons and vice versa.

The publication in 1976 of *Trees and woodland in the British landscape* by Oliver Rackham was a watershed in the development of historical landscape studies in the UK and it has had considerable influence in Europe and North America as well. Twenty years on, the International Conference on

Advances in Forest and Woodland History at Nottingham showed that woodland history studies have become a challenging and dynamic area of research with no shortage of workers willing to contribute their ideas and findings. However, it appears to us that new ideas and concepts are currently perhaps more likely to come from the Continent. In England, historical studies seem to be in a consolidation phase, although recent work in Scotland may provide the next big shift in the understanding of the UK's woodland history.

In *Trees and woodland in the British landscape*, Oliver Rackham was quite pessimistic about the future of the ancient woods that he so assiduously described. We are glad to report that the conservation prospects are now much better (although still not assured). Ancient woodland conservation will continue to face many challenges in the twenty-first century, but ignorance of its existence and significance can no longer be an excuse.

References

Darby, H.C. (1956) The clearing of the woodland in Europe. In: Thomas, W.L. (ed.) *Man's role in changing the face of the earth*. University of Chicago Press, Chicago, pp. 183–216.

Metaille, J.-P. (1992) *Protoindustries et histoire des forests*. Les Cahiers de L'Isard, Toulouse.

Peterken, G.F. (1981) *Woodland conservation and management*. Chapman & Hall, London.

Rackham, O. (1976) *Trees and woodland in the British landscape*. Dent, London.

Russell, E. (1997) *People and land through time. Linking ecology and history*. Yale University Press, New Haven and London.

Salbitano, F. (ed.) (1988) *Human influence on forest ecosystems development in Europe*. Pitagora, Bologna.

Sheail, J. (1980) *Historical ecology: the documentary evidence*. ITE, Abbots Ripton.

Spencer, J.W. and Kirby, K.J. (1992) An inventory of ancient woodland for England and Wales. *Biological Conservation* 62, 77–93.

Tubbs, C.R. (1968) *The New Forest: an ecological history*. David & Charles, Newton Abbot.

Watkins, C. (ed.) (1993) *Ecological effects of afforestation: studies in the history and ecology of afforestation in Western Europe*. CAB International, Wallingford.

Watkins, C. (1998a) *European woods and forests: studies in cultural history*. CAB International, Wallingford.

Savanna in Europe

Oliver Rackham

Corpus Christi College, Cambridge, UK

This chapter is an introduction to the occurrence, history, functioning and conservation of those European tree-lands in which the trees are scattered among non-tree vegetation. Contrary to common belief, these are real ecosystems intermediate between forest and grassland; they result from various combinations of natural factors and human uses of land; they were more abundant in the past than today; and they are not sharply distinct from the savannas of other continents. They are of special importance since they contain most of the ancient trees (other than coppice stools) in Europe.

Do Trees Equal Forests?

Of course they do, say ecologists and foresters: natural vegetation all over Europe is supposed to consist of trees in the form of forests, with forest ground vegetation and forest plants and animals. In practice, trees occur in many non-forest situations. Some are obviously artificial, or largely so: hedgerow trees, trees in fields, street trees and orchard trees. My concern is not with these, but with natural or semi-natural combinations of trees with non-tree vegetation: grassland, heath, shrubs or undershrubs.[1] Non-forest trees and ecosystems in which they occur are regarded by foresters as not proper forest[2] and by grassland specialists as not proper grassland; they tend to be disqualified from study and conservation.

[1] 'Shrubs' and 'undershrubs' are different life-forms. Shrubs (in this chapter) are trees reduced to the stature of shrubs: for example *Quercus coccifera*, *Phillyrea media*, *Arbutus unedo*. Undershrubs are not potential trees, for example species of *Calluna*, *Cistus*, *Phlomis* and *Lygos*. See Rackham and Moody (1996).
[2] For example, a monumental history of Spanish forests (Bauer Manderscheid, 1991) has little to say about the *dehesas* which are such a distinctive feature of Spain.

European Savanna

This chapter has two starting-points. The first is in *Trees and woodland in the British landscape*, where I pointed out two parallel traditions of tree management: woodland and wood-pasture (Rackham, 1976). These appear as separate in the earliest written records; they are well established in the Domesday Book in 1086. They had an unknown, probably lengthy, period of prehistory before being set down in writing. Woodland in England normally consisted of coppice-woods; wood-pasture of tree'd grassland or tree'd heath. From the eleventh century, wood-pasture developed through the introduction of deer husbandry and fallow deer (*Dama dama*). In some wood-pastures it took on a compartmentalized form: some areas were treated as regular coppice-woods within the wood-pasture, and fenced for some years after each felling in order to prevent the deer and other animals from eating the regrowth (e.g. Bernwood Forest (Thomas, Chapter 19, this volume) and Rockingham Forest (Best, Chapter 6, this volume)).

Wood-pastures comprised the English traditions of wooded commons, deer-parks, and wooded Forests.[3] Commons were areas on which certain people other than the landowner exercised rights of pasturage, of cutting wood (seldom timber) or both. Parks were private land on which the owner kept deer. In Forests, by definition, the king (or some other very great magnate) had the right to keep deer and to catch and eat them; this activity was added to, and did not replace, the normal activities of landowners and commoners on the same piece of land. These arrangements were permanent. The 'tragedy of the commons' (Hardin, 1968) seldom happened. Once the thirteenth century had passed, pasturage rarely destroyed the trees. The balance between trees and grassland often remained stable for centuries, and sometimes outlived the social institutions. Hatfield Forest, a supremely complex example of compartmentalization, can still be seen in operation.

These parallel management traditions – coppice-woods and tree'd grassland – exist over most of Europe and beyond. Compartmentalization, however, seems to be a specially English development.

My second starting-point is the observation that, at European conferences on forest history, speakers habitually allude to 'forests' in the past being used as pasture for cattle, sheep and goats. If asked what it was that the animals were eating, they give an evasive answer. Most shade-bearing grasses and herbs are of little nutritional value, if not inedible or poisonous. Livestock love tree leaves, but cannot climb for them. If let into woodland they eat all the leaves within reach, creating a 'browse-line' beneath which nothing edible is allowed to grow (Fig. 1.1). For example, in England, any wood inhabited by moderate numbers of deer (*Cervus elaphus, Dama dama, Capreolus capreolus,*

---

[3] The term 'Forest' in England traditionally means a place of deer, not necessarily a place of trees. It was used for: (i) the area within which deer were protected by the Forest legal system (e.g. Forest of Essex); and (ii) the area – usually much smaller – where the deer lived and where common-rights were exercised (e.g. Epping Forest, Hatfield Forest). It is important not to confuse these meanings, and to separate them from (iii) 'forest' in its modern sense, without a capital F, meaning a place of trees (e.g. Kielder forest).

Muntiacus reevesi) loses its lower foliage: when this has gone, the deer feed mostly outside the wood. When historical documents mention grazing in 'forests' they cannot, therefore, be referring to continuous forest as modern foresters understand it: they imply at least some non-forest vegetation. Many medieval forests were evidently sparse, so that light-demanding plants grew between the trees.

What name are we to give to non-forest tree-land? It must include the whole range: at one extreme there is 'open forest' with only a small non-forest component; at the other extreme there is grassland or heath with only the occasional tree, which it would be perverse to call 'forest' at all. There should be a neutral, descriptive term for the whole formation, not depending on judgements as to its origin or on resolving the often difficult question of how far it is a natural or cultural feature.

The English term 'wood-pasture' has a wider meaning, including areas of woodland within a compartmental wood-pasture system. 'Pasture-woodland' is unsuitable for the sparser formations. 'Parkland' is a term often used by British writers familiar with the husbandry of deer in sparsely-tree'd parks (Fig. 1.2), but parks are defined by having a boundary fence, and the term is inappropriate for unenclosed land.

In other continents, similar formations are called *savanna*. The word comes from a Caribbean language; its original meaning is unknown. From the

Fig. 1.1. Browse-line created by fallow deer. The tree canopy ends abruptly at 1.27 m above ground, and there is little ground vegetation. (Hayley Wood, Cambridgeshire, UK, May 1984.)

sixteenth century onwards it was used by travellers for various kinds of North American and Caribbean vegetation. In the eighteenth century it was often used for grasslands, roughly as a synonym of *prairie*. It later appears among European writers as a general term for trees scattered among other types of vegetation in America, Africa, tropical Asia and Australia (Fig. 1.3). This was the meaning on which it had settled by the early twentieth century, as can be seen from German and Italian encyclopaedias. It is in this structural sense that I write of savanna in Europe: of trees scattered among non-tree vegetation, regardless of their environment or of any theory of how they may have developed.[4]

When forests abut on a region where some factor – cold, drought, fire, browsing – prevents tree growth, there are various possible transitions. The forest may suddenly stop. Or patches of forest may outlie beyond the forest edge in favourable places. Or the trees may gradually get shorter and turn into shrubs (maquis), as happens widely in the Mediterranean. Here we are concerned with where continuous trees give way to scattered, full-sized trees in the form of savanna.

Savannas can make nonsense of official statistics of forest areas. A statement that the forest area of Crete was 4.5% (in 1981) or 33% (in 1992) is meaningless without a definition of how big, and how closely spaced, the trees have to be to constitute forest. Comparisons between countries, or between different years in the same country, are useless unless it is known that the definition remained the same. (I have a map of Portugal and Spain in which the area shown as forest stops at the border. There is no difference on the ground. The Portuguese cartographer regarded *montado* (savanna) as forest, but his Spanish colleague evidently thought of *dehesa* (savanna) as non-forest).[5]

Natural and Cultural Savannas

It is not easy to separate natural and cultural savannas. In Palaeolithic times, mankind had already eliminated elephants and other great browsers, and had presumably tilted the balance from savanna towards forest. Some savannas today turn almost instantly into forest if the human activity that created them ceases; others persist indefinitely.

In damp, cool England, savannas usually occur at a grazing boundary. They mark a concentration of grazing animals which is generally regarded as artificially high – although not enough is known about degrees of browsing in prehistory to be certain that this is so.

---

[4] Some southern European ecologists believe 'savanna' should have a narrower meaning, e.g. particular types of tree'd grassland, or tree'd grassland that does not depend on human activities. These restrictions are not supported by the usage of the word. Many 'classic' savannas, e.g. in Africa, once thought to be wholly natural, are now known to be partly or wholly cultural landscapes (e.g. Sturm, Chapter 2, this volume).

[5] Spanish and modern Greek have no unequivocal word for 'forest': Spanish *monte* can mean anything from 'mountain' to 'heath', and Greek δάσος is almost as wide.

Fig. 1.2. Pollard oaks in medieval deer-park. (Bradgate Park, Leicestershire, UK, March 1995.)

Fig. 1.3. Savanna of *Eucalyptus oleosa*. Note the special 'corymbose' habit which many savanna eucalypts have. (Narrogin, West Australia, December 1996.)

At the cold boundary, forest sometimes turns into non-forest via a savanna-like zone. Examples are the cypress limit in the mountains of west Crete (probably natural (Rackham and Moody, 1996)) and the *Lärchenwiesen* (larch meadows, modified by mowing) in the Alps.

In the Mediterranean there is a drought boundary. In wet parts there is enough moisture to support continuous forests. In dry parts there are no trees except in places where water collects. In between there is a wide zone with enough moisture to support trees but not forests. Either the trees are reduced to the stature of maquis, or they are widely spaced. They may be widely spaced above ground, but their roots may fill all the available space below ground. 'Moisture' includes, besides rainfall, variations in the amount of moisture stored in the soil or rock, and in the degree to which the roots can penetrate the rock to get at the moisture. Especially within the zone of 400–600 mm mean annual rainfall, geology and geomorphology can make the difference between forest, maquis, savanna, and undershrubs or steppe (Rackham and Moody, 1996).

Browsing and fire modify these relations. They interact with moisture and with each other. For example, *Quercus coccifera* can exist as a shrub or a tree, and can grow from one into the other in the face of a certain degree of browsing (Fig. 1.4). The growth in height that it puts on in a year depends on the rainfall. In a wet climate, or in a run of wet years in a generally dry climate, it can more quickly reach the 'get-away point' at which animals can no longer reach the top and it can develop unhindered into a tree. In a wet climate, more browsing is needed to prevent tree growth than in a dry climate.

The typical management of trees in savanna is pollarding or shredding (Fig. 1.5) to yield permanent supplies of wood or leaves used as fodder (fresh or dried). Pollarding and shredding make a tree unsuitable for timber and tend to prolong its life, giving rise to ancient trees (Rackham, 1976). Coppicing, being incompatible with browsing, is seldom practised in savannas; nor do they normally yield timber. Pollarding and shredding involved specialized equipment. The mysterious *houlette*, a kind of sharp-ended spoon on a long, beribboned staff, which is the symbol of the fashionable shepherdess as portrayed in Marie-Antoinette's France, was a real tool (Moreno and Raggio, 1990); it enabled her to detach leaves from a shredded tree without risking her neck.

Development and Stability

Savanna as 'degraded forest'

Those who decline to study savanna often call it 'degraded forest', meaning presumably forest from which some trees have been subtracted. This is more often alleged than demonstrated, but it can happen. In Hatfield Forest, part of the present 'plains' – grassland with scattered pollard trees – formerly comprised three coppices, and is represented on a map of 1757 as intermediate between woodland and savanna (Rackham, 1989).

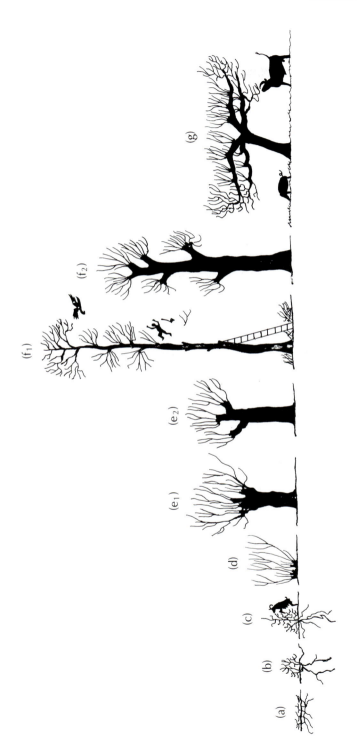

Fig. 1.4. Forms that trees can take. a: ground-oak; b: maquis shrub; c: get-away; d: coppice stool; e1, e2: two styles of pollarding; f1, f2: two styles of shredding; g: modern Spanish style of pollarding. No one tree species assumes all these forms, though *Quercus coccifera* almost does so.

Fig. 1.5. Shredded trees of *Ostrya carpinifolia*. (S. Maria Val di Vara, Liguria, Italy, 27 March, 1996.)

Savannas which are degraded forest ought to bear traces of their forest origin, for example through the trees being tall and narrow, having competed with neighbouring trees in the past. Where, as more often, the trees are low and spreading, they have grown up without above-ground competition: the savanna has existed for at least the lifetimes of the present trees (Fig. 1.6).

Renewal of trees

In most savannas, new generations of trees are not arising all the time. Regeneration is episodic: this is sometimes thought to be a conservation problem. A new generation of trees may arise when browsing falls below a

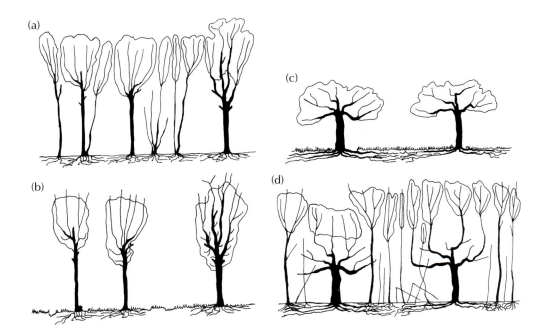

Fig. 1.6. a: forest; b: savanna formed out of this forest by subtracting trees; c: savanna not formed from forest; d: this savanna after infilling.

critical limit, as in the New Forest, England (Tubbs, 1986), or when there is a war and marauding armies have eaten the goats, as happened in the Balkans in World War II.

However, there is an alternative. In many tropical savannas the dominant partners are the grasses. Trees and shrubs arise as a result of too much browsing, not too little. Severe browsing, by wild or domestic ungulates, prevents grasses from taking all the moisture and leaves some over for the trees (Walker *et al.*, 1981). Is this ever so in Europe?

Infilled savanna

Savannas, left unused, often turn into forest. This observation is often cited to support the claim that savanna in Europe is no more than an artefact. Sometimes existing trees multiply and fill in the gaps, producing the unmistakable structure of *infilled savanna*, with big old trees surrounded and overtopped by their crowded children (Fig. 1.6d). The mountain above Sparta in Greece is dotted with scattered big pines (*Pinus nigra*) in a photograph taken in 1906 (Dawkins, 1929, frontispiece). The old pines are now embedded in continuous pine forest (Fig. 1.7).

Savanna can fill in to produce odd types of forest, infilled with different trees from the originals. At Staverton Park, England, the medieval savanna

Fig. 1.7. Savanna infilled since 1906. The old trees and the infill are both *Pinus nigra*. (Taygetos Mountains, Laconia, S Greece, August 1984.)

with its thousands of ancient pollard oaks (*Quercus robur*) has, in part, become infilled with holly (*Ilex aquifolium*), which grows over 20 m high and has overtopped and killed the oaks (Peterken, 1969). Another English example is part of Leigh Woods, Bristol, where the original *Tilia* wood turned into an oak savanna and then, when browsing ceased, became infilled with ash (*Fraxinus*) and elm (*Ulmus glabra*) (Rackham, 1982).

In drier climates, savanna may not easily infill. Each tree may occupy a very specific niche, e.g. where a rock fissure allows roots to penetrate. If it does infill, the result may be a mass of intensely competing, stunted, small stems, which may go dead at the top. This can be seen in the mosaic of deciduous oakwood, savanna and steppe in the Píndhos mountains around

Grevená, northwest Greece. Here I suspect that moisture is the limiting factor and that continuous forest does not represent the natural vegetation.

Stability of cultural savanna

For an example I turn to middle Texas. The commonest trees are an ever-green oak (*Quercus rotundifolia*) and a deciduous elm (*Ulmus crassifolia*). Both are clonal and occur in the form of *motts*, clusters of a few dozen stems. Savanna of motts scattered in grassland is the main type of vegetation extending west from Waco across the Edwards Plateau, ending at the contour of 400 mm annual rainfall.[6]

By American standards this ecosystem (at least the tree component of it) is remarkably stable. Individual stems can be at least 250 years old, which means that the motts themselves must be much older. On 2 November, 1831, James Bowie (inventor of the Bowie-knife) and his cronies had a gun-fight with a scalping party of Tawakoni and Waco (Wilbarger, 1889). The scene as they described it is much the same today, with its prairies and live-oak motts, in one of which the gang holed up (Fig. 1.8). These savannas have lived through four different human cultures: the Indians of earlier centuries, the Indians of the Wild West, the Latin-Americans, and the present, now declining, ranching culture. The grassland component is probably very different, but the trees are the same. However, this is a cultural landscape, and if grazing, the common factor in these cultures, ceases, it soon turns into infilled savanna. The commonest infill is neither oak nor elm but *Juniperus ashei*.

History of Savanna

The earliest written allusion, as far as I know, is in the biblical story of Absalom, who had rebelled against his father King David and lost:

> For the battle was there scattered over the face of all the country. ... And Absalom rode upon a mule, and the mule went under the thick boughs of a great oak,[7] and he was taken up between the heaven and the earth; and the mule that was under him went away. ... And a certain man saw it, and told Joab ... And [Joab] took three darts in his hand, and thrust them through the heart of Absalom, when he was yet alive in the midst of the oak.
>
> 2 Samuel 18: 8–14

This could happen to anyone today in the *Quercus coccifera* savannas of Crete, with low spreading oaks that it seems possible to ride under but which pluck one out of the saddle.

---

[6] Almost the same rainfall at which evergreen oaks give out in the Mediterranean, although the seasonal distribution is very different.
[7] The Hebrew word can mean *Quercus coccifera* (= *calliprinos*) or *Pistacia terebinthus*; either will do.

Fig. 1.8. Savanna with *Quercus fusiformis* motts, the scene of a Wild West fight in 1831. (Near Menard, Texas, January 1996.)

Spain and Portugal

Savanna, known as *montado*, covers at least one-quarter of Portugal; under the academic name of *dehesa* it covers about one-sixth of Spain, typically on sands or thin soils over granite. The trees are cork-oaks (*Quercus suber*) or live-oaks (*Q. rotundifolia*). The grassland is very rich in species; often the grassland under the trees is a different plant community from that between them. As often in Europe, trees tend to grow on thin soil or bare rock, grassland on deep soils.

The trees produce wood and acorns; *Q. suber* also produces cork. They are usually lopped in a very formal style, which is supposed to increase the production of acorns. The acorns feed black Iberian pigs: Spaniards are particular about their hams and sausages, and they also eat the acorns. The grasslands feed merino sheep and cattle (including bullfighting bulls). Evergreen oaks are not thought of as timber trees: the timbers of ancient buildings are either poplar from bottomlands or pine dragged from distant sierras. These practices are often thought to be immemorial; *Don Quixote*, for example, seems to be set in the live-oakeries of Estremadura. However, most savannas lack old trees: the oaks are no more than a century or so old, and there are no stumps to account for their predecessors.

Travellers such as Bowles in *c.*1753, Ponz in 1755–1756, Cavanilles in *c.*1793, Link in 1797–1798, and Borrow in 1835–1836 often followed routes still in use: one can compare their reports with what is there now. Much of what is now savanna was then heath or maquis. The trees are mostly young because they are the first generation of trees on the land. However, there are limited areas with ancient oaks, pollarded in a different style from the nineteenth century trees. These have a documentary record. Scattered references to savanna and pollarding occur in the history of the Mesta, the medieval Spanish guild of transhumant shepherds (Klein, 1920). In the Serra de Mamede, Portugal, where there are ancient trees, there was a cork industry at least as early as 1512 (Videira, 1908).

Alfonso XI, King of Spain in the fourteenth century, was a traveller and a mighty hunter. He stuck fierce boars and slew terrible bears with his royal hand. In his *Book of hunting* (*c.* 1350) he noted topographical details of some 1400 sites where he prowled, such as this:

> La Madroñera [near Trujillo] … it is a flat *monte*, and it is an *encinar hueco* [an 'empty' stand of live-oak, that is one with a clear space under the trees] in which a man can go on horseback because it is raised up, and for those who keep the hounds it is a good *monte* to walk.

This is probably one of the live-oakeries mentioned in the voluminous travel notes of Fernando Colón in *c.* 1510. About that time, Pizarro the Conquistador left his pigs and his oaks at Trujillo and went off to grab an empire. Another four centuries on, there is still a tract of savanna with huge pollard live-oaks, although the arbutus (*madroña*) after which Alfonso named it seems to have gone (Fig. 1.9).

Savanna, I infer, has been a limited practice in dryish parts of Spain and Portugal down the centuries. Originally it was relatively informal. In the latter nineteenth century, when wood-pasture was declining in the rest of Europe, there was a privatization of common-land in Spain and Portugal. Savanna was not destroyed as it was in similar circumstances in, for example, England (Rackham, 1976) and Italy; instead, someone invented a new, formal kind of savanna which became fashionable and gave a new lease of life to wood-pasture (Grove and Rackham, 1998).

Balearic Islands

Majorca has extensive woods of *Q. ilex*, not *rotundifolia* or *suber*. Some have the structure of long-infilled savanna. Savanna itself is now apparently confined to high altitudes; it forms a transition to the alpine zone at the tree-limit, which is probably the lowest in the Mediterranean (*c.* 900 m).

Sardinia

Sardinia has *Quercus suber* and *Q. ilex*. The former is the commonest tree in the island and is the basis for the cork industry. It occurs as savannas and forests, but in forests it lasts for only one generation before being replaced by other oaks.

Cork had many minor pre-industrial uses, but as an industry it is relatively

Fig. 1.9. Savanna with ancient *Quercus rotundifolia*, close to the medieval savanna of La Madroñera. (Jaraicejo, Estremadura, Spain, April 1994.)

late in Sardinia. The cork industry has little control over its raw material: the British entepreneurs who began in the 1830s depended on cork trees already waiting to be harvested.[8] How these trees could have come into existence is still not clear. Nor is it clear what happened to them: very few of the present cork-oaks in the island are as old as this, and there are remarkably few stumps. The few ancient cork-oaks are pollards.

In the mountains there are other kinds of savanna with *Quercus ilex*, ancient deciduous oaks (Fig. 1.10), and pollard yews. These are not necessarily derived from wildwood: sometimes they overlie field systems associated with the Iron Age towers called *nuraghes*.

Sardinia now has extensive and increasing forests; fire plays a large part in the remaining heaths and savannas. Another indication that in the past its ecology may have been very different is the abundance and large size of juniper as a historic building timber. Juniper is a non-forest, often successional, tree, sensitive to both shade and fire; from the thirteenth to eighteenth century the ecology of the island was evidently much more favourable to it than today.

[8] From studies in the archives of Cagliari (Grove and Rackham, 1998).

Fig. 1.10. Savanna with ancient deciduous oaks; the division walls result from nineteenth century privatization. (Fonni, Barbagia Ollolai, Sardinia, April 1992.)

Italy

Savannas were once widespread, and often an intensively managed source of leaf-fodder (Bargioni and Sulli, Chapter 4, this volume). They are now largely replaced by forests; but ancient pollard trees and other remains of savanna can be found in remote places (Moreno, 1990; Moreno and Poggi, 1996).

The most extensive remains are of the chestnut-groves of the Apennines, intermediate between savanna and orchard. *Castanea sativa* dominates whole landscapes. How far it is native is still not clear; the nuts (from trees grafted to special nut-bearing varieties) were a major foodstuff from at least Roman times until this century. The groves, sometimes of big old trees, were carefully maintained for pasture as well as nuts and wood. Chestnut bread is now unfashionable, and few of the groves are still managed. Unmanaged groves have been replaced by coppice-woods or have turned into infilled savanna.

Greece

Greece has many kinds of savanna. In the Píndhos mountains oak savannas exist on a scale comparable to those in Spain. The oaks are deciduous, of many species (e.g. *Quercus pubescens, brachyphylla, cerris, trojana*) and hybrids. They are treated in many ways: oak forests, oak coppices, and oak savannas pollarded or shredded in various styles cover the middle altitudes uninterrupted for tens of kilometres. Pollarding and shredding produce leafy shoots, dried like hay and used to feed sheep; they are important in a climate with barely two months' growing season.

The grassland component is rich in species and includes several endemics. Some species of oak can exist in a ground-oak form: a thicket of shoots, less than 1 m high, which persists indefinitely under browsing and is part of the pasture. Different treatments of oak – timber trees, pollards, coppice stools, ground-oak – are fluid and turn from one into another (Fig. 1.11). A coppice-wood, for example, may be an infilled savanna. Conversely, a coppice may be converted, via browsing, into a combination of ground-oak and shredded trees. Ancient oaks have periods of wide and narrow annual rings corresponding to the shredding cycle. This establishes that some savannas are at least 500 years old. At higher altitudes there are pine savannas, with ancient trees of *Pinus nigra*. In the Peloponnese these may be pollarded; in the north they have no apparent cultural function.

Crete

Crete probably has more varied savannas than anywhere else in Europe (Rackham and Moody, 1996). In east Crete the limestone mountains are

Fig. 1.11. Part-infilled savanna with decidous oaks, some of which are *c.* 500 years old. The ground vegetation is partly ground-oak. (Leipsi, Grevená, NW Greece, May 1988.)

sprinkled with *Quercus coccifera*, often cut into a 'goat-pollard' form: a platform of dense foliage, into which a goat climbs to nibble the shoots. The immense *Q. coccifera* trees of the Lassíthi Mountains, admired by Tournefort in 1700, are still there (Rackham, 1972). Between the trees is a mosaic of maquis, undershrubs and steppe: the maquis is often *Q. coccifera* in its ground-oak form.

West Crete has a chestnut culture rather as in Italy, with giant ancient pollards between which is a distinctive tall-herb vegetation dominated by the umbellifer *Lecockia cretica*. At the alpine limit in the White Mountains, *Cupressus sempervirens* savannas and forests pass into desert via an extraordinary zone of scattered, dwarf, half-dead trees, some of which are at least 1500 years old.

Middle and north Europe

In England, remains and survivals of savanna-like wood-pastures are abundant enough for books to be written on them. Such a wealth of survival probably occurs nowhere else outside southern Europe. They are less common in Wales and Scotland; I have found one infilled savanna in Ireland (Rackham, 1995).

The pollard-meadow type of savanna was once widespread all over Europe (Hæggström, Chapter 3, this volume). Most countries had other forms too. In Germany, pollard savannas survived late enough to have an academic name (*Hudewald*). Ancient pollard *Fagus* and *Carpinus* still exist in parts of Westphalia (Pott, 1981). Among the savannas of the Alps may be mentioned the Großer Ahornboden ('Great Sycamore Flat') in a remote valley of the Karwendel north of Innsbruck, with its giant, apparently non-pollarded *Acer pseudoplatanus*.

In Norway, shredding for leaf-fodder of *Ulmus*, *Tilia* and *Betula*, often scattered among pasture, was once widespread and had its own technical vocabulary (Austad, 1988); it is still occasionally practised. In south Sweden, invaded by recent woodland and modern forestry, such survivals are rare (cf. Berglund, 1991); but in a supposedly ancient *Picea* forest in Småland I was shown a few big, spreading, hemmed-in *Pinus* which were evidently the remains of a former savanna.

Was there Savanna before Civilization?

Palynology

The key to the origin of savanna ought to lie at the drought boundary of tree growth in the Mediterranean. Palynology here, however, encounters the following difficulties:

1. The Mediterranean climate as we know it (hot dry summers, warm wet winters) has been in existence for at most 5000 years – for less time than civilization (Grove and Rackham, 1998). The drought boundary has shifted.

2. Pollen-preserving sites are rare, and represent unusually wet places in a generally arid landscape.
3. In the Mediterranean, trees and shrubs are often of the same species and produce the same pollen.
4. The important non-trees, which might form the vegetation between the trees in a savanna, are often insect-pollinated and shed little pollen.

To summarize numerous studies, glaciations are the normal state of the last 2 million years, and interglacials like the present are relatively brief interludes. Glaciations in the Mediterranean took the form of dry rather than cold periods. The landscape was typically dominated by grasses or *Artemisia*, with some tree pollen but not enough to indicate forest. There seems to be no exact parallel to this anywhere in the world today, but it often suggests savanna, with either patches or scattered individual trees; palynologists use terms like 'forest steppe'. This was either a natural savanna or a Palaeolithic cultural savanna.

As the Holocene advances, tree pollen becomes more abundant. Palynologists interpret this as meaning forest, corresponding to the wildwood of northern Europe; some indeed write of 'dense oak forest'. This may not be true. There is the uncertainty of distinguishing forest from maquis: trees may be of the same species as shrubs and produce the same pollen. There is also the question of pollen representation in savanna, with undershrubs or steppe between the trees.

Palynologists often use the tree-to-non-tree ('AP/NAP') pollen ratio as an index of forest versus non-forest. However, it does not work here, because of the impossibility of knowing what is or is not a tree, and because non-trees vary widely in their pollen production. Many important non-trees shed little pollen and do not flower in shade; for example, *Asphodelus*. Even a few grains of asphodel pollen indicate something in the landscape beside forest. Asphodel is taken by many modern ecologists to be an indicator of a landscape degraded by excessive grazing. Yet it is present already in the early Holocene in south European pollen diagrams, for example at Ayia Galíni in Crete (Bottema, 1980).

Observation suggests that trees in savanna produce at least as much pollen as in forest. Grasses may also be copious producers, but are not significant because pollen may come from local concentrations of wetland grasses around the site of the deposit. What matters is the pollen of three under-producing categories: (i) undershrubs (e.g. *Cistus*, *Labiatae*), mostly poor producers; (ii) wind-pollinated non-shade-bearing herbs (e.g. *Chenopodiaceae*, *Plantago*), moderate producers; and (iii) insect-pollinated non-shade-bearing herbs (e.g. *Asphodelus*, *Cruciferae*), mostly poor producers. The ratios of the pollens of these three categories to those of trees and shrubs should be investigated as a future basis for discriminating between forest and savanna. In Table 1.1 I summarize their values as found in the pre-Neolithic in 12 pollen diagrams from the European Mediterranean. I offer a first suggestion as to what might be the point above which the proportion of these pollens indicates savanna rather than forest.

Table 1.1. Possible savanna indices.

Category	Range encountered	Suggested limit to indicate savanna (%)
(Undershrubs)/(Trees and shrubs)	nil – 4.6% (one site 88%)	> 1%
(Non-shade-bearing wind-pollinated herbs)/(Trees and shrubs)	nil – 11.3% (one site 51%)	> 10%
(Non-shade-bearing insect-pollinated herbs)/(Trees and shrubs)	nil – 80% (one site 104%)	> 1%

Northwest Europe, in contrast, lies well within the climatic limits of tree growth. In England, if anywhere, there ought to have been wildwood from coast to coast. But even here a thin but continuous record of a number of dry-land plants of open ground runs through the Holocene. These are now mostly 'woodland grassland' species such as *Ajuga reptans*, *Lychnis flos-cuculi*, *Succisa pratensis*; plants of permanent, artificially maintained, open areas within woodland. Because they are insect-pollinated, a few grains mean a substantial presence of the plant. It would be rash to claim real savanna in pre-Neolithic England, but there was clearly more to wildwood than just forest (Tubbs, 1996).

There are indications that 'woodland grassland' plants were commoner in earlier interglacials (Godwin, 1975, see for example *Ajuga reptans*). What were the effects of elephants on temperate vegetation? Could they have maintained something like savanna even in the climate of England?

Savanna adaptations

Cork-oak is not a forest tree: it will not stand shade. It is highly adapted to fire, but without being flammable itself. It is adapted to having flammable vegetation around it. It is thus a savanna tree: it grows among grassland or *Cistus* and other undershrubs. These provide fuel and ignite from time to time, burning up cork-oak's competitors. The cork-oaks are saved by their insulating bark; in a fierce fire the foliage is destroyed, but the tree sprouts from the bigger boughs in a self-shredding manner (Fig. 1.12). If there is no fire, cork-oak is replaced by holm-oak (*Quercus ilex*) or other less fire-resistant trees.[9]

Another savanna adaptation is the ground-oak form. There are evidently genes enabling an oak to survive indefinitely under severe browsing in the form of a shrub, and to get back as a tree when browsing slackens. *Quercus coccifera* in the east Mediterranean does this well, and even produces acorns in the ground-oak state. Some Balkan deciduous oaks have a ground-oak

---

[9] An apparent exception is the stable forest of cork-oak, Los Alcornocales, in south Spain (Marañón and Ojeda, Chapter 10, this volume). Dr Marañón tells me that this area lacks any other evergreen oak which might succeed to the cork-oaks.

Fig. 1.12. Cork-oak sprouting from big branches three years after fire. (Portimão, Algarve, Portugal, May 1993.)

form, as to some degree have *Q. rotundifolia* and *suber. Q. ilex*, however, has a ground-oak form only in Majorca and in one locality in east Crete; elsewhere in its range it survives in browsed terrain only out of reach on cliffs. A similar adaptation has evolved in other trees. In Crete, *Phillyrea media* and *Acer sempervirens* have ground-oak-like forms. So has the Greek *Abies cephalonica*, unlike *A. alba*.

Significance and Conservation

Savanna in Europe is not a recent or even a historic development. Its roots go back far into the Pleistocene. It has existed long enough to have left a pollen record, and to have given rise to definite adaptations on the part of the trees. It is difficult to be specific about how wood-pasture functioned before civilization. Its history is bound up with those of fire and of browsing by wild animals. Degree of browsing, and the frequency and intensity of fires, are among the least-known aspects of European wildwood.

Various human activities have affected savanna: extermination of great herbivores, promotion of domestic livestock, alteration of the fire regime, woodcutting, and recently forestry. Later prehistoric and historic activities encouraged and extended it. Savanna is thus partly cultural and partly natural. Here, as often, increasing knowledge fails to define the boundary between these: it becomes more difficult to tell where the natural ends and the cultural begins.

In most countries savanna has been unfashionable for centuries. It has lost ground to the rising ideology of professional forestry: forests are now meant for forestry, and trees are for growing timber, and nothing else. Most foresters disapprove of browsing animals, common-rights and ancient trees. The forests of Europe have been getting denser and younger for over a century. Trees and livestock are separated to a degree that previously happened only with coppice-woods. Where modern forestry took hold early and pervasively – in Germany, France, Switzerland, Austria, Sweden, Italy – little savanna now remains. In Spain, Portugal and Sardinia, savanna took on a new lease of life in the nineteenth century: it is now extensive but often lacks ancient trees. In England, Greece and Crete, where modern forestry arrived late, earlier tree-management often survives, with many ancient trees.

Savanna is an important habitat, especially where it combines grassland and ancient trees. It includes some of the best examples of old grassland. Ancient trees are a specific habitats, or assemblage of habitats, for a multitude of creatures. Every old pollard is an ecosystem in itself, with bats roosting in hollow branches, specific invertebrates in heartwood rotted by different fungi, others in rot-holes in which water collects, and specific lichens on old dry bark under overhangs. Savanna is especially important for organisms requiring more than one habitat, such as hole-nesting birds (e.g. hoopoe, *Upupa epops*), or birds that feed in the open and nest in trees (e.g. cinereous vulture, *Aegypius monachus*), or insects that require dead wood and a nectar source at different stages of their life cycles (Harding and Rose, 1986; Parra, 1990; Rackham, 1990).

The non-tree component loses if savanna turns into forest. This happens most often through infilling, but also with neglect of pollarding. Epping Forest near London was a pasture closely set with *Fagus*, *Carpinus* and *Quercus*, which were kept small by pollarding. Woodcutting ceased in the 1870s, since when the trees have grown to full size: there is no space between them, and little grows beneath them (Rackham, 1978).

Another threat is the changing fire regime. Vegetation increases as savannas fill in; much of this comprises fire-promoting trees such as pines, or flammable undershrubs such as *Cistus ladanifera*. A regime of frequent non-destructive small fires is replaced by infrequent conflagrations which kill the big trees. Foresters may make matters worse by planting eucalyptus and pines which are fire-promoting trees; they then complain when these trees catch fire. They delude themselves that fires can be prevented. Fire prevention is too often a matter of preventing minor fires, and thus allowing fuel to build up and cause a conflagration. Even in non-flammable England, fire becomes a threat to savannas: a fire recently ravaged the pollards of Ashtead Common, Surrey.

The cultural values attached to trees particularly depend on ancient trees. Ancient trees are loved by the British as antiquities and objects of beauty, romance and meaning (Watkins 1998a). They are given names; songs are sung and dances danced in their honour. More recently their special value as habitats has been put on a scientific basis (Alexander, Chapter 7, this volume).

In England, the City of London Corporation, owners of Epping Forest and Burnham Beeches, has long done research into the management of ancient trees (Read, 1991). This initiative has been followed by the founding of the Ancient Tree Forum, followed by English Nature's Veteran Trees Initiative. New sites of ancient trees continue to be discovered.

Conservationists have hitherto devoted much of their energies to forests, especially to those which fit preconceptions of what a wholly natural forest is supposed to look like. Ancient trees, however, are more often in savannas; where they occur in forests, the forest often turns out to be infilled savanna. Savannas, especially those with ancient trees or where the grassland remains intact, deserve to be rescued from misunderstanding and neglect (Fig. 1.13).

Fig. 1.13. Revival of the complex land uses of an English wood-pasture, including new pollards. (Hatfield Forest, June 1994.)

Acknowledgements

This chapter is based on observations made over many years during research activities including the archaeological surveys of Sphakiá (Crete) and Grevená (Greece), and the Mediterranean Desertification and Land Use Project III, commissioned by the European Community. I am especially indebted to A.T. Grove, with whom I have made many visits to southern Europe, and to Jennifer A. Moody, my colleague over many years in Greece and Crete; also to Professors A. Aru and I. Camarda (Sardinia), Dr D. Moreno (Italy), Professors B.E. Berglund and Gunilla Olsen (Sweden), Dr Margaret Atherden (Greece), Lucia Nixon and Professor P.M. Warren (Crete), Jennifer A. Moody and Wick Dossett in Texas, and Professors Paul Adam and Neville Marchant in Australia.

References

Alfonso XI (*c.* 1350) *Libro de la montería.* Seniff, D.P. (ed.), Hispanic seminary of medieval studies (1983), Madison.

Austad, I. (1988) Tree pollarding in western Norway. In: Birks, H.H. and H.J.B., Kaland, P.E. and Moe, D. (eds) *The cultural landscape – past, present and future.* Cambridge University Press, Cambridge, pp. 11–30.

Bauer Manderscheid, E. (1991) *Los montes de España en la historia.* Fundación Conde del Valle de Salazar, Madrid.

Berglund, B.E. (ed.) (1991) *The cultural landscape during 6000 years in southern Sweden – the Ystad Project.* Ecological Bulletins 41, Lund.

Borrow, G. (1843) *The Bible in Spain* (many editions). London.

Bottema, S. (1980) Palynological investigations on Crete. *Review of Palaeobotany and Palynology* 31, 193–217.

Bowles, G. [ie w.] (1783) *Introduzione alla storia naturale e alla geografica fisica di spagna* (trans. f. milizia), Parma.

Cavanilles, A.J. (1795) *Observaciones sobre la historia natural, geografia, agricultura, poblacion y frutos del Reyno de Valencia.* Imprenta Real, Madrid.

Colón, F. [1517–] *Descripción y cosmografia de España.* Patronato de Huérfanos de Administración Militar, 1910, Madrid.

Dawkins, R.M. (1929) *The sanctuary of Artemis Orthias at Sparta.* Macmillan, London.

Godwin, H. (1975) *The history of the British flora.* Cambridge University Press, Cambridge.

Grove, A.T. and Rackham, O. (1998) *Ecological history of Southern Europe* [provisional title]. Yale University Press (in press).

Hardin, G. (1968) The tragedy of the commons. *Science* 162, 1243–1248.

Harding, P.T. and Rose, F. (1986) *Pasture-woodlands in lowland Britain.* Institute of Terrestrial Ecology (ITE), Monks Wood.

Klein, J. (1920) *The Mesta: a study in Spanish economic history 1273–1836.* Harvard University Press, Cambridge, Massachusetts.

Link, H.F. (1801) *Travels in Portugal, and through France and Spain.* Trans. J. Hinckley. Longman, London.

Moreno, D. (1990) *Dal documento al terreno: storia e archeologia dei sistemi agro-silvo-pastorali.* Il Mulino, Bologna.

Moreno, D. and Poggi, G. (1996) Storia delle risorse boschive nelle montagne mediter-ranee: modelli di interpretazione per le produzioni non legnose in regime consuetudinario. In: Cavaciocchi, S. (ed.) *L'Uomo e la Foresta: secc. XIII–XVIII.* Le Monnier, Prato, pp. 635–654.

Moreno, D. and Raggio, O. (1990) The making and fall of an intensive pastoral land-use-system. Eastern Liguria, 16–19th centuries. *Rivista di Studi Liguri* A56, 193–217.

Parra, F. (1990) *La dehesa y el olivar.* Debate Ediciones del Prado, Madrid.

Peterken, G.F. (1969) Development of vegetation in Staverton Park, Suffolk. *Field Studies* 3, 1–39.

Ponz, D.A. (1772–1794) *Viaje de España.* Ibarra, Madrid.

Pott, R. (1981) Anthropogene Einflüsse auf Kalkbüchenwalder am Beispiel der Niederholzwirtschaft und anderer extensiver Bewirtschaftungsformen. *Allgemeine Forstzeitschrift* 23, 569–571.

Rackham, O. (1972) The vegetation of the Myrtos region. In: Warren, P.M. (ed.) *Myrtos: an Early Bronze Age settlement in Crete.* Thames & Hudson, London, pp. 283–298.

Rackham, O. (1976) *Trees and woodland in the British landscape.* Dent, London.

Rackham, O. (1978) Archaeology and land-use history. In: Corke, D. (ed.) *Epping Forest – the natural aspect?* Essex Naturalist N.S.2, pp. 16–57.

Rackham, O. (1982) The Avon Gorge and Leigh Woods. In: Limbrey, S. and Bell, M. (eds) *Archaeological aspects of woodland ecology,* British Archaeological Reports International Series 146, pp. 171–176.

Rackham, O. (1989) *The last forest: the story of Hatfield Forest.* Dent, London.

Rackham, O. (1990) *Trees and woodland in the British landscape,* 2nd edn. Dent, London.

Rackham, O. (1995) Looking for ancient woodland in Ireland. In: Pilcher, J.R. and Mac an tSaoir, S. (eds) *Woods, trees and forests in Ireland.* Royal Irish Academy, Dublin, pp. 1–12.

Rackham, O. and Moody, J.A. (1996) *The making of the Cretan landscape.* Manchester University Press.

Read, H. (ed.) (1991) *Pollard and veteran tree management.* Corporation of London.

Tubbs, C.R. (1986) *The New Forest: a natural history.* Collins, London.

Tubbs, C. (1996) Wilderness or cultural landscapes – conflicting conservation philosophies? *British Wildlife* 7, 290–296.

Videira, C. (1908) *Memoria historica da muito notável villa de Castello de Vide.* Lisboa.

Walker, B.H., Ludwig, D., Holling, C.S. and Peterman, R.M. (1981) Stability of semi-arid savanna grazing systems. *Journal of Ecology* 69, 473–498.

Watkins, C. (1998a) 'A solemn and gloomy umbrage': changing interpretations of the ancient oaks of Sherwood Forest. In Watkins, C. (ed.) *European woods and forests: studies in cultural history.* CAB International, Wallingford, pp. 93–113.

Wilbarger, J.W. (1889) *Indian depredations in Texas.* Reprinted 1985. Eakin Press, Austin.

Development and Dynamics of Agricultural Parks in West Africa

2

Hans-Jürgen Sturm

Special Research Programme 268, Botanisches Institut, Geobotanik und Pflanzenökologie, Siesmayerstr. 70, 60323 Frankfurt/Main, Germany

Agricultural parks are traditional agroforestry systems which are very common in West Africa. They are mainly composed of multipurpose trees and they are undoubtedly an anthropogenic creation. Selection, promotion and introduction of tree species lead to different types of agricultural parks. These parks are strongly related to changes in population density. Based on interdisciplinary historical studies in the West African savanna a model of the development of agricultural parks for the last 100 to 200 years has been established. This is a further step towards an improved understanding of the complex historical processes in the savanna landscape.

Introduction

Agricultural parks or farmed parkland, as Pullan (1974) noticed, cover some millions of square kilometres on the African continent. They can be defined as a landscape where trees occur scattered on farm- or fallow land. Many different tree species varying in age and size may occur on the same field, but more usually a park is made up of just one or two dominant species with similar aged trees.

This chapter tries to answer the following questions:

- Why do agricultural parks exist?
- How are they created?
- How will they develop?

The main zone of interest is the Sudan/Guinea zone of West Africa.

Why do Agricultural Parks Exist?

The answer is quite easy: because they are useful. Most of the trees in agricultural parks are multipurpose trees (Table 2.1). The products of these trees

© CAB INTERNATIONAL 1998. *The Ecological History of European Forests* (eds K.J. Kirby and C. Watkins)

Table 2.1. Examples of multipurpose trees in West Africa.

Species	Human nutrition	Fodder	Medicine	Wood	Others
Adansonia digitata L. (*Bombacaceae*)	fruits, leaves, flowers, seeds	leaves, fruits	every part	construction	–
Balanites aegyptiaca L. Del. (*Balanitaceae*)	fruits	fruits, leaves, sprouts	bark, root, fruit, leaves	construction, tools, fuelwood, charcoal	supplement to peanut butter and soap
Faidherbia albida (Del.) A. Chev. (*Mimosaceae*)	–	leaves, fruits	bark, leaves, gum	tools, construction	soil fertility, soap production
Parkia biglobosa (Jacq.) Benth. (*Mimosaceae*)	fruits, seeds	–	bark, leaves, seeds	construction	dye and soap production
Vitellaria paradoxa C.F. Gaertn. (*Sapotaceae*)	fruits	leaves, sprouts	bark, leaf, root	fuelwood, charcoal	dye production

are mainly used on the subsistence level and are often sold at local markets. Some of them do have a wider economic importance, even occasionally on the international level like *Vitellaria paradoxa* C.F. Gaertn, the sheabutter-tree (Sturm, 1997). They also influence the ecological conditions, mainly microclimatic and soil parameters, for example *Faidherbia albida* (Ouedraogo, 1995).

Much is known about individual species of multipurpose trees, for example the ICRAF (International Council for Research in Agroforestry) special databank on the subject, but much less information exists on the park- or system-level. There we have to look at the interactions between economic, cultural and ecological functions. Most studies on agricultural parks have concentrated on only one of these aspects. An integrated view of these influencing factors is the main target of this research.

How are these Parks Created?

Discussion on the origin of savannas has been going on for several decades, but there is a consensus that savannas are tropical vegetation formations where trees and shrubs occur scattered in grassland (although see Rackham, Chapter 1, this volume). There are climatic and edaphic savannas, but in West Africa most of them are seen as made by humans (Bourlière and Hadley, 1983). The agricultural parks, which develop from the savanna vegetation, are undoubtedly an anthropogenic creation (Fig. 2.1).

Pélissier (1980) presents a scheme of the creation of agricultural parks, which is complemented through our own field studies in Burkina Faso, Nigeria and Benin (see Kerber *et al.*, 1996). In a first step a farmer will clear a piece of land to cultivate. There he will keep some trees which he chooses that are useful to him, while the others are cut down. A first type of park, the so-called 'rest-park', is created. The floristic composition of rest-parks depends mostly on the original savanna vegetation. A typical rest-park of the

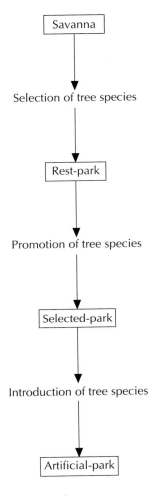

Fig. 2.1. The creation of agricultural parks.

Sudan-Zone is the one with dominance of different *Terminalia* species together with *Vitellaria paradoxa*.

Further promotion of selected tree species leads to a second type of park: 'selected-park', such as those with *Adansonia digitata* and *Parkia biglobosa*. The most important selected-park in the Sudan-Zone has a predominance of *Vitellaria paradoxa*. The fruits of this multipurpose tree offer a very important source of vegetable fat and *V. paradoxa*-parks are therefore seen as characteristic for ethnic groups without livestock because of the lack of animal fat in their diet.

In a last step, these tree stands may be supplemented or replaced by other tree species which are foreign to the local flora (e.g. *Faidherbia albida*), often neophytes like the niem-tree *Azadirachta indica*. This type of park is called an 'artificial-park'. As *Faidherbia albida* provides a valuable fodder for animals in the dry season, this type of park is often seen as typical for livestock-keeping ethnic groups.

How will they Develop?

The floristic composition of a park is not static. Rest-parks change into selected-parks which may change into an artificial-park. The type of park reflects certain climatic and edaphic conditions but also different land use systems, which are influenced through demographic and historical factors.

Historical studies in the West African savanna are unfortunately not easy. There is an enormous lack of information as there are no detailed data on plantations of trees or written information as can be found in Europe. Sources of information are mainly colonial archives, reports of travellers and oral traditions. To exploit these data to a maximum an interdisciplinary team of researchers is necessary. This study on agricultural parks is part of a larger research programme of the University of Frankfurt on the history of the West African savannas (Special Research Programme 268), a collaboration between archaeologists, archaeobotanists, anthropologists, linguists, geographers and botanists.[1]

Our research team tries to link data about the history of agricultural parks and their actual appearance with information on settlement history from anthropologists and linguists as well as with the results from archaeological and geographical dating methods. If it is possible to describe the impact of different types of utilization on the vegetation cover in time and space, it might be feasible to deduce the history of the savanna landscape from the vegetation pattern.

Our model of the history of agricultural parks for the last 100 to 200 years is based on contributions from the different disciplines and is complemented by archives studies and interviews with older peasants and former colonial civil servants (Fig. 2.2). The model is still incomplete.

In precolonial times only very powerful groups could cultivate an extended area of land. Most of the settlements were, for reasons of security, on cleared islands in the savannas, where a permanent cropping system was practised. Within these areas, often several kilometres in diameter, agricultural parks were created. Permanent cropping was not possible without fertilization through manure so the role of fodder trees, especially *Faidherbia albida* became important to feed animals during the dry season. Additionally fruit trees such as *Adansonia digitata*, *Vitellaria paradoxa*, *Parkia biglobosa* and *Tamarindus indica* were also kept on the fields. These trees were of such importance that it was strongly forbidden to cut them down. In former times, and even today, it was forbidden by death to plant a tree. 'If you plant a fruit tree, you will die before eating his fruits', according to the oral tradition of many different ethnic groups in West Africa. A scientific explanation may be

[1] Special Research Programm 268 'History of Culture and Language in the Natural Environment of the West-African Savanna', J.W. Goethe-University, Liebigstr. 41, 60323 Frankfurt, Germany, http://www.informatik.uni-frankfurt.de/~sfb268/welcome.html.

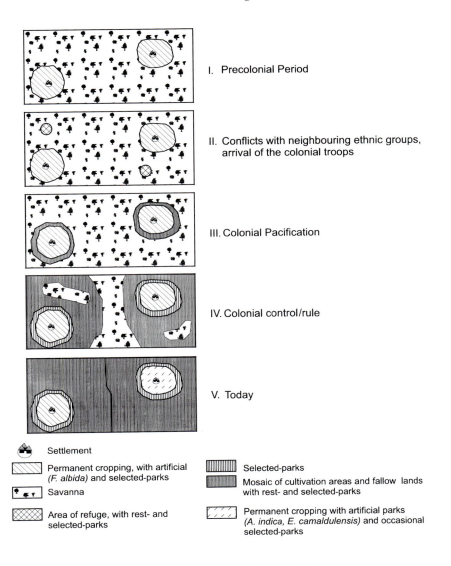

I. Precolonial Period

II. Conflicts with neighbouring ethnic groups, arrival of the colonial troops

III. Colonial Pacification

IV. Colonial control/rule

V. Today

Settlement

Permanent cropping, with artificial (*F. albida*) and selected-parks

Savanna

Area of refuge, with rest- and selected-parks

Selected-parks

Mosaic of cultivation areas and fallow lands with rest- and selected-parks

Permanent cropping with artificial parks (*A. indica, E. camaldulensis*) and occasional selected-parks

Fig. 2.2. Model of the historical development of agricultural parks.

seen in the very slow growth of the trees. The peasant normally dies before the tree produces fruits. *V. paradoxa*, for example, does not produce fruits until it is 20–50 years old. With the first arrival of colonial troops, and mainly through their modern weapons (guns), even these compact settlement areas were no longer safe. The savannas, formerly seen as dangerous areas, became a refuge, a process I would like to call 'Robin-Hood-Effect'.

After the colonial pacification, supplementary circles of agricultural parks – mainly selected-parks with local fruit trees – appeared around the settlements

(Hervouet, 1980). Finally there came the colonial orders to double the cultivation areas especially for cash crops such as cotton. The former permanent
cropping system was abandoned in favour of an extensive farming, today
known as shifting cultivation.

In the beginning of this century we can observe this development for
different ethnic groups in West Africa. The change between cultivation and
fallow gives preference to selected-parks such as the *Vitellaria paradoxa*-
type, which needs fallow to regenerate. Artificial-parks, which require
permanent settlement, are limited to the more densely populated areas,
where shifting cultivation is not practicable. This sequence can still be found
today in the West African savannas.

Succession of Agricultural Parks

As the distribution of the agricultural parks is related to demographic factors,
it will change if population changes. One of the best examples of the
dynamics of parks is the succession from *Vitellaria paradoxa*- to *Faidherbia
albida*-parks (Fig. 2.3).

V. paradoxa-parks are first created as rest-parks and become, through
management, selected-parks, where this tree dominates over other species.
As already mentioned, this species needs a period of fallow to regenerate. So
in older stands no young individuals can be observed. The trees get older
and may be replaced by other species. If the population pressure makes
permanent cropping necessary, *Faidherbia albida*-parks become more
important. This development takes place over a period of 100–200 years.

Local tree species in West Africa are normally very well adapted to the
prevailing ecological conditions such as seasonal drought and are resistant to
insects such as termites, but their productivity is low. The increasing need for
wood due to population growth leads today to a change in species composition. Introductions like *Azadirachta indica* and *Eucalyptus camaldulensis* are
becoming more important especially close to settlement areas. One can
observe this development in densely populated areas, where local species
have often already disappeared (Fig. 2.2). These introduced trees are planted,
in contrast to most of the local species.

Perspective

If one follows this development in West Africa one can see many parallels to
European history:

- Gathering wild fruits.
- Integration of trees in farming lands.
- Plantations of fruit trees.
- Substitution of local trees by foreign ones.
- Industrial plantations of wood trees.

Ouedraogo stated in 1995 that 'man by planting trees becomes the master of
the nature and will no longer be his dependant. The earth is an instrument of

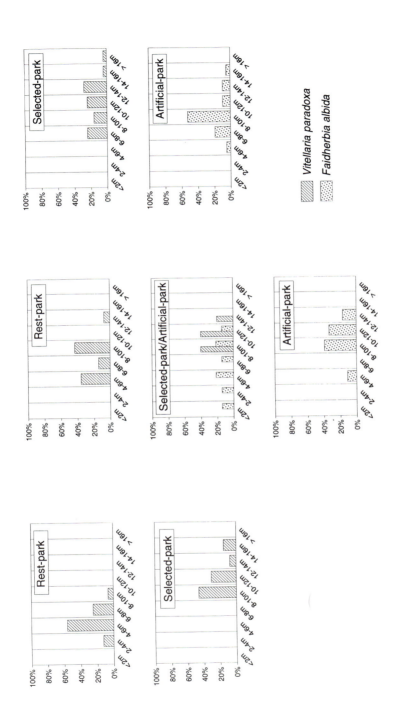

Fig. 2.3. The succession from *Vitellaria paradoxa*- to *Faidherbia albida*-parks in West Africa.

production and no longer the nourishing mother'.[2] I hope that development in West Africa will take all the problems that we have in Europe into account, so that the people can learn from our faults.

Acknowledgements

The study is financed by the German Research Foundation (DFG) and I would like to thank them very much for their support. I also want to thank my colleagues from the Special Research Programme 268 at the J.W. Goethe University, Frankfurt/M.

References

Bourlière, F. and Hadley, M. (1983) Present-day savannas: an overview. In: Bourlière, F. (ed.) *Tropical savannas – ecosystems of the world* 13. Elsevier Scientific Publications, Amsterdam, pp. 1–17.

Hervouet, J.P. (1980) *Du Faidherbia à la brousse. Modifications culturales et dégradation sanitaire.* Orstom, Ouagadougou.

Kerber, D., Reikat, A., Specking, I. and Sturm, H.J. (1996) Les terroirs et la végétation. Paradigmes d'exploitation du sol chez les Mosi et les Bisa dans la province de Bougou, *Berichte des Sonderforschungsbereich* 268 Band 7, Frankfurt am Main, pp. 83–91.

Ouedraogo, S.J. (1995) *Les parcs agroforestiers au Burkina Faso.* Afrena Report No. 79, ICRAF, Nairobi.

Pélissier, P. (1980) L'arbre dans les paysages agraires de l' Afrique tropicale. Cah. *Orstom* 17, 131–136.

Pullan, R.A. (1974) Farmed parkland in West Africa. *Savannas* 3, 119–151.

Sturm, H.-J. (1997) *Nutzbäume in der westafrikanischen Savanne: Der Schibutterbaum* (Vitellaria paradoxa *C.F. Gaertn.) – Charakterbaum der Sudanzone.* Der Palmengarten 61, Frankfurt am Main 41–48.

---

[2] 'L'homme, par la plantation, commence à être le maître du monde naturel et non plus son obligé. La terre est un outil de production et cesse d'être la mère nourricière'. (Ouedraogo, 1995, p. 51).

Pollard Meadows: Multiple Use of Human-made Nature

3

Carl-Adam Hæggström

*Department of Ecology and Systematics, University of Helsinki,
PO Box 7, FIN-00014 Helsinki, Finland*

In many parts of Europe park-like, human-made wooded meadows occur with a vegetation that comprises patches of open meadow alternating with copses of deciduous trees and shrubs, thus forming a mosaic complex. Two kinds of wooded meadows can be discerned, namely *coppice meadows*, the trees of which have multiple stems growing from a common stool, and *pollard meadows*, the trees of which are cut at some height above the ground. The main product of coppice meadows is hay. The trees are cut with an interval of often some decades, producing stakes, poles, firewood and wood for carpentry, at which time also loppings (leafy twigs for fodder) are collected. The main products of pollard meadows are hay and loppings. Hazel nuts, crab apples and other fruits, as well as wooden products are also gathered. Both types of wooded meadows are frequently used for grazing during a short period of the year.

Pollard meadows can be found in several parts of Europe; along the shores of the Baltic Sea (e.g. SW Finland, S Sweden) as well as in the mountains of western Norway, Spain and northern Italy. Today, all of the very few Finnish, and the bulk of the more numerous Swedish pollard meadows, are located in nature reserves and managed more as outdoor museums than as genuine pollard meadows. However, in the mountains of Europe pollard meadows are still used by peasants for hay and loppings.

Introduction

In many parts of Europe, park-like, human-made wooded meadows occur (Hæggström, 1983, 1995). The vegetation consists of two quite different plant communities: patches of open meadow alternating with copses of deciduous trees and shrubs. The copses and meadow glades vary in size and alternate more or less irregularly. Thus a wooded meadow is a vegetation complex of the mosaic complex type (Du Rietz, 1932).

Fig. 3.1. Pollard meadow with newly cut ash pollards. The wooden fence is of the traditional type. Finland, Åland Islands, Eckerö, Kyrkoby Skag. 20 August, 1969. All photos by the author.

Two kinds of wooded meadows have been discerned, namely pollard meadows and coppice meadows (Bergendorff and Emanuelsson, 1990; Emanuelsson and Bergendorff, 1990; Hæggström, 1992a, b). A third slightly different kind ought to be included, namely orchard meadows. The main products of pollard meadows (Fig. 3.1) are meadow hay and the loppings from the pollards (leafy twigs for fodder). Hazel nuts (*Corylus avellana*), crab apples (*Malus sylvestris*) and other fruits, as well as wood products such as poles, are also gathered.

In coppice meadows most of the trees are multi-stemmed (Fig. 3.2), but single stemmed trees are also not uncommon. The coppice meadows resemble coppices, but their main product is hay. The trees are cut at an interval of often some decades, producing stakes, poles, firewood and wood for carpentry, at which time loppings for fodder may also be collected. Sjöbeck (1932) suggested that orchards have their origin in wooded meadows and some types of orchards are regularly used also as hay meadows. In these orchard meadows, fruit trees, such as apple and plum trees, grow scattered or in fairly dense stands (Fig. 3.3).

The different types of wooded meadows are not sharply delimited. For instance, one of the wooded meadows of Nåtö Island in the Åland Islands, SW Finland, has ash (*Fraxinus excelsior*) and black alder (*Alnus glutinosa*) pollards intermingled with multiple stem alder trees. The orchard meadows of the village of Ribaritsa in the Balkan Mountains in Bulgaria are bordered by hazel shrubs and hornbeam (*Carpinus* spp.), ash and large-leaved lime

Fig. 3.2. Birches (*Betula pubescens*) with multiple stems in the coppice meadow of Laelatu in western Estonia. This wooded meadow is extremely rich in species (cf. Kull and Zobel, 1991). 27 June, 1990.

(*Tilia* sp.) pollards. All types of wooded meadows are frequently used for grazing during a short period of the year.

Management of Pollard Meadows

Leaves of deciduous trees have apparently been used as winter fodder for domestic animals since the Neolithic Age (Rasmussen, 1988, 1990a, b, 1991, 1993). New pollards are created on small trees, of between about 5 to 15 cm diameter at breast height (Hæggström, 1983; Mitchell, 1989). Thereafter the trees are pollarded regularly, usually at 3–5 year intervals. Even shorter intervals of 1 or 2 years may occur. Cutting height varies from 1.5 to about 10

Fig. 3.3. Hay making in an orchard meadow with low prune trees. A few hazel shrubs are seen in the background to the left. Bulgaria, Balkan Mountains, Teteven, Ribaritsa. 18 July, 1996.

Fig. 3.4. Shredded ash pollards the year after repollarding. The trees are about 4–8 metres high. *Trifolium repens* and *T. pratense* flower in the foreground. Sweden, Blekinge, Steneryd Nature Reserve. 6 July, 1981.

metres. The lateral branches of high pollards are usually cut near the trunk, thus forming shredded pollards (Fig. 3.4). Shredding is a special type of lopping; only the lateral branches are cut, whereas the top of the tree is left intact (Rackham, 1980; Mitchell, 1989; Christensen and Rasmussen, 1991; Slotte, 1992, 1993).

Fig. 3.5. The ash pollards grow in rows along the banks of the narrow meadow terraces. The meadows are not cut, perhaps due to recent abandonment. Spain, Huesca, Benasque, Cerler at about 1450 m above sea level. 6 August, 1987.

The best leaf fodder is considered to be derived from *Ulmus* spp. and *Fraxinus excelsior.* Other commonly pollarded taxa are *Acer, Alnus, Betula, Carpinus, Juglans regia, Morus, Populus, Quercus, Salix* (especially *S. alba* and *S. caprea*), *Sorbus* and *Tilia* (whose branches are also cut for their bast). In the Mediterranean area, *Melia azedarach, Morus alba, Olea europaea, Platanus orientalis, Tamarix* spp. and even *Eucalyptus* spp. may be pollarded. *Populus nigra* is often only shredded and many of the oak species (*Quercus* spp.) are chiefly cut into shredded pollards. *Fraxinus excelsior* and *Ulmus glabra* are cut into low, medium and shredded pollards.

On flat ground, or on gentle slopes, the pollards grow singly, or in small stands, forming, with the meadow patches, the irregular mosaic pattern of pollard meadows. On steep mountain slopes pollards usually grow along the banks of narrow meadow terraces (Fig. 3.5). Thus the meadows, which are often irrigated, form gently sloping open-meadow strips delimited by rows of pollards.

Pollard Meadows in Practice

Pollard meadows can be found in several parts of Europe; along the shores of the Baltic Sea (e.g. SW Finland, S Sweden) as well as in the mountains of western Norway (Hæggström, 1995), Spain (e.g. the Pyrenees (Gomez and Fillat, 1981, 1984; Creus *et al.,* 1984) and the Cantabrian Mountains, northern Italy (at least around Lake Como; Hæggström, 1983) and Bulgaria. Today, all of the very few Finnish (Hæggström, 1988), and the bulk of the more numerous Swedish pollard meadows are located in nature reserves and managed more as outdoor museums than as genuine pollard meadows (Hæggström, 1987). However, in the mountains of southern Europe pollard meadows are still used by farmers for hay and loppings. Two pollard meadow areas of the Stara Planina (the Balkan Mountains) in Bulgaria may serve as examples of what was once a much more widespread practice.

Ribaritsa pollard meadows

The village of Ribaritsa is located about 10 km southeast of the town of Teteven and 80 km east of Sofia. The meadows of Ribaritsa are located both on plain ground in the valley and as patches on the rather steep slopes at approximately 600–700 metres above sea level. Many of them are bordered by pollards. The species-rich dry and mesic meadows are partly dominated by the grass *Chrysopogon gryllus.* Here and there on the slopes are orchard meadows with quite dense stands of various species but particularly small plum trees. (The area is renowned for its plum brandy.) In July, these orchard meadows, together with other meadows are mown for hay with scythes (Fig. 3.3).

The amount of hay gathered each year depends on the precipitation. A normal yield is about 2000 kg dry hay ha^{-1}. In favourable years, two yields may be gathered and the total hay yield may rise to 5000 kg ha^{-1}. The meadows are usually fertilized with animal manure only, but during the collective farm period up to 200 kg of nitrogen and phosphate fertilizer ha^{-1} was used.

Fig. 3.6. Mr Georgi Sabov carries a few fresh hornbeam leaf sheaves on his back. The meadow has been cut a few days earlier. Bulgaria, Balkan Mountains, Teteven, Ribaritsa. 18 July, 1996.

In dry years, there is not enough hay and the trees are pollarded from mid-July to the end of September. Pollarding only takes place in dry weather conditions. Regularly pollarded trees in Ribaritsa include *Acer campestre*, *A. platanoides*, *Carpinus betulus*, *C. orientalis*, *Fraxinus excelsior*, *Morus nigra*, *Quercus* spp. (with the exception of *Q. cerris* which is not eaten by the animals), *Salix alba*, *Tilia platyphyllos* and *Ulmus* species. *Fagus sylvatica* is used only during years with severe fodder shortage. *Alnus glutinosa* is rarely used. *Corylus avellana* is not used for loppings, but its nuts are collected and the leaves are also used as a remedy against kidney disease and prostate problems in humans.

The trees are pollarded with an interval of 3 or 4 years. A man climbs the pollard and hews the leafy branches with an axe. Leafy twigs about 1–1.5 metre length are cut from these and put in heaps. Either a branch of the same tree, or twigs of *Salix alba* or *Clematis vitalba* are used to bind the twigs into a bundle, a leaf sheaf (Fig. 3.6). The sheaves are preserved either in a barn or on a platform at 2–3 metres height between two trees. Sometimes the leaf sheaves are piled. Two persons can gather 4 m³ of loppings a day.

The normal amount of winter fodder for a sheep or a goat is about 100 kg of dry hay or 1 m³ of leaf fodder. In Ribaritsa, the loppings are used for goats only. Cows and sheep and other domestic animals such as horses, mules and donkeys, are fed with hay. Loppings are fed to animals other than goats only if there is a real shortage of hay.

Fig. 3.7. A small orchard meadow with low apple and plum trees. *Quercus dalechampii* shredded pollards are seen to the left and on the slope in the background. Bulgaria, Balkan Mountains, Milanovo. 15 May, 1991.

Milanovo pollard meadows

The village of Milanovo near Lakatnik in the Iskar Gorge about 45 km north of Sofia is located at about 600–700 m above sea level. The rather steep south and west slopes are mainly covered with *Carpinus orientalis* coppices (Hæggström, 1992b). There are also a few *Carpinus* and *Salix* pollards together with numerous tall *Quercus dalechampii* shredded pollards on the upper parts of the slopes (Fig. 3.7). At the southern outskirts of the village dry and mesic meadows with pollards cover the gently undulating slopes. The bedrock is limestone and the meadow flora is rich in species. The grass *Brachypodium sylvaticum* is one of the main dominant species.

Fig. 3.8. Pollard meadow landscape with low fruit trees and pollards. The high trees are *Quercus dalechampii* shredded pollards. The hay is newly cut and gathered in piles. Bulgaria, Balkan Mountains, Milanovo. 20 July, 1996.

The meadow hay is mown in mid-July and raked into large piles (Fig. 3.8) that are brought to the barns near the farmhouses by lorry. The *Quercus dalechampii* trees are pollarded every third year. The loppings are used both for goats and sheep in Milanovo. At the valley bottom a brook runs from north to south and joins the river Iskar in Lakatnik. This brook is bordered with narrow strips of meadow and pasture land with numerous pollards. Many of these are rather small trees of *Malus* sp. and *Crataegus* spp.

A feature which is probably typical for the Bulgarian pollard meadows is that sheep and goats are allowed to graze and browse in the meadows both before and at the same time as hay making takes place. In Finland and Sweden only milk cows and horses were allowed to graze in the wooded meadows and mostly only after the hay making.

Acknowledgements

Thanks are due to all those persons who have helped me to reach the villages with pollard meadows, particularly the late Professor Stefan Kozuharov and his son Dimitar, Mr Krassimir Apostolov, Drs Spassimir Tonkov, Dolja Pavlova and Mr Georgi Marinov Sabov. The grants and funds awarded by the University of Helsinki and the Ministry of Education are appreciated.

References

Bergendorff, C. and Emanuelsson, U. (1990) Den skånska stubbskottängen. *Nordisk Bygd* 4, 14–19.

Christensen, K. and Rasmussen, P. (1991) Styning af træer. *Eksperimentel Arkæologi, studier i teknologi og kultur* 1, 23–30.

Creus, J., Fillat, F. and Gomez, D. (1984) El fresno de hoja ancha como arbol semi-salvaje en el Pirineo de Huesca (Aragon). *Acta Biologica Montana*, 445–454.

Du Rietz, G.E. (1932) Vegetationsforschung auf soziationsanalytischer Grundlage. *Handbuch biol. Arbeitsmethoden* 11, 293–480.

Emanuelsson, U. and Bergendorff, C. (1990) Löväng, stubbskottäng, skottskog och surskog. *Bebyggelsehistorisk Tidskrift* 19, 109–115.

Gomez, D. and Fillat, F. (1981) La cultura ganadera del fresno. *Revista Pastos* 11, 295–302.

Gomez, D. and Fillat, F. (1984) Utilisation du Frêne comme arbre fourrager dans les Pyrénées de Huesca. *Écologie de Milieux Montagnards et de Haute Altitude. Documents d'Écologie Pryénéenne* 3–4, 481–489.

Hæggström, C.-A. (1983) Vegetation and soil of the wooded meadows in Nåtö, Åland. *Acta Botanica Fennica* 120, 1–66.

Hæggström, C.-A. (1987) Löväng. In: Emanuelsson, U. and Johansson, C.E. (eds), *Biotopvern i Norden. Biotoper i det nordiska kulturlandskapet.* Miljörapport 6, Nordiska ministerrådet, pp. 69–88.

Hæggström, C.-A. (1988) Protection of wooded meadows in Åland – problems, methods and perspectives. *Oulanka Reports* 8, 88–95.

Hæggström, C.-A. (1992a) Wooded meadows and the use of deciduous trees for fodder, fuel, carpentry and building purposes. *Protoindustries et histoire des forêts. Gdr Isard-Cnrs. Les Cahiers de l'Isard* 3, 151–162.

Hæggström, C.-A. (1992b) Skottskogar och skottskogsbruk. *Nordenskiöld-samfundets tidskrift* 51, 81–112.

Hæggström, C.-A. (1995) Lövängar i Norden och Balticum. *Nordenskiöld-samfundets tidskrift* 54, 21–58.

Kull, K. and Zobel, M. (1991) High species richness in an Estonian wooded meadow. *Journal of Vegetation Science* 2, 711–714.

Mitchell, P.L. (1989) Repollarding large neglected pollards: a review of current practice and results. *Arboricultural Journal* 13, 125–142.

Rackham, O. (1980) *Ancient woodland: its history, vegetation and uses in England.* Edward Arnold, London.

Rasmussen, P. (1988) Løvfodring af husdyr i Stenalderen. En 40 år gammel teori vurderet gennem nye undersøgelser. In: Madsen, T. (ed.) *Bag Moesgårds maske. Kultur og samfund i fortid og nutid.* Aarhus Universitetesforlag, pp. 187–192.

Rasmussen, P. (1990a) Pollarding of trees in the Neolithic: often presumed – difficult to prove. In: Robinson, D.E. (ed.) *Experimentation and reconstruction in environmental archaeology. Symposia of the Association for Environmental Archaeology No. 9.* Roskilde, Denmark. Oxbow Books, Oxford, pp. 77–99.

Rasmussen, P. (1990b) Leaf foddering in the earliest Neolithic agriculture. Evidence from Switzerland and Denmark. *Acta Archaeologica* 60, 71–86.

Rasmussen, P. (1991) Leaf foddering of livestock in the Neolithic: archaeobotanical evidence from Weier, Switzerland. *Journal of Danish Archaeology* 8, 51–71.

Rasmussen, P. (1993) Analysis of goat/sheep faeces from Egolzwil 3, Switzerland: evidence for branch and twig foddering of livestock in the Neolithic. *Journal of Archaeological Science* 20, 479–502.

Sjöbeck, M. (1932) *Löväng och trädgård. Några förutsättningar för den svenska trädgårdens utveckling.* Fataburen, Nordiska Mus. Skansens Årsbok, pp. 59–74.

Slotte, H. (1992) Lövtäkt – en viktig faktor i formandet av Ålands grässvålar. *Svensk Botanisk Tidskrift* 86, 63–75.

Slotte, H. (1993) Hamlingsträd på Åland. *Svensk Botanisk Tidskrift* 87, 283–304.

The Production of Fodder Trees in Valdagno, Vicenza, Italy

4

Elena Bargioni and Alessandra Zanzi Sulli
Università Degli Studi, Istituto di Selvicoltura,
Via S.Bonaventura 13, 50145 Firenze, Italy

This case-study focuses on a small livestock farm in northeastern Italy and uses topographic and dendrometrical surveys, interviews of peasants, and quantitative modelling of past cultivation systems and their productivity. From the twenties until 1992, one-third of the fodder supply derived from trees. The results highlight the conflicting needs of timber production and livestock raising. In this sense fodder trees could serve as an important element of conserving the cultural significance of the forests in the region, and contribute to species diversity and richness.

Introduction

Leaf fodder production and collection have been frequently mentioned, described or quoted in laws and technical reports – a good example is Brockmann-Jerosch (1936) – and in several recent forest history papers (Salvi, 1982; Sigaut, 1982; Lachaux *et al.*, 1987; Rasmussen, 1989; Haas and Schweingruber, 1993). However, there seems to have been no analysis of the role of leaf fodder production in cattle farming in prealpine regions, nor of the significance of this practice in the dynamics of a territory economically dependent upon agriculture, forestry and range management. Having found a farm in which, until 1993, cows were fed with leaf fodder, we decided to describe the management of the whole unit.

The farm is located in the commune of Valdagno (NW section of the Vicenza province), in the Agno valley. The valley, 30 km long, cuts, in a north–south direction, the eastern slopes of the Lessini mountains and covers 185 km^2. The local climate is humid subcontinental, the mean annual precipitation is 1489 mm, with cold and long winters, short and hot summers and a large annual temperature range. The farm is situated near the village of Castelvecchio, at 800 m a.s.l. on the side of the Agno river. The rocky parent material is basalt; landslides are rather frequent on steep slopes.

© CAB INTERNATIONAL 1998. *The Ecological History of European Forests* (eds K.J. Kirby and C. Watkins)

By the thirteenth century, the valley bottom already supported both agriculture and manufacturing industries (iron work and weaving), whereas in the hilly areas the only possible activities were agriculture, forestry and animal husbandy (Maccà, 1815). The wool textile industry established itself firmly albeit on a small scale, despite periods of crisis, until Luigi Marzotto started a weaving industry in 1836, with 12 employees. After that the industry kept growing, and currently employs 20,000 workers. Between 1918 and 1940 the town of Nuova Valdagno was created, with the aim of having the factory as a promoter of social life.

Until the 1960s, the Marzotto factories controlled the development of the entire valley. Basic management principles were to employ in the factory only one member per family, and to concentrate all phases of production in its factories. In this way small cattle farms (average area 3.1 ha) run by the owner himself and his family were kept alive. Thus the exploitation of primary resources of the hilly area was a social and economic complement to the extreme industrialization of the valley, resulting in a very stable structure for the territory (Visonà *et al.*, 1994).

The Study Farm and Methods

The farm we have been studying is a typical example of the majority of the farms in the Agno valley controlled by Marzotto as far as ownership, management, size, fragmentation and choice of crops are concerned. The total area changed from the 5.0 ha of the thirties to the 5.9 ha of today, of which 3.1 ha are arable land and hay meadows. The tenancy was, and still is, held by a family which between 1918 and 1940 was comprised of parents and six children, one of them a full-time employee of the Marzotto factory.

Until recently the farm production was planned as follows:

- Arable fields for wheat, corn, potatoes, beans, turnips, for consumption on the farm.
- Hay meadows, mainly clover, with two crops kept for winter and a third one grazed by cattle.
- Pastures with trees, either singly, in groups or in rows. The tree species were mainly ash, some black alder, black poplar, cherry, maple and elm. Most of the trees were kept for fodder production but some of them were green-pruned to produce timber. Grazing, supervised by children, lasted from the middle of May until the end of October for three hours in the morning and three hours in the afternoon. Meadows were neither sown nor cultivated; however, when trees were cut and the stumps uprooted, the soil uncovered was seeded with mixture of seeds collected from the hay barn.
- A woodlot covered by hazel, ash and alder, usually producing fuelwood and timber, but also leaf fodder during meagre years.
- A beech coppice with standards, now turned to high forest, producing fuelwood and some timber (both for sale), and litter and grass for domestic animals kept in the barn.

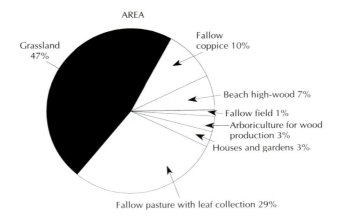

AREA

Grassland 47%

Fallow coppice 10%

Beach high-wood 7%

Fallow field 1%

Arboriculture for wood production 3%

Houses and gardens 3%

Fallow pasture with leaf collection 29%

Today's land use	Area (m^2)
Fallow coppice	5721
Grassland	27,666
Fallow pasture with leaf collection	17,028
Houses and gardens	1670
Arboriculture for wood production	2000
Fallow field	700
Beech high-wood	4233
Total farm's area	59,018

Fig. 4.1. Today's land use.

There were 4–5 milk cows of the local Burlina variety, each producing 10 l of milk per day. These cows were fed in the barn from November to May with hay produced in the meadows, leaf fodder (*frascari*) and grass from the woods collected day by day according to need. Between May and October cattle were grazing in the wooded pastures and in the meadows after the second mowing. Sometimes fodder was insufficient because of drought. Green leaves shredded from the trees were then added to the food. The other livestock were chickens (25–30), a pig and 2–3 sheep for wool.

The technical analysis of the farm and its history since the thirties has been based on various maps: IGMI topographic map 1:25,000 edition 1955 (but based on 1935 records) and 1965; 1:50,000 edition 1969; technical regional map 1:5000 (based on aerial photographs 1981); and survey maps (cadaster, NCT-UTE of Vicenza) of 1936 and subsequent updating until the early nineties. In addition the land survey and planning documents of the *Comunità Montana*, a bureau of territory administration in mountain areas, were used. Field visits were undertaken to check the current condition of individual land survey plots and to record site morphology and vegetation cover. The owner provided further information on past land use of these plots.

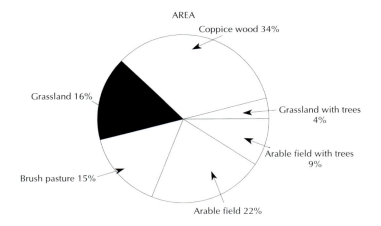

Fig. 4.2. Cadaster land use.

Land use	Area (m^2)
Coppice wood	20,395
Grassland	9338
Brush pasture	8557
Arable field	12,834
Arable field with trees	5405
Grassland with trees	2459
Total farm's area	58,988

Results

The data collected have been treated both quantitatively and cartographically. The plot size was below 5000 m^2, but this does not stem from inheritance fragmentation but is linked to soil morphology and aspect. Fragmentation, therefore, is a function of the farm's organization. Ownership borders coincide with walls erected to create terraces and rows of trees. There is also a correlation between the aspect of the plots and their agricultural use: soils were tilled on southeastern slopes, meadows always face southeast and are on less fertile sites (bad drainage or lower temperatures) and pastures with trees and woodlands cover the northwestern slopes.

Results show that wooded plots have been more stable than the plots devoted to agriculture and animal husbandry (Figs 4.1–4.4). The land survey records are more precise than those of the *Comunità Montana* since maps produced by this organization do not differentiate tilled fields and meadows, with or without tree cover. The current plan wipes away all land use forms, typical of past integrated land use, that are intermediate between meadows interspersed with trees, and woods, and establishes a clear cut, but only theoretical, typology of agricultural land and forest land. A further complication is that plots used as rangeland with trees for fodder were classified as

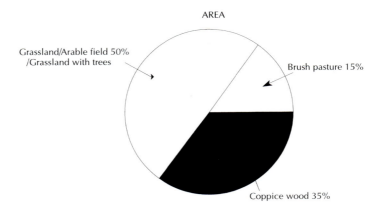

Fig. 4.3. Land use according to the document of 'Comunità Montana'.

Land use	Area (m²)
Coppice wood	20,395
Grassland/Arable field/Grassland with trees	30,036
Brush pasture	8587
Total farm's area	59,018

coppice on the land use records if trees were dense or as shrubby pasture if trees were sparse.

Cattle breeding was the main source of revenue for the farm, despite more than 40% of the farm's surface being occupied by scattered trees, and the absence of grazing in the woods. Hence there is a need to focus on fodder-producing trees, their cultivation, productivity, past role in the farm, and their future role in the evolution of a landscape of decreasing agricultural activity. Formerly grazed shrubby pastures and coppices, and the rows of fodder trees all have a different structure.

Cutting the fodder trees

There are two kinds of fodder: *broco*, consisting of leaves shredded directly from the tree (more precisely from the thin branches on the crown and along the stem) and used directly as fodder, and *frascari*, which are faggots – branches with leaves – collected and preserved for winter nourishment in the stable. The most important species for this purpose is the ash (*Fraxinus* sp.); other species such as alder (*Alnus* sp.), poplar (*Populus* sp.) and hazel (*Corylus* sp.) are all used mainly as fresh fodder, whereas beech (*Fagus* sp.) is most valuable during spring, since its new shoots appear when grass is still lacking under the forest cover.

When trees reach a diameter of 8–10 cm and are approximately 7–8 years old, the top of the tree's crown is cut, so that the stem bifurcates. Also all the

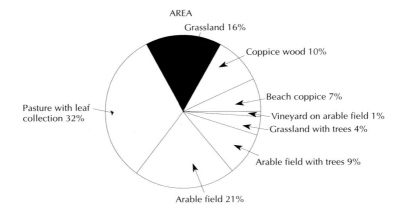

Fig. 4.4. Past land use.

Past land use	Area (m²)
Coppice wood	5721
Grassland	9338
Pasture with leaf collection	19,028
Arable field	12,134
Arable field with trees	5405
Grassland with trees	2459
Vineyard on arable field	700
Beech coppice	4233
Total farm's area	59,018

branches along the stem are cut at 15–20 cm from the main stem leaving a stub (*gropo*): this acts as a small stump producing *frascari*, prevents the rotting of the main stem, and can be used as a support for tree-climbing. *Frascari* are cut from the tree with a billhook (*cortelon*), and are then left on the ground to dry for a whole day. Finally they are collected to form faggots, 25–30 cm in diameter, bound with shoots of *Laburnum anagyroides*. These faggots are left standing in the field for 2–3 days before they are brought into the hay barn and piled flat in criss-cross layers.

While climbing the tree to cut the *frascari*, the farmer shreds all the leaves from the branches that form the top of the crown: these are used directly as fresh fodder for the cows. *Frascari* are collected during the second half of the month of August. The collection is done every 3 years, when the shoots are approximately 1.5 m long. During the intermediate 2 years, only the leaves are picked. The time necessary for the fodder collection is 3–4 h per tree and the farm we have analysed had 250–300 fodder trees. Annual production from 85–100 trees was 700–1000 faggots, with each tree producing approximately 0.5 m³ of fresh leaves and 8–10 faggots.

When the top of the tree stops producing regularly (usually when the trees have a diameter of 25–30 cm), the tree is cut. If it grows in a pasture, the stump is dug out from the ground to be replaced by a seedling, whereas if the tree belongs to a row, its place will be taken by a shoot from the stump.

The faggots are brought from the hay barn to the stable for feeding where they are kept all morning to absorb moisture. The branches are then shredded and the cows fed with the leaves, the twigs being used as fuel. Normally one faggot is used for each cow.

Distribution of fodder trees on the farm

In the Valdagno farm, fodder trees are organized in two kinds of structures: as scattered groups in pastures (grazing land), or as rows bordering the estate or dividing two lots of differently managed land. In 1995 we made a survey of two rows: one has been defined as 'productive', having been exploited in 1993, and the other one has been defined as 'abandoned' since it has not been exploited for the past 10 years.

Highest aerial productivity and lowest space occupancy is obtained by creating row structures of trees, interspersed by shrubs. This structure minimizes shade competition for neighbouring crops because tree branches are cut every 3 years and shrubs are coppiced every 2 years, strongly reducing the light interception.

The 'productive' row is formed by ash trees, hazel, hornbeam (*Carpinus* sp.) and *Laburnum anagyroides*. Ash is used for *frascari* production whereas shrub species are coppiced for fuelwood and, during dry summers, for fresh fodder which is browsed directly in the field. There are also dead stumps, which shows that maintenance is lacking, since dead trees were once substituted with seedlings, so as to avoid gaps in the row. The *frascari*-producing ashes have been pollarded at 4.5–5.0 m from the ground, have stubs (*gropi*) at 2.0 m from the ground, and are spaced approximately every 50 cm. The distribution of the stubs around the stem shows the desire to develop the branches parallel to the row and, along property borders, on the inner side of the farm land.

The 'abandoned' row is quite similar to the other one as far as species composition is concerned but there are different tree shapes. Trees which had been pruned for fodder production in the past are recognizable by the pollarding and the stubs similar to those of the 'productive' row, but they are taller (12 m, as against the 8 m of the other row) and show a greater vertical development instead of the normal sequence of crown layer and trimmed branches. There are several trees, apparently derived from seedlings and not coppice shoots, which have not been pollarded or trimmed in *gropi*. These have evidently been kept for the production of timber. Some other trees have been pollarded at 8–9 m above the ground, but do not show stubs along the stem. Possibly these trees were reserved for the production of fuelwood bigger than that offered by shrubs. The 'abandoned' row is only abandoned therefore insofar as branch and twig production are concerned. The various tree types represent a voluntary choice by the farm manager for a different kind of tree exploitation leading to different techniques being used to shape the stem. With no change in tree species, the structure has changed from forage production to wood production.

Trees in groups

Scattered groups of trees on pastures cover slightly more than 1 ha on slopes of north–northeast aspect. The slope is irregular, sometimes quite steep, drainage is locally poor and small landslides appear here and there. Site conditions did not permit tillage since crops would not reach maturity; while use as meadow would not have provided enough hay for the farm because the area is small and the aspect is such that it would be difficult to get two crops of hay. Therefore the area was used as pasture with small groups of trees for the production of fodder. As a result, fodder production could be increased without excessive shading of the grass cover; on the contrary, some shade was beneficial both to the grass and to the grazing animals. The result should have been the creation of an 'aerial meadow' (Sereni, 1961) with the advantages of a forest climate.

This structure has not actually been described and mapped since, according to the owner, the current situation does not represent the original density and distribution. However, ash, maple (*Acer* sp.), elms and cherries (*Prunus* sp.) used to be distributed in groups and nearly all of them were cultivated for twigs. Sometimes the finest specimens, mostly cherry trees, were managed for timber production. Regeneration was provided either by spontaneous seeding or by planting. Natural regeneration was never sufficient because pruned trees did not produce enough seed and also because seedlings were browsed by grazing animals. When shepherds spotted a seedling they used to surround it with thorny branches as protection. Planting of seedlings was usually done in rows to reduce shading. The arrangement of trees growing in pasture is thus quite similar to that of trees growing in the 'abandoned' row. Here too, utilization ceased some 10 years ago but abandonment seems to be more radical since the only cultivation activity has been the felling of over-mature individuals.

The twig collection method and productivity were the same as for the trees in rows. The only differences are in the greater height of the pollarding – 7–8 m from the ground – because there is no concern about the shading of neighbouring crops, and in the arrangement of the stubs (unless trees were growing near a border).

Analysis of the farm productivity in forage units and an estimate of animals which could be supported show that, hectare for hectare, this system of fodder trees supports one cow more than by hay feeding alone.

Discussion

All methods that have been used – interviews, field research, comparison of data collected with the existing bibliography – have added something to the research. Interviews are the best system to describe the details of a technique in a specific geographical and economic frame which is also tied to its historical development. Obviously the fruitfulness of the interview depends partly on the memory of the person providing the information and his or her ability to communicate (in our case we have been lucky), and partly on the interviewer who must be able to listen and ask useful questions. A basic knowledge of techniques and economy of the past is thus needed, much of which came from the literature review.

The field research gave us the opportunity to document, through the maps we made and a database that can be updated, past tree structures which, once abandoned, are developing towards more natural structures. In addition we could quantify and put into context a complex historical land use organization based on agriculture, forestry and animal husbandry that was not always adequately described in the available literature. Through our research, however, and the literature review we can now provide a reasonably accurate general model of the former role that fodder trees had in the forage resources budget of a farm.

The Valdagno farm is a good example of how apparent limitations on primary resource productivity have been overcome through adopting techniques different from those currently accepted as rational for particular specialized land use types. The main limitations are:

- the extent of the farm (4–5 ha), which is too small for the needs of the farmer's family, and it is not possible to buy more land;
- the aspect and slope of the land which limited to some 2 ha the land available for tillage; and
- the need to devote part of the land to crops for the domestic use of the family.

The limitations have been overcome as follows:

- The land situated in the warmest and least steep slopes has been tilled and seeded with wheat and corn; less fertile land, but still favourably oriented, has been devoted to meadows; the remnant has been left to pastures with trees and woodland.
- Animal husbandry has been used to transform resources otherwise useless to humans, such as pastures, meadows and trees.
- To make up for the limited amount of ground available for husbandry, the vertical dimension has been exploited for energy fixation and fodder production. Trees were therefore used primarily for forage production and secondarily for wood production.

The last choice is only possible if a culture of the tree exists, if its architecture and processes of growth and reproduction are known. It is hard to imagine such knowledge, common heritage in a mountain society, among today's specialized farmers.

The traditional forest worker sees the cultivation of fodder trees as something that does not pertain to his job, or as some kind of insult to trees! However, we must examine the Valdagno case according to our modern knowledge of forest ecosystems and to the new concepts of resource conservation.

Part of the territory has been used for animal husbandry by means of twig collection and pastures with trees, instead of keeping it only as forest or only as pasture, despite a tendency by foresters to keep apart the two kinds of land use. This integration offered the possibility of:

- preserving valuable hardwood species such as ash, elm and maple;
- devoting individual trees or groups of trees for timber production, since fuelwood necessary to domestic needs was largely supplied by the left-over faggots, branches and by the fodder trees at the end of the production cycle; and
- maintaining forest tree species and their culture in an agriculturally productive system.

All this helped to maintain diversity in the Italian forest flora. The mature hardwood individuals acted as seed sources for colonization of abandoned pastures and meadows. Finally, together with knowledge of leaf fodder production, it was also possible to preserve knowledge of hardwood silviculture, otherwise very limited among Italian foresters.

References

Brockmann-Jerosch, H. (1936) Futterlaubbaume und Speiselaubbaume. *Berichte der Schweizerischen Botanischen Gesellschaft* 46, 594–613.

Haas, J.N. and Schweingruber, F.H. (1993) Wood anatomical evidence of pollarding in ash stems from the Valais, Switzerland. *Dendrochronologia* 11, 35–43.

Lachaux, M., de Bonneval, L. and Delabraze, P. (1987) Pratiques anciennes et perspectives d'utilisation fourragere des arbres. *Fourragere* (special issue on *L'animal, les friches et la foret: la foret et l'elevage en region mediterraneenne francaise*), 83–104.

Maccà, G. (1815) *Storia del territorio vicentino*, Anastatic edition. Libreria Alpina Degli Espositi, Bologna.

Rasmussen, P. (1989) Leaf foddering in the earliest neolithic agriculture; evidence from Switzerland and Denmark. *Acta Archaeologica* 60, 71–86.

Salvi, G. (1982) La scalvatura della cerreta nell'alta valle del Trebbia. *Quaderni storici* 49, 148–156.

Sereni, E. (1961) *Storia del paesaggio agrario italiano*. Laterza, Bari.

Sigaut, F. (1982) Gli alberi da foraggio in Europa: significato tecnico ed economico. *Quaderni storici* 49, 49–55.

Visonà, A., Vigolo, G.T. and Cornale, P. (1994) *Valle dell 'Agno: guida alle risorse naturali ed ambientali*. Litovald, Valdagno.

Wood-pasture in Dutch Common Woodlands and the Deforestation of the Dutch Landscape

5

G.H.P. Dirkx

DLO Winand Staring Centre for Integrated Land, Soil and Water Research, Department of Historical Geography, PO Box 125, 6700 AC Wageningen, the Netherlands

Forest grazing has been reintroduced in parts of the Netherlands to improve the nature conservation value of woodland. We carried out a small historical–ecological study to gain an insight into the long-term effects of grazing on the development of vegetation. Late medieval regulations drawn up by commoners demonstrate some of the problems that grazing animals caused. We conclude that in general the amount of grazing led to the destruction of woodland and that it is unlikely that heavy grazing led to high nature conservation values. Open glades in woodland eventually became so extended that no woodland survived. The regulations studied provide little information about the size of herds of grazing animals. They do show, however, the great importance placed on the time of the year when it was allowable to drive herds into the woods.

Introduction

In the Netherlands there is not a rich tradition of research on forest history. Publications are scarce, they usually concern relatively new forests, and often have a strong legal, economic and silvicultural bias (e.g. Schaars, 1974; Buis, 1985; De Rijk, 1989). The landscape and ecological aspects of old forests, as discussed for instance by Rackham (1976, 1980), are hardly described in Dutch literature on forest history. This might be because ancient woods, according to the definition of Rackham (1976), are extremely rare in the Netherlands. Most of the forests that survived the initial agricultural clearances were badly degraded due to wood-cutting, litter extraction, sod-cutting and grazing, and eventually disappeared.

By about 1800 only 4% of the Dutch landscape was covered by forest. Vast areas of woodland had turned into treeless heaths, although since the beginning of the sixteenth century some new woods have also been planted (Buis, 1985). This new planting increased until in the second half of the nineteenth century heaths were being reclaimed and afforested on a large scale. These young heathland plantings are, however, of little natural and scientific value. The age composition of the tree layer, usually consisting of coniferous species, has little variation; there is no shrub layer and the ground flora is species-poor. The transition between woodland and the surrounding area is distinct: transition zones and edge vegetations are absent.

The few ancient woods that survive, by contrast, have a high nature conservation value, partly because they contain species that have become rare in the Netherlands. These woods are regarded as nuclei from which rare species can spread into younger woods (Al *et al.*, 1995). Historical forest management has also been receiving increasing attention since it became clear that the conservation of natural values is often connected with the continuation of the historical management. This is true, for example, for the orchid-rich coppices in the southern Netherlands (De Kroon, 1986).

Grazing as a Woodland Management Tool

Forest grazing, which is a form of historical forest management, has received little attention so far, although it is frequently practised in woods and nature reserves. However, this is mainly in the relatively uniform, younger woods. The grazing animals are expected to increase the variation in the structure of the wood with species-rich transition and edge vegetation and herb-rich glades. Grazing is also applied in new nature reserves created by nature development. Here too, grazing is aimed at creating a more varied vegetation and at preventing development of dense thickets of trees and shrubs. The effects of grazing on the vegetation are being investigated, as well as the suitability of different grazers for a specific terrain, the health of the animals in relation to the terrain condition, etc. (Van der Bilt, 1986; Bekker and Bakker, 1989; Bokdam and Gleichman, 1989; Bokdam and Meurs, 1991; Bokdam and Wallis de Vries, 1992).

Nevertheless, more research is necessary to gain insight into the long-term effects of grazing on vegetation development (Van Wieren *et al.*, 1989). The short period over which the effects of grazing can be studied now is still a major limitation (During and Joosten, 1992), for grazing has been applied in nature management only since the seventies. Therefore it is difficult to predict its long-term effects.

Historical information could be used to bridge some of this gap, since forest grazing has been part of agriculture in the Netherlands, for the last 7000 years. Historical–ecological research may yield information that can improve the application of different management regimes (Dirkx *et al.*, 1992). With respect to grazing this might focus on the size of the herd, the animal species, the grazed area and the changes that occur in the vegetation. Information on the size of the herd and the animal species can be derived from archaeozoological research and from written sources; the grazed area might be derived from the size of the village territory; changes in vegetation can be studied by palaeoecological research, but also from the measures that were taken to protect woods and from data on timber sales.

In order to test the usefulness of historical information in grazing research, we carried out a short study (Dirkx, 1997) concentrating on the

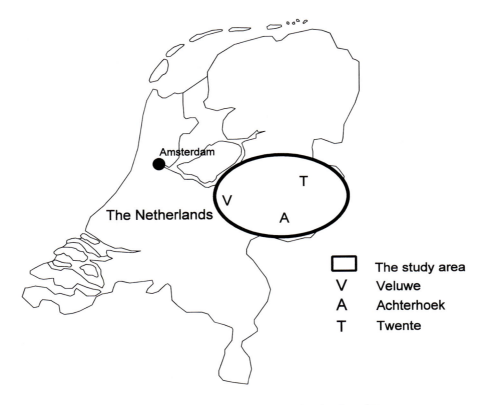

Fig. 5.1. The study area with the regions of the Veluwe, Achterhoek and Twente.

practice of grazing common village pastures. These consisted of both wood-land and open heathland. For this chapter we confined our work to the so-called *marke boeken*. These were books in which the users of the commons, in Dutch *marken*, laid down the regulations for the use of the commons. The regulations supply a lot of information on the management and use of the commons.

Woodland in our Study Area

The study was focused on the central and eastern parts of the Netherlands (Fig. 5.1). The landscape mainly consists of nutrient-poor sandy soils. The forest that grew here in the Middle Ages was secondary, differing greatly from the woods that had developed during the Atlantic Period. These latter woods had much elm (*Ulmus* sp.) and lime (*Tilia* sp.) but had been changed during prehistory and in some places had disappeared completely. With the introduction of agriculture in the Neolithic Period, a process of soil nutrient depletion and podzolization was started on the nutrient-poor sandy soils (Iversen, 1973; Behre, 1988). The competitive balance between the various

tree species in the forest changed and oak (*Quercus* spp.) could gain ground.
Livestock farming also contributed to the changes in forest vegetation.
Livestock was pastured in the woodland adjacent to the settlements. The
grazing livestock damaged trees and shrubs in the wood, some tree species
being affected more than others. The collection of leaves as winter fodder has
also been claimed as part of the reason for the decline of elm. Regeneration
was suppressed because saplings were eaten. Thus, glades – both former
fields and storm holes – were kept open by the livestock. Openings formed
in the canopy so that photophilous shrubs, herbage and grasses found their
way into the wood (Behre, 1988; Pott and Hüppe, 1991). In the course of
time, more and more wood disappeared from the Dutch landscape and the
Veluwe must already have had a rather open character in Roman Times (Van
Geel and Groenman-van Waateringe, 1987).

After the Roman Empire collapsed, the population in large parts of the
Netherlands was reduced greatly. Large parts of the cultivated land were
abandoned and woodland spread back in many places. Settlement names
indicate that the Netherlands must have been a densely wooded land in the
Early Middle Ages, a picture that is corroborated by palaeoecological research
(Vervloet, 1977; Koster, 1978; Theuws, 1988). Because of soil changes, however, complete restoration of the former woodland was not possible. Elm and
lime did not recover the lost ground, and woods with much oak and birch
(*Betula* sp.) were formed (cf. Barker, Chapter 15, this volume). On the ice-
pushed ridge of the Veluwe, with sandy soils in which cambic podzols have
developed, *Fago–Quercetum* woods were probably found. In the eastern part
of the Netherlands (Achterhoek and Twenthe), where acid carbic and gleyic
podzols had developed, *Betulo–Quercetum* woods were found.

In the course of the Early Middle Ages (500–1000) the population started
to grow again. In many places new settlements were founded. In the High
Middle Ages (1000–1350) the population growth was accelerated. The
pressure on unreclaimed woodland and heathland strongly increased. In the
second half of the thirteenth century the people with an interest in such
unreclaimed land, organized in the *marken,* started to set down in writing the
regulations for the use of this ground.

The *Marke Boeken*

In our study area the first *marke boeken* with regulations on the use of the
commons were published in the fourteenth century. The oldest *marke
boeken,* however, are not always available. Some of them have been lost in
wars.[1] (See below for manuscript sources.)

Other books got lost because they were not passed on when a new
judge was appointed. Sometimes, agreements on the use of the common
were not laid down at all, but were kept and passed on verbally, in
accordance with old custom.[2] Most of the regulations that we have at our
disposal thus date from the seventeenth century. This is a major limitation
when we want to study the use of woods. Towards the end of the Middle
Ages the process of degradation had already proceeded so far that the

woods were in a deplorable state (Buis, 1985). In the seventeenth century the commons mostly consisted of open heathland. In the earliest regulations we can already read a great concern about the continued existence of the wood. Indeed it was sometimes explicitly mentioned that the *marke boek* had been written up to stop further loss of the wood. This was the case in the Niersenerbos in the Veluwe area, where the book was written in the late sixteenth century.[3] Elsewhere in Europe the use of woods had also become increasingly regulated since the Middle Ages (Rackham, 1976). The rules in the *marke boeken* are especially concerned with wood-cutting and grazing livestock (Buis, 1985). Wood-pastures appeared to be important, because in nearly all the commons with woodland the regulations include provisions on wood-pastures (Pott and Hüppe, 1991). In the Veluwe area, for instance, 20 commons still had woodland within their boundaries and 18 of the *marke boeken* included provisions on wood-pastures (Buis, 1985).

Feeding Pigs with Acorns

The utilization of woodland for pasture dates further back than the fourteenth century (Heringa, 1982, 1985). Ninth century sources describe the habit of driving pigs into the wood (ten Cate, 1972), particularly to areas with acorns and beech-mast. In many places the size of a forest was measured in the number of pigs that could be driven into it (Buis, 1985).

This acorn-feeding could cause damage to the wood. Pigs root up the earth searching for acorns, beech-mast, wild apples, plant roots and insects. Seedlings of young trees are thus destroyed. On the other hand, acorns and beech-mast that had not been eaten had a clean seed-bed. The most damage to the wood, however, was probably caused not so much by the pigs themselves as by the habit of beating, shaking or throwing unripe acorns out of the tree, so that the pigs could eat these too. Unripe acorns that fell to the ground but were not eaten could not sprout, and therefore the habit was later forbidden in many places.[4]

Limitations with respect to the number of pigs to be acorn-fed are rarely found in the *marke boeken*. This was probably not necessary, because the number of pigs that was allowed to be driven into the woods was determined each year on the basis of the amount of acorn mast.[5] When this number had been decided on, the farmers herded their pigs together. Each pig was branded[6] and any unbranded pigs that were driven into the wood illegally could be easily recognized and confiscated.[7] To prevent tampering with or misuse of the brand, the commoners of the Eder bos in the Veluwe area decided in 1512 that it should be put away in a box with two locks.[8] The herd was pastured by a herdsman, for example in Didam, where a special *hierd* was appointed (ten Cate, 1972). Pigs were allowed in the wood only in autumn and winter, when they could feed on the acorn mast (Fig. 5.2). During the rest of the year they could have been let in to feed on roots, insects and herbage but the damage that this would cause was considered unacceptable.

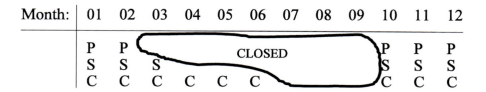

Month:	01	02	03	04	05	06	07	08	09	10	11	12
	P	P				CLOSED				P	P	P
	S	S	S							S	S	S
	C	C	C	C	C	C				C	C	C

Fig. 5.2. Summary of regulations about the months in which the woods could be grazed by pigs (P), sheep (S) and cattle (C).

Cattle and Sheep in the Wood

Although medieval sources mainly mention swineherding in woods, we should not overrate its significance. Archaeozoological research has shown that the livestock in sandy areas consisted of about 50% cattle (Groenman-van Waateringe and van Wijngaarden-Bakker, 1990). Cattle forage in the wood in quite a different way, eating mainly grass and herbage in the undergrowth. They are considered typical grazers (Van de Veen and Van Wieren, 1980). Sheep are also primarily grazers, although they do eat away the woody heather shrubs and saplings (Bottema and Clason, 1979).

If cattle are to find food in the wood, there must be glades with a lusher vegetation (see Rackham, Chapter 1, this volume). Around the Early Medieval Kootwijk (Veluwe) a rather open and lush vegetation was present according to the palaeoecological research (Van Geel and Groenman-van Waateringe, 1987).

The historical sources record that the cattle pastures in the wood also needed regulation because it was feared that cattle would affect the saplings.[9] In our study area parts of the wood that had been recently cut were closed to livestock (Buis, 1985) for three years,[10] and this was also the case, for example, in the t'Ename forest in Belgium (Tack *et al.*, 1993) and in Britain (e.g. Best, Chapter 6, this volume). During the growing season the wood was also closed to sheep and cattle. This was apparently intended to allow the twigs and saplings, which develop in spring, to grow until they were no longer attractive for the animals. Sheep were feared more than cattle. In a forest in the Veluwe area sheep were kept out from February to September, whereas cattle were kept out only from the end of June to September.[11] In other regions sheep were also kept out of the wood longer than cattle, for sheep typically from April until October, but for cattle only from July until October (Fig. 5.2).

Protection of the Wood

Besides the regulations for forest grazing, other rules were issued to protect the wood. Severe penalties were imposed on illegal wood-cutting. Even the amount of wood that farmers were legally allowed to cut became increasingly curtailed. In the first half of the fifteenth century, for example, the farmers in a common in Twente were allowed to fell an oak and a beech each year; in

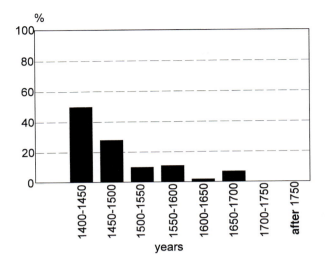

Fig. 5.3. Regulations about pigs driven in the woods as a percentage of the total number of regulations.

the second half of that century they were allowed to fell only one tree. A bit later still, farmers had to demonstrate the necessity before they were allowed to fell a tree, besides which they were obliged to plant a sapling when a tree was felled. In the seventeenth century each farmer had to plant 15 saplings. Nevertheless, there were still complaints about a shortage of timber (Bakker and Groot, in prep.). In other regions the permission to fell trees was also connected with an obligation to plant saplings (Buis, 1985), as was sometimes the right of wood pasture.[12] The planting material was obtained from nurseries and stewards had to see to it that farmers looked after their nurseries well (Ter Kuile, 1908). There was not always enough planting material available however. In that case, a shrubbery was purchased, which sometimes involved several thousands of trees.[13] The newly planted saplings had to be protected from the grazing livestock, usually by putting thorny branches around them, so that the animals could not touch them.[14]

All the attempts to protect the woodland from complete destruction, however, were of little avail. By the end of the Middle Ages the area of woodland had been strongly reduced (Buis, 1985). None was left, for example, around the village of Groenlo (Achterhoek), whose name refers to a green wood (Künzel *et al.*, 1988). A sixteenth century map of this village shows fields, grassland and heathland with only a few scattered trees. The loss of woodland in the medieval period also reduced the significance of acorn-feeding of pigs. Fifteenth century bills from Twente show that acorn-feeding of pigs did occur but only rarely (Dirkx, 1994). While in the fifteenth century many provisions on acorn-feeding were still included in the *marke boeken*, from the sixteenth century onward their significance decreased (Fig. 5.3). The year 1500 is therefore a clear turning-point.

The many rules in the *marke boeken* concerning the use of the woodland had apparently not been able to prevent the loss of woodland, unlike the position at Hatfield Forest (Rackham, Chapter 1, this volume). The commoners were unable to turn the tide. We believe that pasturing played a key role in the loss of woodland because woods that were not utilized as common pastures did survive. One of these forests is the Speulderbos in the Veluwe area. It was closed to pasturing in 1547.[15] Now it is one of the few old forests in the Netherlands. Similarly, private woods, where no common pasturing rights were asserted, were also saved. An example of this is the wood of Espelo manor in Twenthe, which is now a nature reserve.

Conclusions

Our study has shown that pasturing livestock in woods has been an important factor in the deforestation of the Dutch landscape. It remains to be seen whether the late medieval woods on poor sandy soils did have the natural values that we now want to create by grazing. Transition and edge vegetation types seem unlikely to have survived under the grazing pressures present. Glades, which contribute to the structure variation in woods, did occur and, as a result of grazing, were extended so much that the wood eventually disappeared.

It appears, therefore, that the historical grazing density in Dutch forests was too great. Unfortunately, we have not yet got an impression of what that density was. The *marke boeken* do not provide information about the area of woodland in the commons, nor about the head of livestock pastured there. It seems that the commoners were more concerned about the time of year in which the herd was driven into the wood, than about the size of the herd.

Manuscript Sources

1. Mark van Weenum, Geldersche Markerechten I: 4.
2. Buurschap van Emst en Westendorp, Geldersche Markerechten I: 376.
3. Maalschap het Niersenerbosch, Geldersche Markerechten I: 407.
4. Markeregt van Losser, Overijsselsche Stad-, Dijk- en Markeregten 25; Lierder en Speldermarken, Geldersche Markerechten I: 111.
5. Mark van Rekken, Geldersche Markerechten II: 135; Mark van Baak, Geldersche Markerechten II: 325.
6. Mark van Dieren, Spankeren en Soeren, Geldersche Markerechten I: 509.
7. Marken van Waverlo, Loel en Milsterio, Geldersche Markerechten II: 94.
8. Maalschap het Eder bos, Geldersche Markerechten I: 324.
9. Markeregt van Leussen, Overijsselsche Stad-, Dijk- en Markeregeten: 1.
10. Lierder en Speldermarken, Geldersche Markerechten I: 111; Mark van Loenen en Sylven, Geldersche Markerechten I: 157.
11. Maalschap het Niersenerbosch, Geldersche Markerechten I: 407.
12. Mark van Mallem, Geldersche Markerechten II: 142.
13. Markeregt van Bathmen, Overijsselsche Stad-, Dijk- en Markeregten: 19; Mark van Mallem, Geldersche Markerechten II: 142.

14. Markeregt van Losser, Overijsselsche Stad-, Dijk- en Markeregten: 25.
15. Maalschap het Speulderbos, Geldersche Markerechten I: 418.

References

Al, E.J., Koop, H. and Meeuwissen, T. (1995) *Natuur in bossen. Ecosysteemvisie bos.* IKC – Natuurbeheer, Wageningen.

Anonymous (1873–1967) *Overijsselsche Stad-, Dijk- en Markeregren*, Derde deel, *Vereeniging tot beoefening van Overijsselsch regt en geschiedenis.* Zwolle, 1–23.

Bakker, M. and Groot, L. (in prep.) *Historische ecologie van de bossen van Twente.* DLO – Staring Centrum, Wageningen.

Behre, K.E. (1988) The role of man in European vegetation history. In: Huntley, B. and Webb, T. III (eds) *Vegetation history* (Handbook of vegetation science Volume 7). Kluwer, Dordrecht, pp. 633–672.

Bekker, D.L. and Bakker, J.P. (1989) Het Westerholt IX: veranderingen in vegegatiesamenstelling en -patronen na 15 jaar beweiden. *De Levende Natuur* 90, 114–119.

Bilt, E. van der (1986) Begrazing van natuurreservaten in Drenthe. *Huid en Haar* 5, 169–178.

Bokdam, J. and Gleichman, J.M. (1989) De invloed van runderbegrazing op de ontwikkeling van Struikheide en Bochtige smele. *De Levende Natuur* 90, 7–14.

Bokdam, J. and Meurs, C.B.H. (1991) Is bijvoedering van heideschapen nodig? *De Levende Natuur* 92, 110–116.

Bokdam, J. and Wallis de Vries, M.F. (1992) Forage quality as a limiting factor for cattle grazing in isolated Dutch nature reserves. *Conservation Biology* 6, 399–408.

Bottema, S. and Clason, A.T. (1979) *Het schaap in Nederland.* Thieme, Zutphen.

Buis, J. (1985) *Historia forestis; Nederlandse bosgeschiedenis.* HES, Utrecht.

Cate, C.L. ten (1972) *Wan god mast gift ... Bilder aus der Geschichte der Schweinezucht im Walde.* Pudoc, Wageningen.

Dirkx, G.H.P. (1994) *Enschede-Noord; een historish-geografisch onderzoek voor landinrichting.* DLO-Staring Centrum (Rapport 32), Wageningen.

Dirkx, G.H.P. (1997) ' ... ende men sal van een erve ende goedt niet meer dan een trop schaepe holden ...' *Historishe begrazing van gemeenschappelijke weidegronden in Gelderland en Overijssel.* DLO-Staring Centrum (Rapport 499),Wageningen.

Dirkx, G.H.P., Hommel, P.W.F.M. and Vervloet, J.A.J. (1992) Historische ecologie. Een overzicht van achtergronden en mogelijke toepassingen in Nederland. *Landschap* 9, 39–51.

During, R. and Joosten, J.H.J. (1992) Referentiebeelden en duurzaamheid. Tijd voor beleid. *Landschap* 9, 285–295.

Geel, B. van and Groenman-van Waateringe, W. (1987) Palynological investigations. In: Groenman-van Waateringe, W. and van Wijngaarden-Bakker, L.H. (eds) *Farmlife in an Carolingian village.* Van Gorcum, Assen/Maastricht, pp. 6–30.

Groenman-van Waateringe, W. and van Wijngaarden-Bakker, L.H. (1990) Medieval archaeology and environmental research in the Netherlands. In: Besteman, J.C., Bos, J.M. and Heidinga, H.A. (eds) *Medieval archeology in the Netherlands.* Van Gorcum, Assen/Maastricht, pp. 283–297.

Heringa, J. (1982) *De buurschap en haar marke.* Drentse Historische Studien 5, Assen.

Heringa, J. (1985) Lijnen en stippelijnen in de geschiedenis van de buurschap. *Nieuwe Drentse Volksalmanak* 102, 69–93.

Iversen, J. (1973) *The development of Denmark's nature since the last Glacial.* Geology of Denmark III. Danmarks Geologiske Undersogelse V. Raekke 7c.

Koster, E.A. (1978) *De stuifzanden van de Veluwe; een fysisch-geografiche studie*. Publicaties van het Fysisch geografish en Bodemkundig Laboratorium van de Universiteit van Amsterdam nr. 27, Amsterdam.

Kroon, H. de (1986) De vegetaties van Zuidlimburgse hellingbossen in relatie tot het hakhoutbeheer. Een rijke flora met een onzekere toekomst. *Natuurhistorische Maandblad* 75, 167–192.

Kuile, G.J. ter (1908) *Geschiedenis van den hof Espelo, zijne eigenaren en bewoners*. Tijl, Zwolle.

Künzel, R.E., Blok, D.P. and Verhoeff, J.M. (1988) *Lexicon van Nederlandse toponiemen tot 1200*. Publicaties van het P.J. Meertens-Instituut, Amsterdam.

Pott, R. and Hüppe, J. (1991) *Die Hudelandschaften Nordwestdeutschlands*. Westfalisches Museum Fr Naturkunde, Munster.

Rackham, O. (1976) *Trees and woodland in the British landscape*. Dent, London.

Rackham, O. (1980) *Ancient woodland. Its history, vegetation and uses in England*. Edward Arnold, London.

Rijk, J.H. de (1989) De economische geschiedenis van het Edese bos. *Nederlands Bosbouw Tijdschrift* 61, 106–112.

Schaars, A.H.J. (1974) *De bosbouw van het "Entel" in de tweede helft van de achtiende eeuw*. Gelderse Historische Reeks no. 5, Zutphen.

Sloet, J.J.S. (1911) *Geldersche Markerechten I*, Werken der vereeniging tot uitgaaf der bronnen van het oud-vaderlandsche recht. Tweede reeks no. 12 's-Gravenhage.

Sloet, J.J.S. (1913) *Geldersche Markerechten II*, Werken der vereeniging tot uitgaaf der bronnen van het oud-vaderlandsche recht. Tweede reeks no. 15 's-Gravenhage.

Tack, G.P., van den Bremt, P. and Hermy, M. (1993) *Bossen van Vlaanderen. Een historische ecologie*. Davidsfonds, Leuven.

Theuws, F.C.W.J. (1988) *De archeologie van de periferie*. Studies naar de ontwikkeling van bewoning en samenleving in het Maas-Demer-Schelde gebied in de Vroege Middeleeuwen. Universiteit van Amsterdam, Amsterdam.

Veen, H.E. van de and van Wieren, S.E. (1980) *Van grote grazers, kieskeurige fijnproevers en opportunistische gelegenheidsvreters*. I.V.M., Amsterdam.

Vervloet, J.A.J. (1977) Cultuurhistorie. In: ten Houten de Lange, S.M. (ed.) *Rapport van het Veluwe-onderzoek*. Pudoc, Wageningen, pp. 76–78.

Wieren, S.E. van et al. (1989) *Begrazingsonderzoek in Nederland: samenhang, prioriteiten en samenwerking*. NRLO-rapport 89/31, 's-Gravenhage.

Persistent Outcomes of Coppice Grazing in Rockingham Forest, Northamptonshire, UK

6

Jeffrey A. Best
Nene College, Boughton Green Road, Northampton NN2 7AL, UK

This paper argues that to understand properly the present-day character of coppice-woods once managed as elements of Forest, it is necessary to appreciate not only the consequences of deer husbandry, but also the long-term impact of grazing livestock and other small game. Initial observations from the Forest of Rockingham suggest that, in comparison with non-forest woods, there are differences in the composition of underwood and field-layer vegetation today that can be attributed to the former presence of such animals. More detailed research, in this and other remnant forests with a known history of coppice grazing, is called for.

Introduction

Many ancient woodlands in the British Isles have also been regularly grazed by domestic stock; for example sheep and cattle turned out to graze in wood pastures. Indeed, much of the structure and character of some of the ancient woodlands we treasure today has developed as a direct consequence of a long history of livestock grazing. Patterns and rates of regeneration in them have been influenced by grazing pressure, profoundly altering the age structure of the woodland trees and also significantly modifying the species composition of the ground flora.

(Putnam, 1994b)

Grazing as a factor affecting coppice-woods has been acknowledged by both Rackham (1976, 1980) and Peterken (1981), but research has largely been directed towards particular circumstances. Examples include the impact of large herbivores in upland woods (Mitchell and Kirby, 1990) and the modern impact of deer upon coppice elsewhere (Ratcliffe, 1992; Kay, 1993; Cooke, 1994; Putman, 1994a). Studies of the longer-term relationships between livestock and coppice have mostly been confined to classic lowland sites, particularly the New Forest (Peterken and Tubbs, 1965; Tubbs, 1968, 1986;

© CAB INTERNATIONAL 1998. *The Ecological History of European Forests* (eds K.J. Kirby and C. Watkins)

63

Flower, 1980; Putnam, 1986). This has, however, been followed by a call for the explicit recognition of 'grazed coppice' and for further investigation of its geographical extent (Chatters and Sanderson, 1994). Meanwhile, Evans and Barkham (1992), and especially Barkham (1992), have argued that more work should be undertaken because of the importance of grazing mammals with respect to the composition of the field-layer plant communities of coppice-woods.

This chapter outlines opportunities for consideration of grazed coppice via what is already known about the character and historical experience of less well-known examples in the midlands Forest of Rockingham.

Grazing in Rockingham Forest Woodlands

The contribution of grazing in this particular ancient Forest has been recognized by Peterken (1976), Rackham (1989) and – with regard to the detailed land use and ecological history of constituent woods – Peterken and Welch (1975), Best (1983) and Bellamy (1986). 'Grazed coppice' may be a characteristic but insufficiently appreciated feature of Rockingham past and present (Best, 1994). *Foresta Regis de Rockingham* was established during the twelfth century, upon potentially productive but intractable chalky boulder clay-mantled terrain. Remarkably, the 'legal Forest' persisted well into the nineteenth century, although much reduced in influence from the late sixteenth century. Today, the 'physical forest' is shrunken, fragmented and much-altered, but still represents a locus of ancient, semi-natural woodland, the majority of which was ostensibly managed for the greater part of the time, as coppice.

The woods are classified as W8 *Fraxinus excelsior–Acer campestre–Mercurialis perennis* with substantial representation of W22 *Prunus spinosa–Rubus fruticosus* and W21 *Crataegus monogyna–Hedera helix* scrub (Rodwell, 1991), or 'type 2A, 2Aa, Ab Ash – Maple', plus scrub 'type b' (Peterken, 1981), or 'type VI/VII Ash – Hazel – Maple' (Rackham, 1980). The old Forest woods are, however, subtly different from non-Forest woods of these kinds nearby, with (in comparison to the Purlieu Woods) fewer species; the woodland species are of more restricted occurrence; and those present are evidently those most tolerant of grazing (Rixon, 1975).

The situation is complicated by the fact that many of the Purlieu Woods are apparently primary, whereas many of the old Forest woods are of ancient-secondary origin (Rixon, 1975; Best, 1983; Bellamy, 1986). Nevertheless, there remains the possibility of a differential effect resulting from 800 years of livestock impact, where grazing is known to have been a major use of the Forest woods.

Coppice Grazing in the Medieval Forest System

For background, reference should be made to Young (1979) and particularly for Rockingham to Pettit (1968), Steane (1973, 1974), Best (1983, 1989) and Bellamy (1986). Maps from the 1580s show that a large proportion of

Rockingham Forest was structured and managed as a 'compartmented Forest'. There were spacious, enclosed lawns, reserved for support of the king's deer, extensive, unenclosed grassy plains, accompanying groups of woods – a proportion of which were, at any time, enclosed by temporary fences or hedges – and interstitial ridings, together with meadows, assarted pastures and open fields. Accompanying these were complex patterns of ownership, privileges, rights and claimed rights on the part of the various land-using interests alongside the crown, which were regulated and interpreted under the prescribed, regulatory framework of the Forest administration modified by the actual behaviour of its officials. The wooded tracts and non-wooded parts of the Forest were not separate, but an integrated whole.

Within the woods, apart from sheltering the venison, production of underwood was also expected. In addition, notwithstanding the supposedly limited resource value of the field-layer growth, this element of the Forest coppice-woods was evidently deemed to be of some value by potential users and regulators. Common grazing rights existed, held by inhabitants of the Forest townships, which predated establishment of the legal Forest. Any grass and herb production not required by the deer could be released by agistment, for the benefit of the local economy and exchequer coffers.

There are comparatively few documentary references to coppice grazing in Rockingham, although there is mention during both the thirteenth and fourteenth centuries. Better preserved are records of the formal regulations and procedures that were introduced to mitigate the resultant impact of grazing upon the woods. According to the *Act for inclosing of woods in the Forests, Chases and purlieus* of 1482, it was declared that,

> divers Subjects having Woods growing in their own ground within the Forest of Rockingham … might not before Time, cut nor inclose their said ground, to save the young Spring of their Wood so cut, any longer Time than for Three Years

and later

> the same young Spring hath been in Times past and daily is destroyed with beasts and Cattle of the same Forest … to the great Hindrance, as well of his said Subjects, as of his (Edward IV) Deer, Vert, and Venison in their covert, and otherwise to the likely destruction of the Same Forests …

Instead, it was proposed that henceforth,

> … it shall be lawful … immediately after the Wood so cut, to cope and inclose the same ground with sufficient Hedges, able to keep out all Manner of Beasts and Cattle forth of the same ground, for the preserving of their young Spring; and the same Hedges so made the said Subjects may keep them continuously by the Space of Seven Years after the same inclosing, and repair and sustain the same as often as shall need within the same Seven Years …

This provision for exclusion of cattle for seven years is also found in a Statute of 1543. Risk of damage was not thereby entirely removed though, for

> The proprietors of the underwood, in the forest woods, are empowered, by the ancient laws and customs of the forest, to fence in each part or sale, so soon as it is cut, and to keep it in band, as it is here termed, for seven years, except against

> the deer, which are let in at the expiration of four years … so that there are
> always seven parts, or sales, constantly in band, and in which the cattle of the
> commoners are not permitted at any time to de-pasture …
>
> (cited in Pitt, 1809)

Restrictions were placed on the kinds of beasts and the times of year when commoners might enjoy their privileges within the Forest coppices. Only cattle, horses and young pigs were permitted between 25 April and 11 November – the winter 'Heyning' preserving such forage as remained for the deer alone – but excluding 'fence month' when the deer were giving birth (notionally the two weeks before and after Midsummer Day). Mature pigs could only enter the woods between 14 September and 11 November to fatten on acorns and (unlikely in Rockingham) beech-mast, according to rights of pannage (Bellamy, 1986). In all events, the number of animals was not supposed to exceed what could be kept on other land when excluded from the forest woods.

Abuse and Neglect of the Forest Coppices

In practice, the intricate Forest system was poorly regulated, with many breaches and disputes between the various interests. Probably before, and certainly during and after the sixteenth century, the enclosed coppices suffered unwarranted entry by livestock, overstocking, and untimely pressure upon regenerating coppice and field-layer growth, over and above the persistent impact of the comparatively few deer. Entitled usage, agistment and commoning evidently exceeded originally intended limits, to the detriment of other village husbandry let alone the integrity of the underwood. By the middle of the sixteenth century it would appear that the Forest coppices were in a deplorable state, yielding little profit to the crown, with the commoners profiting more than the monarchy (Pettit, 1968). Contemporary accounts and surveys point to the limited extent of coppice in saleable condition, neglected woods cut sporadically, and many reduced to, or containing a high proportion of, thorns, as a consequence of grazing abuse (Peterken, 1976).

With piecemeal alienation of crown woods the situation became even worse as better maintained private underwood outsold the royal residue. By the seventeenth century the keepers were actively involved in taking advantage of the protracted demise of the Forest institution, putting all manner of beasts into their poorly fenced coppices. Seeing this, the local population followed suit, even claiming greater entitlement to do so. Agistment had been ousted by claims to common of pasture, regulations were often disregarded and cattle allowed without stint. Even sheep joined the numerous cattle 'by pretence of common' (Rackham, 1976). And so the situation continued into the eighteenth century, from which there is evidence of outrageous manipulation of the commoning system, and, not surprisingly, many complaints from private owners of the effects of both the deer and unfettered livestock upon their young trees and underwood (Wake and Webster, 1971).

'The whole of what are now considered to be forest woods are subject to the depasturage of the deer, and, at a stated time of the year, to the depasturage also of the cattle belonging to those who reside in the adjoining townships, and who claim to be possessed of a right of commonage on these accounts, the profit arising to the proprietors of these woods, from the cutting of the timber and underwood, is small, compared with that arising from regular well-managed purlieu woods, which are not subject to the annoyance of the deer and cattle'.

From the same source, with particular reference to Geddington Chase:
    'The injury sustained, by the deer being admitted into the young spring wood, in the first instance, is very considerable; but that injury is small indeed, when compared to the destructive havoc made by the devouring jaws of a herd of hungry cattle, admitted into the young coppice just as the leaves have begun to appear, and at a season of the year when it sometimes happens they have just survived a state of famine, the consequence of a want of sufficient fodder, in a hard and severe winter . . . and . . . are reduced to an extreme state of leanness and poverty at the time they are turned into the woods, when whole herds of them rush forward like a torrent, and every thing that is vegetable and within their reach inevitably falls a sacrifice to their voracious and devouring appetite'.

    'The Woods from various causes have much decreased in value. Amongst others, the increased price of labour, and the general use of coal for fuel, have had great effect - the early admission of the deer into the Woods - overstocking the commons with large beasts - putting horses into the Woods whilst inclosed - felling large quantities of timber on the underwood - cutting it in summer to draw the timber out of the woods, and the general depredations of the neighbouring Towns, are causes which have all materially contributed to bring them into their present state. So little demand is there now for underwood, composed of thorns chiefly, that Twenty Acres a year can scarcely be disposed of . . .'

Complaint from Sir Richard Brooke upon the occasion of being asked to pay a fee to Mr Hatton, the hereditary Keeper nay Warden of the Bailiwick of Rockingham, in Rockingham Forest, 1815.

Fig. 6.1. Complaints regarding the impact of grazing of coppice-woods in Rockingham Forest, late eighteenth/early nineteenth century (Pitt, 1809; Best, 1983).

From Forest to Recent Times

The descendants of Sir Christopher Hatton, recipient of many Forest properties and rights from Elizabeth I, upheld the archaic Forest with zeal, well into the nineteenth century. This caused further revealing complaints (see Fig. 6.1), and was commented upon unfavourably in the Report to the Board of Agriculture (Pitt, 1809; Best, 1983). Disputes delayed disafforestation and enclosure until the time when other values allowed at least some of the woods to escape conversion to pasture or tillage (Kirby, 1992). Many survived as game preserves (Edlin, 1970). Fox hunting and pheasants were by then well established, but it was the deliberate encouragement of rabbits, *Oryctolagus cuniculus*, for sport which promised to sustain grazing impact for another century. Rabbits were apparently uncommon and noteworthy in the early eighteenth century (Wake and Webster, 1971) but may have been kept in warrens, from which they later spread. Rabbits were specifically indicted as an additional reason for coppice failure at the beginning of the nineteenth century, and again a hundred years later (Nisbet, 1906; Best, 1983). Detailed records from one estate reveal that they were maintained at high density within a couple of its woods until the late 1940s (Best, 1983). It is only for the most recent half century, then, that some of the old Forest woods have been free of any kind of grazing impact.

Persistent Outcomes of Coppice Grazing

Persistent, heavy grazing can result in changes to the composition and relative abundance of species of trees, shrubs and plants of the woodland floor, with some being disadvantaged but others favoured (Barkham, 1992; Peterken, 1996; Sanderson, 1996). Preliminary work suggests that field maple, *Acer campestre*, common hawthorn, *Crataegus monogyna*, blackthorn, *Prunus spinosa*, crab apple, *Malus sylvestris* and elder, *Sambucus nigra*, tend to be more frequent in the grazed coppices whereas ash, *Fraxinus excelsior*, hazel, *Corylus avellana*, and small-leaved lime, *Tilia cordata*, are less frequent.

Acer campestre is selectively favoured by the preferential grazing of more palatable species. For Rackham (1980), it is 'unusually and significantly abundant in the coppices of those few Forests that are on calcareous soils' of which Rockingham is given as an example. *Crataegus monogyna* is seemingly more frequent in compartmented Forests, such as Hatfield (Rackham, 1980, 1989). The association of *Prunus spinosa* with old Forests has been noted with regard to the presence and larval dependence of the localized black hairstreak, *Strymonidia pruni*, butterfly (Marren, 1990). *Malus sylvestris* is understood to have been utilized as a source of valuable winter food for deer, and *C. monogyna*, *P. spinosa* and *M. sylvestris* were identified in the Forest of Epping, as not to be cut from the vert, because deer feed on them (Rackham, 1976). The presence of *Sambucus nigra* in abnormal frequency and abundance in King's Wood might, in part, be attributed to the activities of rabbits there, as it is one of the few woody species they avoid (Sumption and Flowerdew, 1988).

Large, above-ground coppice stools of ash, *Fraxinus excelsior,* appear to be markedly less frequent in grazed woods although there are many young maiden trees and saplings. Ash is preferentially grazed by deer and other livestock, but is known to return following the exclusion of rabbits (Rackham, 1980). In woods that have been grazed, the frequency of palatable *Corylus* is lower and its distribution patchy. In some Rockingham woods spared the attention of modern deer, there is currently regeneration of new hazel plants. The status of *Tilia* is less clearly related to grazing because it is confined to the most likely primary woodland sites in Rockingham and so might never have occurred in the post-Roman woods that were later incorporated within the regime of the medieval Forest. Peterken (1976) suggests a change in Rockingham Forest since the sixteenth century, with an increase in *Corylus* and *Fraxinus*, a marked decrease in 'thorn', and a decline in sallow, *Salix caprea/S. cinerea*, and oak, *Quercus robur*. This could be interpreted as a progressive response to the gradual release from grazing pressure.

Field-layer species that are apparently more frequent in grazed coppices include tufted hair grass, *Deschampsia cespitosa,* wood sorrel, *Oxalis acetosella*, sanicle, *Sanicula europaea*, barren strawberry, *Potentilla sterilis*, burdock, *Arctium minus* and ground ivy, *Glechoma hederacea*. Those species apparently absent or less frequent in grazed coppices include herb Paris, *Paris quadrifolia*, wood speedwell, *Veronica montana*, broad-leaved helleborine, *Epipactis helleborine*, greater butterfly orchid, *Platanthera chlorantha*, bird's-nest orchid, *Neottia nidus-avis*, pignut, *Conopodium majus*, wood anemone, *Anemone nemorosa* and bluebell, *Hyacinthoides non-scripta*.

With allowance for soil character, the pattern of field-layer species in coppice may be determined by four variables: (i) frequency, scale and type of canopy disturbance; (ii) length of time taken for canopy closure following large-scale disturbance; (iii) demography of the species; and (iv) grazing (Barkham, 1992). In the case of Rockingham, a determinant of species absence could be the secondary origin of some of the woods concerned, a number of 'primary' woodland indicators appearing in the list above. *Anemone* and *Hyacinthoides* are present, the latter sometimes abundant, but displaying a decidedly patchy distribution within the woods. Both are sensitive to, and suppressed by, grazing (Sanderson, 1996). This, together with reduction of the other dominant, dogs mercury (*Mercurialis perennis*), allows such species as *Sanicula europaea* to increase in frequency (Chatters and Sanderson, 1994; Sanderson, 1996).

The old Forest coppices are distinctively 'grassy', with *Deschampsia* frequent throughout. Favoured by local soil conditions (with a tendency to seasonal waterlogging) it could well have been encouraged by grazing in the past. Although not appetizing, it can be eaten by deer, and it has been suggested it was probably the chief component of the *herbagium* of woods (Rackham, 1976). Livestock treading might also have resulted in damage to the soil structure that would increase its competitive advantage. Defence mechanisms as well as palatability may be important (Putman, 1994b), which could account for the otherwise abnormal frequency of *Oxalis* in these woods. Other small herbs such as *Potentilla sterilis,* which are early developing, but of limited bulk, may be intrinsically well adapted to avoid grazing pressure.

The reinstatement of coppice management in one of the woods that is now a Local Nature Reserve has revealed an interesting emergence of species from the seed bank (Brown and Warr, 1992). As evidence of past livestock impact, the observed high frequency and abundance now of *Arctium* and *Glechoma* is in accordance with the erstwhile presence of rabbits there. Both species are disliked and avoided, and decline in the frequency of *Glechoma* following the impact of myxomatosis upon rabbits has been recorded elsewhere (Sumption and Flowerdew, 1988). The status and dynamics of bramble, *Rubus fruticosus*, warrants further scrutiny, now reappearing as a result of conservation coppicing, but known in other circumstances to be avidly browsed by deer, horses and cattle.

Other Persistent Outcomes of the Forest Grazing System

Chatters and Sanderson (1994) have called for reappraisal of the character and conservation value of both the woodland and grassland elements of grazed commons because their studies contradict the widespread assumption that grazing always results in impoverishment. For Rockingham Forest, the distinctive character of associated grasslands, particularly the old lawns and ridings, has been acknowledged (Peterken and Harding, 1974; Peterken, 1976), the latter noting that such grasslands were floristically rich and that this was an effect of long-sustained grazing within the Forest. Of the once extensive array of lawns, plains and ridings, most have been lost to cultivation, afforestation or improvement, with few instances remaining. However, Peterken (1976) conjectured that fragments could still survive in later modified, open habitats, of which a major example has recently been discovered in the unlikely setting of Corby New Town (Best and Logue, 1991). This includes a small but characteristic tract of MG4 *Alopecurus pratensis–Sanguisorba officinalis* (Rodwell, 1992) which is almost certainly an *in situ* remnant of old Forest hay meadow. Other swards nearby include a mosaic of dry–wet, calcareous–mildly acid grassland communities; representation of a number of old woodland species, including goldilocks, *Ranunculus auricomus*, barren strawberry, *Potentilla sterilis*, bluebell, *Hyacinthoides non-scripta*, and wood sedge, *Carex sylvatica*, growing in open swards; a number of county rarities once characteristic of unimproved pastures and meadows including oval sedge, *Carex ovalis*, common sedge, *Carex nigra* and adder's tongue, *Ophioglossum vulgatum*; plus notable representation of three species, betony, *Stachys officinalis*, lady's mantle, *Alchemilla vulgaris* and devil's bit scabious, *Succisa pratensis*, that were deemed, over 50 years ago, to be characteristically associated with old Forest ridings (Druce, 1930). Interestingly, in this particular context, such grassland appears to have increased in extent, at the expense of neighbouring old Forest coppice-woodland in recent years, as a consequence of conversion to urban parkland. The relict grassland communities now flourish once more thanks to a surrogate regime of low intensity grass cutting.

Discussion

Many more issues are raised than resolved by this foray into the relationships between grazing and the character of Rockingham Forest woodlands. There are considerable difficulties in isolating the effects of past grazing from the other processes and dynamics of woodland (Best, 1983; Barkham, 1992; Sanderson, 1996). There are few directly comparable instances where grazing can be observed today while the responses to release from grazing may mask previous impacts. Past grazing effects must be separated out from exploitation for underwood, physical fragmentation of the sites, replanting, selective abstraction of timber, a recent episode of neglect, and the reappearance of unregulated deer numbers. Grazing impact is likely to have varied from place to place and time to time, according to particular combinations of animals and sites. However, the requisite details of the kinds, numbers, husbandry and performance in particular places and years, of the grazing/browsing ruminants and non-ruminants involved, as well as the extent and capability of the precise grounds where grazing occurred, are almost entirely lacking for all but the modern period. Bellamy (1986) furnishes a rare set of such information, derived from early eighteenth century Drift accounts. Many more histories of individual woods are called for, to test Forest-wide generalizations, bringing together the historic profile of grazing and appropriate information about the subsequent structure and species dispositions within the site. Whilst there is the possibility of persistence of any vestige of grazing artefact, mapping of underwood relics and field-layer species mosaics ought to be undertaken as an urgent priority.

Other ancient Forests need to be examined for equivalent opportunities for investigating the consequences of grazing impact through time on a much broader basis, to determine whether there are universally shared characteristics and any consistent spatial variations in terms, for example, of base richness of the terrain across the total spectrum of Forests. It has been said that Rockingham is atypical in this sense (Peterken, 1976), but insufficient is known of the conditions of other Forests to be sure about this.

References

Barkham, J.P. (1992) The effects of coppicing and neglect on the performance of the perennial ground flora. In: Buckley, G.P. (ed.) *Ecology and management of coppice woodlands*. Chapman & Hall, London, pp. 115–196.

Bellamy, B. (1986) *Geddington Chase – the history of a wood*. Bellamy, Geddington.

Best, J.A. (1983) *King's Wood, Corby – description, history, explanation of habitats and wildlife*. Nene College, Northampton.

Best, J.A. (1989) The changing vegetation. In: Colston, A. and Perring, F. (eds) *The nature of Northamptonshire*. Barracuda, Buckingham, pp. 40–50.

Best, J.A. (1994) *The past, present and future of Rockingham Forest*. The Wildlife Trust for Northamptonshire, Northampton.

Best, J.A. and Logue, W.L. (1991) *The grasslands of Hazel and Thoroughsale Woods, Corby*. The Northamptonshire Wildlife Trust, Corby.

Brown, A.H.F. and Warr, S.J. (1992) The effects of changing management on seed banks in ancient coppices. In: Buckley, G.P. (ed.) *Ecology and management of coppice woodlands*. Chapman & Hall, London, pp. 147–166.

Chatters, C. and Sanderson, N. (1994) Grazing lowland pasture woods. *British Wildlife* 6, 78–88.

Cooke, A.S. (1994) Colonisation by muntjac deer and their impact on vegetation. In: Massey, M. and Welch, R.C. (eds) *Monks Wood National Nature Reserve: the experience of forty years 1953–1993*. English Nature, Peterborough, pp. 45–61.

Druce, G.C. (1930) *The flora of Northamptonshire*. T. Buncle & Co, Arbroath.

Edlin, H.L. (1970) *Trees, woods and man*, 3rd edn. Collins, London.

Evans, M.N. and Barkham, J.P. (1992) Coppicing and natural disturbance in temperate woodlands – a review. In: Buckley, G.P. (ed.) *Ecology and management of coppice woodlands*. Chapman & Hall, London, pp. 79–98.

Flower, N. (1980) The management history and structure of unenclosed woods in the New Forest, Hampshire. *Journal of Biogeography* 7, 311–328.

Kay, S. (1993) Factors affecting severity of deer browsing damage within coppiced woodlands in the south of England. *Biological Conservation* 63, 217–222.

Kirby, K. (1992) *Woodland and wildlife*. Whittet, London.

Marren, P. (1990) *The wild woods*. David & Charles, Newton Abbot.

Mitchell, F.J.G. and Kirby, K.J. (1990) The impact of large herbivores on the conservation of semi-natural woods in the British uplands. *Forestry* 63, 333–353.

Nisbet, J. (1906) Forestry. In: Serjeantson, R.M. and Adkins, W.R.D. (eds) *The Victorian History of the counties of England – Northamptonshire II*. Constable, Westminster, pp. 341–352.

Peterken, G.F. (1976) Long-term changes in the woodlands of Rockingham Forest and other areas. *Journal of Ecology* 64, 123–246.

Peterken, G.F. (1981) *Woodland conservation and management.* Chapman & Hall, London.

Peterken, G.F. (1996) *Natural woodland – ecology and conservation in northern temperate regions.* Cambridge University Press, Cambridge.

Peterken, G.F. and Harding, P.T. (1974) Recent changes in the conservation value of woodlands in Rockingham Forest. *Forestry* 47, 109–128.

Peterken, C.P. and Tubbs, C.R. (1965) Woodland regeneration in the New Forest, Hampshire, since 1650. *Journal of Applied Ecology* 2, 159–170.

Peterken, G.F. and Welch, R.C. (eds) (1975) *Bedford Purlieus: its history, ecology and management.* Monks Wood Experimental Station Symposium 7, Huntingdon.

Pettit, P.A.J. (1968) *The royal forests of Northamptonshire: a study of the economy 1558–1714.* Northamptonshire Record Society, 23, Northampton.

Pitt, W. (1809) *General view of the agriculture of the County of Northampton.* Richard Phillips, Bridge Street, London.

Putnam, R.J. (1986) *Grazing in temperate ecosystems: large herbivores and their effects on the ecology of the New Forest.* Croom Helm/Chapman & Hall, London.

Putnam, R.J. (1994a) Severity of damage by deer in coppice woodlands: an analysis of factors affecting damage and options for management. *Quarterly Journal of Forestry* 88, 45–54.

Putnam, R.J. (1994b) Effects of grazing and browsing by mammals on woodlands. *British Wildlife* 5, 205–213.

Rackham, O. (1976) *Trees and woodland in the British landscape.* Dent, London.

Rackham, O. (1980) *Ancient woodland.* Arnold, London.

Rackham, O. (1989) *The last forest. The story of Hatfield Forest.* Dent, London.

Ratcliffe, P.R. (1992) The interaction of deer and vegetation in coppice woods. In: Buckley, G.P. (ed.) *Ecology and management of coppice woodlands.* Chapman & Hall, London, pp. 233–246.

Rixon, P. (1975) History and former woodland management. In: Peterken, G.F. and Welch, R.C. (eds) *Bedford Purlieus: its history, ecology and management.* Monks Wood Experimental Station Symposium, 7, Huntingdon, pp. 15–38.

Rodwell, J.S. (1991) *British plant communities Vol. 1 Woodlands and scrub.* Cambridge University Press, Cambridge.

Rodwell, J.S. (1992) *British plant communities Vol. 3 Grasslands and montane vegetation.* Cambridge University Press, Cambridge.

Sanderson, N. (1996) The role of grazing in the ecology of lowland pasture woodlands with special reference to the New Forest. In: Read, H.J. (ed.) *Pollard and veteran tree management II.* Corporation of London, Burnham Beeches, pp. 111–117.

Steane, J.M. (1973) The forests of Northamptonshire in the early Middle Ages. *Northamptonshire Past and Present* 5, 7–17.

Steane, J.M. (1974) *The Northamptonshire landscape.* Hodder & Stoughton, London.

Sumption, K.J. and Flowerdew, J.R. (1988) The ecological effects of the decline in rabbits *Oryctolagus cuniculus* due to myxomatosis. *Mammal Review* 5, 151–186.

Tubbs, C.R. (1968) *The New Forest: an ecological history.* David & Charles, Newton Abbot.

Tubbs, C.R. (1986) *The New Forest.* Collins, London.

Wake, J. and Webster, D.C. (eds) (1971) *The letters of Daniel Eaton to the Third Earl of Cardigan, 1725–1732.* Northamptonshire Record Society, Northampton.

Young, C.R. (1979) *The Royal Forests of Medieval England.* Leicester University Press, Leicester.

7

The Links between Forest History and Biodiversity: the Invertebrate Fauna of Ancient Pasture-woodlands in Britain and its Conservation

K.N.A. Alexander

National Trust, 33 Sheep Street, Cirencester, Gloucestershire GL7 1RQ, UK

Relict areas of ancient pasture-woodland have been found to support relatively rich assemblages of species of invertebrate, fungi and lichen which have a strong association with old forest conditions in some parts of their European ranges. This identified link between forest history and biodiversity appears to depend on historical continuity of the post-mature generations of tree species. Older trees have been selectively removed over much of the British Isles for a variety of reasons: the trees are 'past their best', they are dangerous, they harbour pathogens, and so forth. Nevertheless, the UK currently holds possibly the largest reserves of old tree habitat in northwest Europe. This chapter will concentrate on saproxylic invertebrates and review the present state of knowledge of these wood-decay communities and their conservation management needs.

Introduction

The association of many species with ancient forests and woodland is well known (e.g. Boycott, 1934; Peterken, 1974; Rose, 1976, 1993; Kerney and Stubbs, 1980; Harding and Rose, 1986; Stubbs and Falk, 1987; Speight, 1989; Alexander, 1996). In this Chapter I will draw on experience with the dead wood invertebrates of old trees, especially Coleoptera.

Ancient woodland – in the conventional narrow sense of dense stands of trees – has attracted considerable interest amongst British ecologists (e.g. Peterken, 1981; Spencer and Kirby, 1992) but other types of ancient wooded countryside such as pasture-woodland have been more neglected (Harding and Rose, 1986; Rackham, 1986).

Types of Pasture-woodland

Pasture-woodlands are types of land management regime which combine grazing livestock with growing trees (Rackham, 1996). They encompass a whole spectrum of landscape features, from conventional woods, where live-stock have access for part of the year at least, down to individual trees within pastures. No particular density of trees is implied. British examples come in a wide variety of categories:

- medieval forests, as at Hatfield Forest in Essex (Rackham, 1989);
- medieval and later deer parks, as at Whiddon Park in Devon;
- landscape parks with rough grazing and deer, as at Petworth Park in Sussex;
- landscape parks with more conventional livestock grazing, as at Kedleston Park in Derbyshire;
- old wooded commons, as at Felbrigg Beeches in Norfolk;
- rough hillside pastures with trees, as in the Cotswold Hills;
- pollard willows and other trees along watercourses;
- long-established orchards;
- field and hedgerow trees.

The last three are often not included as true pasture-woodlands in the formal sense, but as they combine trees and pastures they are effectively the same so far as the associated plant and animal communities are concerned. What is the difference between a historic park on the one hand and an area of farm-land where the hedges or fences have been removed leaving only the former hedgerow trees? The two land use patterns have been switched from one to the other in many cases, as at Ickworth Park in Suffolk (Rackham, 1986).

The Range of Relict Old Forest Invertebrates

A wide diversity of invertebrates have been recognized as characteristic of relict old forest in the UK. Research emphasis has mainly focused on the particularly diverse communities associated with decaying timber (Harding and Rose, 1986; Kirby and Drake, 1993), but many characteristic species exist amongst the epiphytic communities (not just the lichens and bryophytes themselves), in the canopy foliage of the trees, in damp shady situations such as the leaf litter, on the ground flora, and probably elsewhere. These have been reviewed elsewhere (Kirby *et al.*, 1995; Alexander, 1996).

Relationship between History and Species-richness of Deadwood Fauna

The relationship between history and species-richness is conventionally demonstrated by the close association between the richer sites and those with well-documented histories, especially in the cases of surviving areas of forest with medieval and perhaps earlier origins, and in the medieval deer parks (Rose, 1976; Harding and Rose, 1986; Rackham, 1986; Harding and Alexander, 1994). Historical documentation often fails us, however, and individual sites which are rich in 'relict old forest' species may not be well documented. The more informal pasture-woodland systems are very much a problem in this respect as the land management system was often not described in the same way as forests and deer parks.

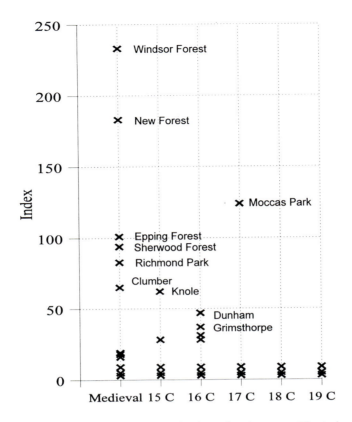

Fig. 7.1. Forest history and site quality for deadwood Coleoptera. (The index is described in Alexander (1988) and Harding and Alexander (1994).)

In Fig. 7.1 the richest sites are compared with the available documentation of their historical development as pasture-woodlands. Moccas Park (Herefordshire) is very prominent as badly fitting the pattern. This reflects poor documentation of the early history of the site rather than an exception to the general rule that older sites are richer in old forest species. As at Ickworth Park in Suffolk, the historical development of pasture-woodland may be complex. Calke Park in Derbyshire was developed in the 1600s from an area which included a block of ancient woodland and some pasture-woodland in the form of old common land. It now exhibits two very different landscape types: the old wooded area now has the appearance of an old deer park, with ancient trees, bracken and unimproved pastures, while other areas, developed from adjacent conventional field systems, combine younger trees with short-sward, herb-poor pastures (Table 7.1).

If we examine the richness of different types of wooded landscape for dead wood Coleoptera, a general pattern emerges; the richest sites are the more intact areas of relict old forest while – of the wooded habitats – the old coppices (the conventional view of woodland) are the least rich in these

Table 7.1. The development history of the area now Calke Park SSSI, Derbyshire, showing how ancient woodland and pasture-woodland were incorporated into a historic park.

c. 1150	Calke landscape referred to as 'the wood' in which the inhabitants of Calke Monastery live
1590	Roche Wood and Bowley Wood (ancient woodlands)
	Castle Close (pasture-woodland?)
	Derby Hills Common (common land including pasture-woodland trees)
1639–1666	present park created by enclosing areas listed above
1712–1713	some landscaping – avenues planted
1760–1830	area of park enlarged

communities (Table 7.2). Using Suffolk sites as an example, the medieval deer park of Staverton (Peterken, 1969) is particularly rich in dead wood beetles, followed by the rough grazed, wooded, heathy pastures of Icklingham Plains, while the classic ancient woodland of Bradfield Woods is notably species-poor (Table 7.3).

There is a need for greater recognition of ancient and over-mature trees as the major reservoirs of these particular old forest communities rather than historic woodland, old forests or ancient deer parks *per se*. A long and unbroken history of over-mature generations of trees is the key feature involved in the survival of relict old forest communities. While historical site documentation helps us to understand how the species in question have persisted, it may be misleading in situations where the land management was informal and so not documented or where documentation has been lost over time. The conservation of the communities of interest needs to focus on the trees, not the historic sites.

The discovery of a large area of southern Worcestershire which supports an extremely rich dead wood fauna in what appears to be fairly ordinary

Table 7.2. Types of situation with continuity of old timber, ordered by species richness of the associated deadwood communities.

Situation	Examples of sites
Old Forests and chases	Windsor, Sherwood, Epping
Historic parks	Moccas, Richmond and Staverton
Wooded commons	Burnham Beeches, Ashtead Common, Wytham Woods
Old rough pastures with trees:	Fens such as Woodwalton and Chippenham
winter fodder situations	Hill country: Cotswold scarp, e.g. Bredon Hill NNR
	Lake District valley sides
	Breck-Icklingham and Stanford
River floodplains: willow and other pollards	Upper Thames
Ancient woodlands	Monks Wood NNR, Cirencester Park Woods

Table 7.3. Beetle index values for a range of sites in Suffolk.

Site name	Type of pasture-woodland	Beetle index
Staverton Park	medieval deer park	49
Icklingham Plains	informal rough pasture-woodland	34
Shrubland Park	landscape park	27
Ickworth Park	landscape park with mixed history involving periods when reverted to conventional farmland	6
Bradfield Woods	ancient semi-natural woodland	4

The beetle index is described in Alexander (1988) and Harding and Alexander (1994).

countryside is an exciting recent development (Whitehead, 1996) and is shaking established views of the sort of places these species require and their mobility in modern farmed landscapes.

Studying the Ecology of Decaying Wood

There have been considerable advances in our understanding of the ecology of wood-decay communities in recent years, especially the role of fungi – see particularly Rayner and Boddy (1988), Rayner (1993) and Boddy (1994) – and of tree mechanics (Mattheck and Breloer, 1994). Improved methodologies for investigating the fauna have also been developed (Hammond and Harding, 1991), especially the invention of the Owen extraction trap (Owen, 1989, 1992; Alexander, 1994). This knowledge has even been applied in an attempt to recreate the breeding conditions for particular species or guilds, notably the rebuilding of a tree as a suitable breeding site for the rare and endangered click beetle, *Limoniscus violaceus* (Müller, P.W.J.) (Green, 1995).

Conservation Issues

The older generations of trees are essential in order to maintain a species-rich wood-decay fauna, and yet it is these very trees which have been selectively removed over much of the UK and Europe for a variety of reasons:

- Timber and wood production involves the removal of trees before they begin to develop decay.
- Trees within pastures have been removed as they shade the sward and reduce productivity.
- Older trees are 'past their best'.
- Old trees 'harbour pathogens', although Winter (1993) has demonstrated how little truth there is in this commonly heard statement.
- Old trees are 'dangerous'. Ancient trees may offer fewer hazards than trees in the transitional stages between young trees and ancient, as they have generally already lost their major boughs and the trunk is short and squat, with a low centre of gravity. Nevertheless, limbs are shed, and for a variety of reasons, and whole trees may collapse or be windthrown.

Mattheck and Breloer (1994) have recently produced an excellent review of assessing tree failure and this should help to reduce the incidence of unnecessary fellings.

Because we now have so few old trees, active intervention may be necessary in order to prolong their lives and to provide time for new generations to be brought on. Crown reduction work may be needed to reduce the danger from windthrow or collapse leading to premature death. Action may need to be taken to reduce losses through neglect or ignorance. Car-parking and mature trees, for example, do not go well together; the combination leads to tree decline from root compaction, and boughs may be cut off due to perceptions of danger to cars and motorists, which all combine to remove the actual trees which were the attraction for the parking in the first place. Grazing animals at commercial levels and trees do not readily coexist; bark-stripping can be a severe problem leading to tree decline and premature death.

The detail of the conservation needs of populations of over-mature trees has recently been reviewed by Alexander *et al.* (1996). Species-rich communities are now rare and fragmented. There may now be little or no contact between populations and this must ultimately be a serious loss. There is an urgent need to promote older trees as desirable everywhere, not just in 'old tree museums' - the historic parks and old forest relics. There is a need to protect all examples wherever they are, to promote the retention of younger generations into old age, and to fill the gaps within today's fragmented forest landscape by new plantings. New plantings should consider the long-term need for varied age structures.

Conclusions

Old forest invertebrate species are not all inherently immobile. Their mobility may be relatively low but it is not zero. Today's reservoirs of these species are the results of historical accidents. We must find ways of allowing them to spread so that they can become part of 'everyday nature' again. They are part of natural nutrient cycling, breaking down dead wood, recycling materials and releasing them for other organisms to use. History has passed these communities on to us through pasture-woodland management systems. What about the future? What is the best way of keeping them? Maintaining old-style pasture-woodland systems is the obvious approach and one which is being promoted to a considerable degree in the UK at the moment, but, as was stressed earlier, it is the old trees, not the sites or the system of management, that are important.

Their future lies in a wider acceptance of old trees as a desirable feature of the countryside in general, not just in special sites; of significance to everyone, not just ecologists. To this end an Ancient Tree Forum has been established in the UK, to promote the conservation of ancient trees, and the statutory nature conservation agency in England, English Nature, has launched a Veteran Tree Initiative in partnership with English Heritage (the statutory agency for the cultural aspects of the countryside), the National

Trust (a major private charity which owns land for conservation objectives), the Ancient Tree Forum and others, to actively pursue many of the ideas presented here.

References

Alexander, K.N.A. (1988) The development of an index of ecological continuity for deadwood associated beetles. *Antenna* 12, 69–70.

Alexander, K.N.A. (1994) The use of freshly downed timber by insects following the 1987 storm. In: Kirby, K.J. and Buckley, G.P. (eds) *Ecological responses to the 1987 Great Storm in the woods of south-east England*. English Nature (Science Report 23), Peterborough, pp. 134–150.

Alexander, K.N.A. (1996) The value of invertebrates as indicators of ancient woodland and especially pasture woodland. *Transactions of the Suffolk Naturalists Society* 32, 129–137.

Alexander, K.N.A., Green, E.E. and Key, R. (1996) The management of over mature tree populations for nature conservation – the basic guidelines. In: Read, H.J. (ed.) *Pollard and veteran tree management II*. Corporation of London, Burnham Beeches, pp. 122–135.

Boddy, L. (1994) Wood decomposition and the role of fungi: implications for woodland conservation and amenity tree management. In: Spencer, J. and Feest, A. (eds) *The rehabilitation of storm damaged woods*. University of Bristol, Bristol, pp. 7–30.

Boycott, A.E. (1934) The habitats of land mollusca in Britain. *Journal of Ecology* 22, 1–138.

Green, E.E. (1995) Creating decaying trees. *British Wildlife* 6, 310–311.

Hammond, P.M. and Harding, P.T. (1991) Saproxylic invertebrate assemblages in British woodlands: their conservation significance and its evaluation. In: Read, H.J. (ed.) *Pollard and veteran tree management*. Corporation of London, Burnham Beeches, pp. 30–37.

Harding, P.T. and Alexander, K.N.A. (1994) The use of saproxylic invertebrates in the selection and evaluation of areas of relic forest in pasture-woodlands. *British Journal of Entomology and Natural History* 7 (Suppl. 1), 21–26.

Harding, P.T. and Rose, F. (1986) *Pasture-woodlands in lowland Britain: a review of their importance for wildlife conservation*. Institute of Terrestrial Ecology (NERC), Huntingdon.

Kerney, M. and Stubbs, A. (1980) *The conservation of snails, slugs and freshwater mussels*. Nature Conservancy Council, Shrewsbury.

Kirby, K.J. and Drake, C.M. (1993) *Deadwood matters: the ecology and conservation of saproxylic invertebrates in Britain*. English Nature (Science Report 7), Peterborough.

Kirby, K.J., Thomas, R.C., Key, R.S., McLean, I.F.G. and Hodgetts, N.G. (1995) Pasture-woodland and its conservation in Britain. *Biological Journal of the Linnaean Society* 56 (suppl.), 135–153.

Mattheck, C. and Breloer, H. (1994) *The body language of trees. A handbook for failure analysis*. HMSO, London.

Owen, J.A. (1989) An emergence trap for insects breeding in dead wood. *British Journal of Entomology and Natural History* 2, 65–67.

Owen, J.A. (1992) Experience with an emergence trap for insects breeding in dead wood. *British Journal of Entomology and Natural History* 5, 17–20.

Peterken, G.F. (1969) Development of vegetation in Staverton Park, Suffolk. *Field Studies* 3, 1–39.

Peterken, G.F. (1974) A method for assessing woodland flora for conservation using indicator species. *Biological Conservation* 6, 239–245.

Peterken, G.F. (1981) *Woodland conservation and management.* Chapman & Hall, London.

Rackham, O. (1986) *The history of the countryside.* Dent, London.

Rackham, O. (1989) *The last forest: the story of Hatfield Forest.* Dent, London.

Rackham, O. (1996) History of woodland and wood-pasture. *Transactions of the Suffolk Naturalists Society* 32, 116–128.

Rayner, A.D.M. (1993) The fundamental importance of fungi in woodlands. *British Wildlife* 4, 205–215.

Rayner, A.D.M. and Boddy, L. (1988) *Fungal decomposition of wood – its biology and ecology.* Wiley, London.

Rose, F. (1976) Lichenological indicators of age and environmental continuity in woodlands. In: Brown, D.H., Hawksworth, D.L. and Bailey, R.H. (eds) *Lichenology: progress and problems.* Academic Press, London, pp. 279–307.

Rose, F. (1993) Ancient British woodlands and their epiphytes. *British Wildlife* 5, 83–93.

Speight, M.C.D. (1989) *Saproxylic invertebrates and their conservation.* Nature & Environment Series No. 42. Council of Europe, Strasbourg.

Spencer, J.W. and Kirby, K.J. (1992) An inventory of ancient woodland for England and Wales. *Biological Conservation* 62, 77–93.

Stubbs, A.E. and Falk, S. (1987) Hoverflies as indicator species. *Sorby Record* (Special Series) 6, 46–49.

Whitehead, P.F. (1996) The notable arboreal Coleoptera of Bredon Hill, Worcestershire, England. *The Coleopterist* 5, 45–53.

Winter, T. (1993) Deadwood – is it a threat to commercial forestry? In: Kirby, K.J. and Drake, C.M. (eds) *Deadwood matters: the ecology and conservation of saproxylic invertebrates in Britain.* English Nature (Science Report 7), Peterborough, pp. 58–73.

8

Interactions between Humans and Woodland in Prehistoric and Medieval Drenthe (the Netherlands): an Interdisciplinary Approach

Theo Spek

DLO Winand Staring Centre, PO Box 125, 6700 AC, Wageningen, the Netherlands

Deforestation on the sandy soils of the Dutch province of Drenthe was studied through a combination of archaeological, historical, palaeoecological and pedological data. The investigations showed that the spatial distribution of deforestation was closely connected to soil distribution and prehistoric settlement history. Poor coversand soils were densely inhabited in the Neolithic Period and the Bronze Age and as a result of this largely deforested. On the richer soils of the boulder clay plateaux deforestation did not start earlier than the Early and Middle Iron Age. In this period human habitation moved from the coversand landscape to the boulder clay landscape, resulting in a large-scale transposition of centre and periphery on a local scale. Deforestation at the boulder clay plateaux accelerated with the large-scale reclamations of the High and Late Medieval Period. The small area of woodland that was left received its final blow between AD 1450 and 1650 when the Drenthian rural economy specialized in wool and meat production and large herds of sheep and oxen resulted in a grazing density that largely exceeded the carrying capacity of the natural vegetation.

Introduction

The Dutch landscape is especially famous in foreign countries for its wetness and flatness. According to many traditional textbooks, we – the Dutch – have been plodding along for ages through a flat and low-lying landscape, full of peat, clay and mud, fighting against the water in order to find a toilsome existence. The major activities of the Dutch are supposed to have been digging canals and ditches, constructing dikes and building windmills. 'God created the earth, except for Holland, which was created by the Dutch themselves' is an age-old saying. This chapter deals with a less conspicuous activity of our Dutch ancestors: the devastation of thousands of square kilometres of woodland. By the end of the nineteenth century the Netherlands

Fig. 8.1. The province of Drenthe in the northeastern part of the Netherlands.

was by far the most sparsely wooded country of the continent. Only 1% of the Netherlands was covered by woodland in the year 1900. The Dutch historical landscape is thus a very open one but, between 1900 and the present, the area of woodland has increased to about 10%.

This deforestation process occurred in the course of prehistory and history in the Pleistocene sandy landscape in the northeastern part of the country, the Province of Drenthe (Fig. 8.1) (Spek, 1993, 1997). In the past centuries, the cultural landscape of this province was dominated by open arable fields, open stream valleys and wide, open heathlands (Fig. 8.2). Between the vast primeval forest of the Atlantic Period (8000–5000 years ago) and the open plains of the late nineteenth century landscape lies a long process of deforestation. When, where and how intensively did deforestation take place?

Methods

The lack of written sources

Hardly any ancient woodland has been preserved in the study area. All that is left of the original woodland are a few, strongly modified patches. These last remnants were useful to our research, but for the vast majority of the woodland that had been lost in the course of time the only information was in written archives and in the soil.

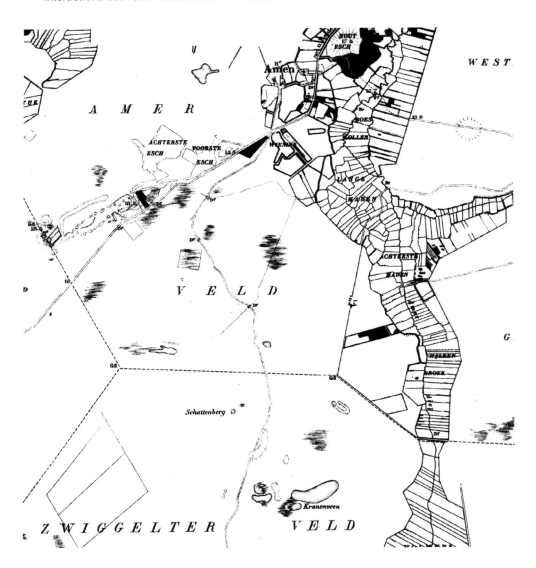

Fig. 8.2. Around 1900 the Drenthian landscape was totally dominated by open heathland, open arable and open grassland. This topographical map from the year 1900 shows the outlines of this nineteenth century landscape. At the top you see a small hamlet: Amer. The open spaces around it mark the medieval open fields. The meadows along the brook have a characteristic strip parcellation. By far the larger part of the landscape was covered by heathland (Amer Veld, Zwiggelter Veld, Grolloer Veld). Woodland is hardly found on this map.

Although the study of medieval archives might be very useful in many areas of Europe, Drenthe was a very remote, poor and marginal area that was not very attractive to the medieval and early modern elite and, therefore, very little was written about landscape management (including about woodland). The exception to this rule is the large number of historical field and place names containing woodland elements. For example, seventeenth century tax maps of my study area appeared to contain many field names with woodland elements. The most frequent are those names ending in *-loo, -holt, -wold* and *-laar*, but there are many others. A thorough mapping of these names could give a useful insight into the distribution of ancient woodland in the medieval period.

The interdisciplinary solution

Even where available, written sources only tell part of the story. For a good view of the prehistoric and historical deforestation process we have to use other methods, especially those that reveal the information that is hidden in the soil. Therefore, we developed an interdisciplinary working approach, involving four scientific disciplines: (i) historical geography; (ii) soil science; (iii) palaeoecology; and (iv) archaeology. During the last few years we have tried out our methods in a study of prehistoric and medieval field systems (Van Smeerdijk *et al.,* 1995) and on the reconstruction of the landscape dynamics of a Dutch wetland area in Mesolithic times (Spek *et al.*, 1997).

Historical geography

Although there are few medieval written sources dealing primarily with woodland and no maps, there are medieval village bylaws, legal and lease contracts that provide some insight into the woodland management of the period. Extrapolation back from seventeenth century tax maps can also give information about the situation and topographic relationships of medieval woodland. Field names may also be suggestive.

Soil science

Our soil research includes extensive testing of soil profiles in the field by boring to determine the spatial distribution of old woodland soils, as well as pedogenetic research on several characteristic woodland soils. An important tool in this pedogenetic research was micromorphology, the microscopic study of thin soil sections (Jongerius and Heintzberger, 1975). During the field work metal boxes were pressed into a profile wall of an ancient woodland soil in order to get an undisturbed soil sample. The boxes were taken to the laboratory and moulded in a polyester resin. Thus, we could 'petrify' a piece of woodland soil. Then, a very thin slice of 25 µm is cut off for microscopical observation. From these thin sections, specialists can derive a lot of information on humus ecology and former biological activity in the ancient woodland soil. The investigation of former root channels, faunal channels and excreta from various depths can help us to gain insight into prehistoric and historical soil formation and woodland ecology, especially if this information is combined with palaeoecological research data.

Palaeoecology

Palaeoecological techniques using pollen analysis, charcoal determination and radio-carbon dating form a third part of our project. During the last few decades several long pollen sections have been analysed from lake sediments and raised bogs in our study area (Casparie, 1972; Dupont, 1985; Bohnke, 1991). They formed the chronological background for our own pollen research in which we sampled and analysed several ancient woodland soils. Although pollen is poorly preserved in sandy soils, the ancient woodland soils appear to have a rather clear zonation of pollen because of the soil acidification processes that have taken place during the centuries. If a brown forest soil changes to a podzol, the biological homogenization often does not reach as deep as previously, so that old pollen spectra are often fossilized in the subsoil (Havinga, 1984).

Archaeology

The fourth element has been the integration of the archaeological data, the spatial distribution of prehistoric settlements and other traces that were found in the study area. The linking of archaeological and soil data yielded interesting new views about the changes in settlement site selection in the past. This selection in turn had a strong relationship with the history of the ancient woodland because prehistoric people used the woodland very intensively (see also Spek, 1997).

Results and Discussion

The natural landscapes of Drenthe

In the province of Drenthe we distinguish four natural landscape types:

1. A boulder clay landscape, dominated by sandy loam deposits from the Saalian ice age.
2. A coversand landscape, dominated by poor sandy soils from the Weichselian ice age.
3. A landscape of brook and stream valleys.
4. A landscape of raised bogs or moorlands.

The first two landscapes together form the high, inhabitable soils of Drenthe. People have lived here continuously for 5500 years, but the differences between the two must be emphasized. Soils in the boulder clay landscape are much more loamy and fertile than those of the coversand landscape.

What did the Drenthe landscape look like 5500 years ago? Pollen diagrams show that the mixed deciduous woodlands of the Atlantic Period were at their height at that time. The boulder clay landscape was covered with mixed deciduous forests of oak (*Quercus*), lime (*Tilia*), elm (*Ulmus*), holly (*Ilex*), and hazel (*Corylus*). In the coversand landscape we found oak, beech (*Fagus*), pine (*Pinus*), and probably also lime and hazel. The stream valleys were covered with alder (*Alnus*) and ash (*Fraxinus*). The percentages of arboreal pollen that were found in pollen spectra from the higher soils of those days were very high. They range from 70 to 90%. This gives us an indication that the forest was still very dense at that time. Mesolithic and Early

Fig. 8.3. Megalithic tomb of the Funnel Beaker Culture (3400–2900 BC). Most of these prehistoric monuments were laid out on coversand soils in the neighbourhood of boulder clay plateaux. The big stones were derived from eroded boulder clay. The picture was taken in 1920 in the neighbourhood of Havelte (SE-Drenthe).

Neolithic people surely had a slight influence on the natural landscape, but the landscape was still covered by dense woodland.

The Middle Neolithic Period

This changed somewhat when the first farmers settled on the Drenthe plateau in the Middle Neolithic Period. These farmers belonged to the Funnel Beaker culture and are especially famous for their megalithic tombs (Fig. 8.3). They applied a shifting-cultivation system, which means that they continually moved within a certain territory, which is expressed in the archaeological findings. A group of farmers who are the first to colonize an area and therefore are free to choose the sites for their fields would be expected to decide upon the most fertile soils, here the loamy soils of the boulder clay landscape, but this was not the case.

From the pedological information on the location of the settlements discovered so far, the first farmers sought the poorer soils of the coversand landscape (Wieringa, 1958; Bakker, 1982). Some 90% of the settlements are found in the coversand landscape, and not in the boulder clay landscape. The ancient principle that the most fertile soils are always reclaimed first appeared not to hold true. How should we explain this preference for the poorer soils? It may be due to the state of technology at that time. The woods on the poorer coversands were easier to reclaim with flint axes. The tree canopy and the understoreys were less dense, more light shone upon the ground and the trees were probably less thick than those on the loamy soils. For arable farming too, the sandy soils were easier to till than the loamy soils and corn will sprout better in light sandy soils. In addition the coversands which we call poor now, were not so poor in the Middle Neolithic Period. The strong eluviation that we find in these podzols certainly dates from later periods (Waterbolk, 1964).

Fig. 8.4. Neolithic and Bronze Age villages were situated on poor sandy soils in the coversand landscape. They were reclaimed from rather open woodlands containing oak, birch and hazel. The removal of the woodland and intensive grazing resulted in acidification and podzolation of the soil. Nowadays the prehistoric habitation areas are covered by poor heath (foreground). The richer soils of the boulder clay plateau have not been podzolized and have kept their woodland vegetation up to now (background).

Furthermore, the site selection of the Middle Neolithic farmers was influenced by the availability of water. The settlements are often close to a lake or brook. A third characteristic is the nearness of other landscape types. Although the settlements were situated on coversand, other landscape types were often quite near. Most settlements were situated on small open patches, not far from the rich woodlands on the boulder clay plateau (Fig. 8.4). Thus, one could optimally take advantage of each landscape type. This means that the prehistoric settlements and ecological transition zones were closely connected.

The Late Neolithic and Bronze Age

For a long time after the Middle Neolithic Period, the inhabitants of Drenthe continued to choose the coversands for their settlements and fields. Settlements from the Late Neolithic Period and the Bronze Age are almost exclusively situated on coversand. Pollen spectra from these periods show that the woodland on these coversands was strongly thinned out. The percentages of the trees of the primary woodland like oak, lime and elm decreased, whereas the percentages of secondary species like birch (*Betula*) and hazel increased, as well as the species of grassland and heathland.

Around the settlements, larger and larger spaces were formed, covered with grass and heath. The decrease of woodland also strongly affected the mineral cycle and water regime of the soil. Because of soil acidification, minerals increasingly leached from the topsoil of the coversands (Spek *et al.*, 1997). After some time, podzol soils were formed, covered with heath. By the end of the Bronze Age, landscape with roughly equal amounts of woodland and heathland had formed. Most of the coversands had been deforested and had turned into grassy heath. Massive deciduous forests were only to be found on the boulder clay plateaux.

Changes in the Iron Age

During the Late Bronze Age and the Early and Middle Iron Age (1100–250 BC) the population of Drenthe increased rapidly. The deforestation and impoverishment of the coversand landscape, which had started in the preceding periods, were then accelerated. We can see this in the pollen spectra (Casparie and Groenman-Van Waateringe, 1980). The percentage of arboreal pollen strongly decreases in the diagram, that of heathland pollen strongly increases. Soil degradation also increased in this period. This made the coversand landscape a less attractive area for living and working. The ecological limits were in sight and, not uncommonly, exceeded. Excavations of settlements from this period show that the Bronze and Iron Age people were much troubled by sand-drifts (Van Gijn and Waterbolk, 1984). The vulnerable coversand landscape could not meet the rapid growth of the population. Two thousand years of occupation had completely exhausted the old habitation area.

How did Iron Age people react to this ecological crisis? One of their solutions was to make agricultural improvements. Arable and livestock farming became more intensively connected and fertilization was intensified. Migration to more suitable, more fertile areas was also needed. Archaeological research has shown that around 500–300 BC people left their old villages and settled elsewhere (Van Gijn and Waterbolk, 1984). We can distinguish three types of migration:

1. *Interregional migration* to other regions, especially to the coastal region, where many new clay areas had developed by accretion.
2. *Regional migration* to formerly uninhabited parts of Drenthe, especially to the edges of the Drenthe plateau.
3. *Local migration* of people within their own habitation area.

The third type is especially important to us. The Iron Age people left the old settlement areas and increasingly settled on the boulder clay plateau, within their own territories. They had lived in the coversand landscape for over 3000 years, but now they migrated to the loamy soils. The massive woods on these soils were gradually cleared and converted to arable land. In the pollen diagram the percentage of primary woodland species like oak decreases very strongly.

The occupation of the boulder clay landscape marked a new phase in the deforestation of Drenthe. The field complexes from the Iron Age are easy to recognize on aerial photographs and in the field by their chessboard pattern. Small, square fields were enclosed by earth-walls (Fig. 8.5). Up to now about

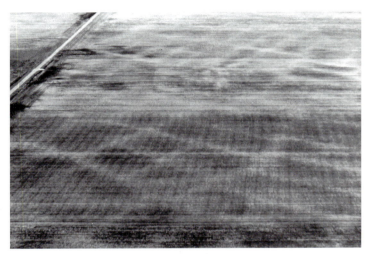

Fig. 8.5. Aerial photograph showing the outlines of an arable field dating from the Middle Iron Age. These fields had a kind of honeycomb-structure: small square plots (15–40 m) surrounded by small embankments. Ironically, these fields are called *Celtic fields* in the Netherlands, although they are not at all related to the Celts.

Fig. 8.6. In 1993 our team from the Winand Staring Centre excavated a part of an Iron Age Celtic field in Zeijen (Drenthe). The picture shows a cross-section of one of the embankments.

Fig. 8.7. Ancient woodland of the Norger Holt (northern Drenthe), situated on a boulder clay soil. The tree layer mainly consists of oak (*Quercus robur* and *Quercus petraea*) and holly (*Ilex aquifolium*).

75 of these Iron Age fields have been found in Drenthe (Brongers, 1976). Our pedological studies (Fig. 8.6) have shown that the larger part of these fields were reclaimed from forest and laid out on boulder clay soils, not in the old coversand landscape (Spek, 1993). The Middle Iron Age appears to have been a turning point. People were forced to reclaim new parts of the landscape because they had exhausted the environment in their former habitation areas.

Transposition of core and periphery

The Iron Age really was a watershed in settlement history, for the great majority of the settlements and fields of Roman and medieval times were also laid out in the forests on the boulder clay plateaux. Our extensive micro-morphological and palaeoecological research of the subsoil of medieval fields and settlements has proved that they were reclaimed from deciduous forests of oak, beech, holly, hazel and birch (Fig. 8.7). The many medieval field and place names containing a woodland element are another indication for the widespread reclamation of boulder clay woodlands. Consequently, the centre of the medieval cultural landscape, and of the present landscape, became situated in the boulder clay landscape, and not in the coversand landscape, as was the case before.

Fig. 8.8. Intensively grazed and coppiced oak woodland of the Schipborger Strubben (central Drenthe). Besides a lot of bracken (*Pteridium aquilinum*), the shrub and herb layer of the Drenthian 'strubben woodlands' contains *Anemone nemorosa, Polygonatum multiflorum, Cornus suecica, Lonicera periclymenum, Stellaria holostea, Oxalis acetosella* and *Polypodium vulgare*.

So we find an important transposition in the landscape in the course of pre-history. Before the Iron Age, the sandy soils of the coversand landscape were the centre of habitation. In that time the woodlands on the boulder clay soils were situated on the periphery. Since the Iron Age the boulder clay landscape has been the centre, and the exhausted coversands, covered with heathland, turned into a marginal area. So, centre and periphery are relative concepts; even as is the concept of marginality. Landscapes that are now being used intensively may have been marginal in prehistory, while landscapes that are now considered marginal areas may have been used intensively in prehistory.

Changes in the woodland

The transposition of core and periphery within the Drenthian landscape is not without significance to our view on the origins of ancient woodland. The areas that were covered by ancient woodland in historical times appeared to have functioned as a place of residence and work for numerous generations of prehistoric people. The soils of these ancient woodland areas are not so undisturbed as many people are inclined to think. This also has implications for our modern nature management. Many areas that biologists consider as undisturbed, wild nature reserves contain many cultural elements and are a kind of cultural landscape.

Middle Ages and Early Modern Period

In this chapter I will not discuss the medieval and early modern deforestation process of the study area in detail. In the pollen diagrams we can see that this

Fig. 8.9. Intensive sheep grazing could result in large drift-sands. The picture was taken around 1900 in the vicinity of Kraloo (SE-Drenthe).

process has accelerated from the tenth century onwards. Large parts of the woodlands of the boulder clay plateau were converted to open fields, meadows and grazing land during the High Middle Ages. The small area of woodland that survived these reclamations received its final blow in the Late Medieval Period and the Early Modern Period, when agriculture was being intensified. The expanding markets of the Dutch cities during the fifteenth, sixteenth and seventeenth centuries had a strong influence on the agricultural economy of all parts of the Netherlands, and also on peripheral areas like Drenthe. Livestock farming especially was intensified. The archives indicate that from the fifteenth century onwards, large herds of sheep and oxen were kept in Drenthe in order to produce wool and meat for the city markets. This resulted in much higher grazing densities on the common woodlands (Fig. 8.8), heathlands and grasslands and, as a result of this, very strict common regulation for these areas, especially in the seventeenth century. In many cases the grazing intensity exceeded the carrying capacity of the natural vegetation (Fig. 8.9). This could lead to extensive sand-drifts (Castel, 1991). Further developments in these areas are covered in papers by Dirkx (Chapter 5, this volume) and van Laar and den Ouden (Chapter 9, this volume).

Conclusions

1. Deforestation in Drenthe started in the Middle Neolithic Period, accelerated in the Late Bronze Age and Early and Middle Iron Age and had a second acceleration in the High and Late Medieval Period.
2. The spatial distribution of the deforestation process was closely connected to geological and pedological aspects. The poor coversand landscapes were deforested in the early prehistoric period; the richer soils of the boulder clay plateaux were largely deforested after the Early Iron Age.
3. In landscape history, centre and periphery are relative concepts. So is the concept of marginality of land.

References

Bakker, J.A. (1982) TRB settlement patterns on Dutch sandy soils. *Analecta Praehistorica Leidensia* 15, 87–124.

Bohnke, S.J.P. (1991) Palaeohydrological changes in the Netherlands during the last 13,000 years. PhD Thesis. Free University of Amsterdam.

Brongers, J.A. (1976) *Air photography and Celtic field research in the Netherlands.* ROB, Amersfoort.

Casparie, W.A. (1972) Bog development in southeastern Drenthe, the Netherlands. *Vegetatio* 4, 1–272.

Casparie, W.A. and Groenman-Van Waateringe, W. (1980) Palynological analysis of Dutch barrows. *Palaeohistoria* 22, 7–65.

Castel, Y. (1991) Late Holocene eolian drift sands in Drenthe (the Netherlands). PhD Thesis. University of Amsterdam.

Dupont, L.M. (1985) Temperature and rainfall variation in a raised bog ecosystem. A palaeoecological and isotope-geological study. PhD Thesis. University of Amsterdam.

Havinga, A.J. (1984) Pollen analysis op podzols. In: Buurman, P. (ed.) *Podzols.* Van Nostrand Reinhold Soil Science Series 3, New York, pp. 313–323.

Jongerius, A. and Heintzberger, G. (1975) *Methods in soil micromorphology. A technique for the preparation of large thin sections.* Soil Survey Papers, No 10. Netherlands Soil Survey Institute, Wageningen.

Spek, Th. (1993) Milieudynamiek en lokatiekeuze op het Drents Plateau (3400 v Chr-1850 na Chr.). In: Elerie, J.N.H. (ed.) *Landschapsgeschiedenis van de Strubben/Kniphorstbos. Archeologische en historisch-ecologische studies van een natuurgebied op de Hondsrug.* Van Dijk and Foorthuis RegioProjekt, Groningen, pp. 167–236.

Spek, Th. (1997) Die bodenkundliche und landschaftliche Lage von Siedlungen, Äckern und Gräberfeldern in Drenthe (nördliche Niederlande). Eine Studie zur Standortwahl in Vor- und Frühgeschichte (3400 v. Chr. – 1000 n. Chr.). *Siedlungsforschung* 14, 200–292.

Spek, Th., Bisdom, E.B.A. and van Smeerdijk, D.G. (1997) *Verdronken dekzandgronden in Zuidelijk Flevoland. Een interdisciplinair onderzoek naar de veranderingen van bodem en landschap in het Mesolithicum en Vroeg-Neolithicum.* Rapport 472, DLO-Staring Centrum, Wageningen.

Van Gijn, A.L. and Waterbolk, H.T. (1984) The colonization of the salt marshes of Friesland and Groningen: The possibility of a transhumance prelude. *Palaeohistoria* 26, 101–122.

Van Smeerdijk, G.G., Spek, Th. and Kooistra, M. (1995) Anthropogenic soil formation and agricultural history of the open fields of Valthe (Drenthe, the Netherlands) in medieval and early modern times. *Mededelingen Rijks Geologische Dienst* 52, 451–479.

Waterbolk, H.T. (1964) Podsolierungserscheinungen bei Grabhügeln. *Palaeohistoria* 10, 87–101.

Wieringa, J. (1958) Opmerkingen over het vergand tussen bodemgesteldheid en oudheidkundige verschijnselen naar aanleiding van de NEBO-kartering in Drenthe. *Boor en Spade* 9, 97–113.

9

Forest History of the Dutch Province of Drenthe and its Ancient Woodland: a Survey

J.N. van Laar and J.B. den Ouden

Department of Forestry, Wageningen Agricultural University, Wageningen, the Netherlands

In the Province of Drenthe most of the woodland had disappeared by the beginning of the nineteenth century, mainly through human activities. Old maps show a number of woodland remnants in a mainly degraded land use system. Reforestation activities, privately initiated, started after 1850 and culminated in the first half of this century as a State activity. Parts of the ancient woodland were incorporated in the new forests. Nowadays, these remaining old woods are valued for cultural, historical, botanical and ecological reasons. Two ancient woods located in the National Forest 'Emmen' are discussed in terms of these values and an inventory of ancient woods is presented as a basis for further research and conservation.

Introduction

Since the beginning of the 1980s there has been a growing scientific interest in land use history and, more specifically, in forest history in the Province of Drenthe (Buis, 1985; van Laar and De Vries, 1985; Bieleman, 1987; Elerie *et al.*, 1993; Den Ouden and Roosenschoon, 1994; Dirkx, Chapter 5, this volume; Spek, Chapter 8, this volume). These studies contribute to a better understanding of the changes in vegetation and landscape patterns, and of socio-economic developments on a local and regional scale in the past. This knowledge may be useful for preserving, restoring and managing vegetation communities and valuable landscape features, as well as helping to awaken a consciousness in both the public and politicians of their cultural–historical environment.

This chapter provides a general description of forest history in Drenthe from the time that written documents became available; a discussion on ancient woods that are of historical interest; and some conclusions. The research is mainly based on literature review and study of topographical maps.

Present Physical and Cultural Setting

The Province of Drenthe is situated in the northeast of the Netherlands and covers 2653.54 square kilometres of land. The human population is relatively small – about 457,500 – and is for the larger part concentrated in five major, expanding urban areas: Assen, Emmen, Meppel, Hoogeveen and Coevorden. Only a small part of the population lives in rural settlements or is dispersed in the countryside. Agriculture is an important land use; forest covers 11.5% of the province – more than the national average – and areas with important landscape values or destined for natural development as reserves are increasing both in numbers and hectares.

In geographical terms, the province is a rather flat loam plateau, crossed by stream valleys, alternating with reclaimed heath in agricultural use, vast blocks of afforestation, small woods, remnants of heathfields and moors and a small number of estates. In the peripheral zones there are former peat areas, which are now mainly in agricultural use. Altitudes range from minus 0.9 metre to plus 23 metres New Amsterdam Watermark.

Woodland and Forest History

Up until the second half of the sixteenth century, very few written references to our study area can be found. Extensive woods have existed in the area (e.g. Waterbolk, 1984; Spek, Chapter 8, this volume) and the authors of *Hedendaagsche Historie of Tegenwoordige Staat van het Landschap Drenthe*, first published in 1792, refer to large woody areas in 'former times'. They indicate that a number of relics of these primeval woods could be recognized in the field at that time (van Lier *et al.*, 1975). This document also provides reasons for the reduction in woodland area.

In the Late Middle Ages, around 1300, a more or less autonomous administration in the so-called 'Lantschap Drenthe' was established, with its own regional laws, jurisdiction, customs and privileges. On a local scale an administration evolved in which prosperous peasants together with entitled farmers from local communities made decisions about the use of communal grounds, (wood) pasturage, the cutting of turves and harvesting timber and firewood. Often meetings took place in the communal woods. 'Sacred groves' existed, as they do today in India. These groups belonged to the so-called 'marke'-organization, which can be defined as a system of local authorities representing the farmers, non-farmers and cottagers that lived within particular boundaries, or a 'marked' area (Dirkx, Chapter 5, this volume). This administrative structure lasted for many centuries. Similar land use systems existed elsewhere in the Pleistocene northwestern European plain until the middle of the nineteenth century.

Local regulations on the use of communal lands and woods could not prevent them degenerating. Even drifting sand areas – small-sized deserts, sometimes threatening settlements – developed. During the many unstable years of political struggle between the sixteenth and eighteenth centuries, looting of the local rural societies caused much destruction, not only of buildings and crops, but of forests as well. From the forests that were left, large quantities of wood were needed for rebuilding. Common oak, *Quercus robur* L., used to be an abundant and favoured species for construction purposes. Even now, Drenthe is well known because of its oaks in and around villages.

So, excessive exploitation – cutting, public sales of timber of common oak, wood-pasturage – wars, and clearing of woods in order to create more agricultural land, resulted in a depleted and deforested region at the end of the eighteenth century. Woods that had been left or had been newly established in some cases, can be recognized on military topographical maps from around 1800 and the Napoleonic maps of 1811–1813.

Some small and local systematic afforestation took place in the period before AD 1800. This happened within the framework of laws that prescribed replanting of common oak after cutting. In 1612 two officials, forest guards or 'holtvoogden', were appointed to monitor the activities and report misuse of woods. In some cases when an estate was built, a park forest was laid out with it. Afforestation could also be found around the few monasteries. Early efforts to stabilize drift sands by means of planting common oak, birch or Scots pine were made in the eighteenth century. To keep out cattle and sheep from the old agricultural lands and to meet firewood needs, coppice of common oak was established at the transition zone of private farm land (so-called 'es') and common wasteland. Remnants of these either private or communal afforestation activities may now be considered ancient woodlands.

Nineteenth century changes

After 1800 afforestation increased. In 1806 Drenthe became a part of the Kingdom of Holland and consequently the region was incorporated in the French Empire in 1810. Provincial authorities and citizens had to deal with French politics and laws until 1813. Among the new legislation a few laws ordered the division of communal lands in order to stimulate reclamation and afforestation. At least 75% of the area was still 'wasteland' - heathland, peat-land and drift sand areas – at the beginning of the nineteenth century (Blink, 1929) but most of this had become divided after 1850. Much heavy timber was extracted from the remaining woods. The Weerdingerholt, Emmerholt, Drouwenerholt and Westdorperholt are examples of woods that disappeared for this reason. Nevertheless there was a substantial increase of forested area in the last century. Most of this systematic afforestation was private, but some-times it was a communal matter. In some cases municipalities initiated afforestation as a means of work relief.

In the nineteenth century, several types of afforestation can be distinguished. Afforestation of drift sands was subject to provincial legislation and was carried out by local communities (e.g. Emmerzand, Mepperzand, Odoornerzand) or private landowners (e.g. Dieverzand). A suitable tree species appeared to be the Scots pine, *Pinus sylvestris* L. Sometimes common oak was planted. To protect these new plantations, lines of birch, *Betula pendula*, were planted. Afforestation of deep peat soil was a private matter for the prosperous peat-moor proprietors and was mainly concentrated near Hoogeveen. A lack of manure meant that the soil was not suitable for agriculture after peat-cutting and one of the methods to meet fertility requirements appeared to be afforestation. Consequently, some of these new forests were only temporary and were converted to agricultural land after four or five decades. Scots pine and common oak, especially for coppicing, were favourite species. The plantings were often combined with crops of cereals, potatoes or tick beans during the first years, a local agroforestry system. By 1900 Hoogeveen had become the most densely forested municipality. This afforestation was also important because it provided much work to the peat-cutters in wintertime.

Heath afforestation increased in the second half of the last century. The importance of the common heath fields in the land use system was diminishing and gradually they became private property. Private persons and societies initiated afforestation projects on a small scale. The easiest method was to burn the heath, sow seeds of Scots pine and cover them with sand from small ditches. Sometimes more intensive site preparation was put into practice by digging about half a metre deep. These new plantations were usually surrounded by walls and were planted with common oak and birch.

The establishment of coppices was of major importance in the late eighteenth and nineteenth centuries and contributed considerably to the increase of the forested area of Drenthe. For farmers, coppice provided a variety of products in a relatively short period of time. Some of the oldest planted coppices are now considered as ancient woodland. Generally an intensive site preparation of digging and manuring was carried out. During the agricultural depression of 1877 to 1895, in particular, arable lands were converted into coppices of common oak. In this century many outgrown coppices have been transformed into more productive forests of coniferous tree species.

Twentieth century changes

In the twentieth century a huge increase of forested area occurred. Afforestation in this era happened mainly on account of the Dutch government and the 'Dutch Moorland Reclamation Society', founded in 1888, or its regional equivalents. The Forest Service, founded in 1899, was considered to have the task of carrying out the government's goals on afforestation. At first, until 1922, the activities in Drenthe were limited to the fixing of extended drift sands at locations that became threatened. After 1922, vast heathlands were bought by the State in order to have these 'waste lands' afforested. Afforestation of heathlands owned by municipalities was also supported by the State. The State-owned forest area had its fastest growth in the twenties and thirties and many National Forests were founded in that period. Since 1960, there have been very few State-initiated heath afforestations (Vos, 1980).

Several stages in afforestation techniques developed. Next to common oak and Scots pine, less common tree species appeared such as Japanese larch, *Larix kaempferi*, black pine, *Pinus nigra*, Norway spruce, *Picea abies*, Sitka spruce, *Picea sitchensis*, Douglas fir, *Pseudotsuga menziesii* and American red oak, *Quercus rubra*. New ways of site preparation, fertilizing,

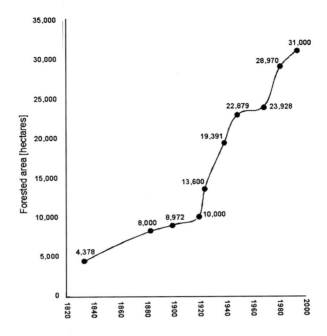

Fig. 9.1. Increase of forested area in Drenthe from AD 1833 to 1996 (adapted from Kalb, 1984).

machines and more sophisticated mixed planting designs came into practice. The increase in planting is closely related to unemployment during the Economic Depression. Afforestation was used as a means of work relief (Vos, 1980). Site preparation, road building and digging ditches were carried out by unskilled labour from all over the Netherlands.

In Fig. 9.1 the increase in forest cover from 1833, the year the first Cadastral Survey was published, until the present, is demonstrated, although definitions of the concept of 'forest' may have slightly changed in the different surveys. Recent years have shown a new afforestation wave on to marginal agricultural lands, especially on reclaimed peat. Every year, hundreds of hectares of agricultural land are expected to be converted into new forests, sometimes temporarily, but mostly as permanent forest. Even new estates have come into vogue. A considerable amount of national and provincial policy making has developed in the last decade. Between all these new forests the few ancient woods surviving have become a valuable type of wood, where local provenances of tree species and herbs, typical for old forest sites, may have survived many centuries.

Ancient Woodland in Drenthe

Ancient woodland can be defined as 'woodland that has existed continuously since before a certain date' (Rackham, 1980). For the United Kingdom,

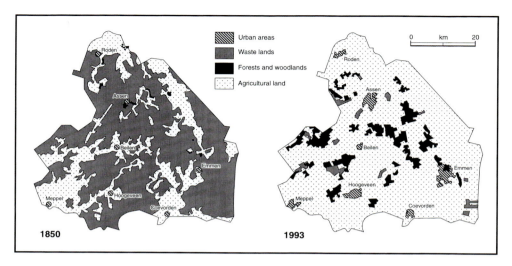

Fig. 9.2. Changes in land use in Drenthe between 1850 and 1970 (adapted from P.P.D. Drenthe (Kettner, 1997)).

Rackham proposes the year 1700. Given the circumstances in the Netherlands, AD 1800 would be a better option, since there are very few data on woods of earlier times available. Moreover, the deforestation process had reached its zenith, and therefore the forest cover was at its lowest point at the end of the eighteenth century.

When we apply this criterion, only a relatively small number of ancient woods, mostly small in size, can be identified in Drenthe (Fig. 9.2). While a wood may have existed for a very long time, its appearance and condition can vary. A high forest can be considered as an ancient woodland (holt), but this may also be the case with former coppiced wood and scrub forest ('strubben') or woodland that has been transformed into another forest type. In Drenthe these wood types all occur. All of them have been exploited to a greater or a lesser degree. Ancient woods have in common that they are at least two centuries old, their tree and other plant species are mainly indigenous and the soil is (relatively) undisturbed. The following classification has been made:

1. Ancient woods that are (partly) surrounding old arable fields ('esrandstrubben') or are located at transition zones between villages and drift sand areas. They are characterized by a scrubby appearance due to overexploitation and grazing. These woods may have been planted in the period between 1550 and 1700. Good examples are at Zeyerstrubben, Anloërstrubben and Schoonloërstrubben.

2. Ancient woods consisting of old outgrown planted coppices, either established on heathland or transformed from degenerated primeval woodland. Site preparation usually happened. Valtherspaan near Emmen represents such a woodland. Old outgrown coppices can also be found on wet sites with *Alnus* sp.

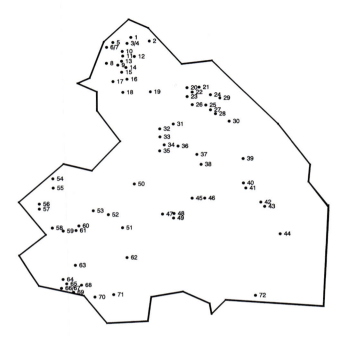

Fig. 9.3. Locations of woodlands in Drenthe that are most likely to be ancient woodlands. Numbers in figure refer to Table 9.1.

3. A few high forests are considered ancient woodland. This type is characterized by mature stands of mainly one tree species, mostly common oak and sometimes with a sub-tree layer, called a 'holt'. The extension '-holt' is often found on old topographical maps indicating such a forest type. One of the best examples of this type of ancient woodland is the 'Norgerholt', which might even be a 'primary woodland' relic, that is, it has had forest cover since the last Ice Age.

4. Finally, ancient woodland that has been gradually converted towards mixed and uneven aged forests can be identified. The Asserbos belongs to this type and was already known about in the fourteenth century. Now it is an urban forest incorporated in the provincial capital of Assen. Park forests near old estates, established before 1800, and the typical old oak plantations in the villages ('brinken') are not considered as ancient woods.

A major source for tracing ancient woodland is the Napoleonic maps of 1811–1813. They cover the major part of Drenthe (Koeman, 1970). Other topographical maps dating from the late eighteenth or the beginning of the nineteenth century are also very useful, such as the 'Topographische Kaart der Provintiën Groningen en Vriesland', dated 1820–1824.

In the field, ancient woods can be recognized by specific plant species that are limited to (or more abundant in) ancient woodland. Where these species are found at non-wooded sites, this may suggest that there was an

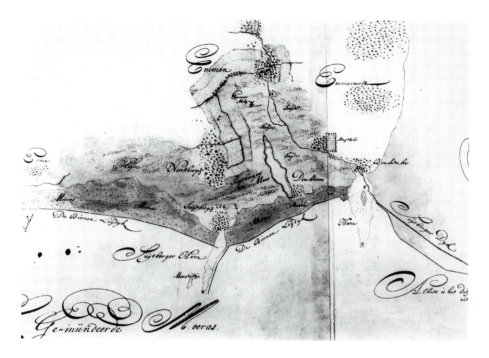

Fig. 9.4. Fragment of a map drawn by cartographer J. van Alberdingh in 1688. It shows wooded areas near the farmers' village of Emmen. Some woods have disappeared but some can be recognized as ancient woodland, which still exists.

ancient woodland there in former times. Ancient woods are locations where indigenous tree, shrub and plant species can be found (Rövekamp and Maes, 1995). Examples are *Malus sylvestris, Prunus avium, Prunus spinosa, Pyrus pyraster, Quercus petraea, Fagus sylvatica, Rhamnus catharticus, Rosa rubiginosa, Rhamnus frangula, Ilex aquifolium* and *Viburnum opulus*. Ancient woods are sparsely located all over Drenthe, giving in total a few hundred hectares (Fig. 9.3, Table 9.1).

The Ancient Woodland near Emmen

Two examples of ancient woodland near the city of Emmen in the south-eastern part of Drenthe are now described in more detail (Fig. 9.4). They are remnants of a more extensive wooded area that can be traced on a map dating from 1725 ('Kaart met betrekking tot geschillen tussen de markegenoten van Valthe en Weerdinge'). This map shows woods named 'Weerdingerholt', 'Valtherholt', 'Spaan', 'Meerbos' and 'Willem Stee' (Den Ouden and Roosenschoon, 1994). The city of Emmen has developed in less than one century from a little village to a major urban centre with a lot of industry and administrative services. The forested area in the peri-urban zone has grown rapidly as a result of heathland afforestation in the 1920s and

1930s. So, within the last century, the two ancient woods have faced dramatic changes in their surroundings.

Both woods belong to the State, are administered by the State Forest Service and are less than 10 ha in size. A few years ago they were carefully surveyed in the framework of local forest history research. The ancient woodland 'Meerbos' is 4.95 hectares and is situated west of Weerdinge. It is a high forest of common oak, *Quercus robur*, locally mixed with beech, *Fagus sylvatica*. There is a sub-tree layer of *Ilex aquifolium* and a shrub layer of *Sorbus aucuparia, Corylus avellana, Frangula alnus* and *Sambucus nigra*, in varying densities. In the herb layer, characteristic ancient woodland species are found including *Oxalis acetosella, Polygonatum multiflorum, Moehringia trinervia* and *Stellaria holostea*. A small part of the soil points to ancient agricultural use, the other part is a heath podzolic soil with loam within 1 m depth. The first time 'Meerbos' can be traced is on a map of 1725 (Den Ouden and Roosenschoon, 1994). Since 1788, the limits of Meerbos have hardly changed. Meerbos is a 'type 3' forest, part of the surrounding woods of the Weerdinger infield. On the northern edge, a little protective embankment, in former days carrying a hedge, can still be seen. Until the end of the nineteenth century topographical maps are the only information source, but planting of oaks is recorded in 1874 and in 1895.

The ancient woodland Valtherspaan, approximately 7 ha, is now embedded in twentieth century heath afforestations, north of Emmen. The limits of Valtherspaan have hardly changed since it appeared on a French map of 1811, the latter being a very accurate source for determining forest history. The Valtherspaan has been exploited as common oak coppice and was last cut in 1927. The soil has been disturbed in places, which is common in planted coppice. A small patch of the wood, however, shows the typical morphology of century-old coppice stools, rising one to two feet (0.31–0.61 m) above the forest floor. The Valtherspaan can be characterized as a 'type 2' ancient woodland.

Conclusions

Some conclusions can be drawn from the provisional research done on the ancient woodland in Drenthe and a number of recommendations made:

1. The area of ancient woodland in Drenthe is very small (about 1%) compared to the total forested area.
2. Most forests in Drenthe are only one or two tree-generations old, which is very young in terms of forest ecosystem development.
3. Four types of ancient woodland can be distinguished in Drenthe.
4. More detailed historical–geographical and ecological research is needed to complete their characterization. We believe this knowledge will be useful in policy making and management concerning these ancient woods and in protecting other possible ancient woodland. In particular, ancient woods hold indigenous genetic tree and shrub species of local genetic stock which are of interest to landscape managers.

Table 9.1. Provisional list of ancient woodlands according to old topographical maps and based on Scheper (1989), Meeuwissen (1991) and own observations, with their sizes, owners, specific features and classification.

	Name or toponym	Size* (ha)	Owner†	Main characteristics	Classif.
1	Het Waal	7	DrLa	Outgrown coppice of *Alnus glutinosa* on bog-peat	2
2	Kluivinghsbosch	13	NatMon	High forest and coppice with standards of *Quercus robur*	3
3	De Kleibosch	7	DrLa	Outgrown coppice of *Q. robur*, *A. glutinosa* and *F. excelsior* on clay	2
4	De Dobben	7	DrLa	Outgrown coppice of *Q. robur*, *A. glutinosa* and *Betula* sp. on clay	2
5	Elzenbroekje Nietap	3	State	Outgrown coppice of *Quercus robur* and *Alnus glutinosa* on moist soil	2
6	Bitse	2	State	Outgrown coppice of *Quercus robur* on clay	2
7	Toutenburgsingel	15	Private	Outgrown coppice and high forest of *Quercus robur* and exotic species	–
8	De Hullen	4	Private	Outgrown coppice and coppice with standards of *Quercus robur*	2
9	Roderesch	4	Private	Coppice of *Quercus robur*, *Betula* sp. and *Prunus serotina*; degraded	2
10	Hooghout	4	Private	Outgrown coppice of *Quercus robur* and *Betula* sp.	2
11	Sterrebosch		State	Estate forest of mixed deciduous and coniferous species	–
12	Tolnerbosch (two parts)	4	Private	Outgrown coppice with standards of *Quercus robur*, *Betula* sp.	2
13	Lieverder Noordbosch	26	State	Outgrown coppice of *Quercus robur*	2
14	Hullig	5	State	Outgrown coppice of *Q. robur*, *A. glutinosa*, *B.* sp.; old *P. sylvestris*	2
15	De Tip (two parts)	5	State	Outgrown coppice of *Quercus robur*, some *Ilex aquifolium* in tree layer	2
16	Achter het Hout	20	State/Private	Outgrown coppice of *Quercus robur* and *Betula* sp.	2
17	Eenderbrug (two parts)	5	NatMon/Private	Outgrown coppice of *Quercus robur*	2
18	Norgerholt	25	NatMon	High forest of *Quercus robur* with sub tree layer of *Ilex aquifolium*	3
19	Zeyerstrubben	45	State	Outgrown scrubby *Quercus robur* along old arable field	1
20	Schapedrift near Zeegse	5	Private	Outgrown scrubby *Quercus robur* on drift sand dunes	1
21	De Strubben/Kniphorstbos (partly)	35	State	Outgrown scrubby *Quercus robur*	1
22	Schipborg	2	State	Outgrown coppice of *Betula* sp., some outgrown *Q. robur* bordering	2
23	Burgvallen	6	State/Private	Outgrown coppice of *Quercus robur* and *Alnus glutinosa* on bog-peat	2
24	Groot Blok/Zuidesch Annen	7	State	Outgrown scrubby *Quercus robur* along old arable field	1
25	Anloër Strubben	35	State	Outgrown coppice of *Quercus robur*	1,2
26	Gastersche Holt/De Stobben	6	DrLa	High forest of *Q. robus*, *Betula* sp. and *A. glutinosa* with rich herb layer	3,2
27	Grootesch Eext	2	Private	High forest of *Quercus robur*	3
28	Binnenesch Eext	1	Private	High forest of *Quercus robur*	3
29	Zwaanmeer (partly)	20	Private	Outgrown coppice of *Quercus robur* and a few coniferous trees	2,1
30	Varik (four parts)/Kleine Houten	10	Private	Outgrown coppice of *Quercus robur*	2
31	Kamps	3	DrLa	Outgrown coppice of *Quercus robur* and high mixed forest	4
32	Asserbosch (east part) + Amelterbosch	80/25	Municipality	High forest of deciduous (*Quercus robur*) and coniferous tree species	3,4/4
33	N.N. near N.A.M.	1	Private	Outgrown coppice of *Q. robur*, *A. glutinosa*; planted *F. excelsior*	2,4
34	Geelbroek	2	State	Outgrown coppice of *Alnus glutinosa* on wet soil	2
35	Amerboschstuk	2	State	Outgrown coppice of *Alnus glutinosa* and *Betula* sp. on wet soil	2
36	Houtesch	4	State	High forest and outgrown coppice of *Quercus robur* and *Alnus glutinosa*	2,3
37	Grolloërholt	5	Private?	High forest of *Quercus robur*	3

	Name	Size	Owner	Description	
38	Schoonloërstrubben	25	State	Outgrown scrubby Q. robur on drift sand dunes along old arable fields	1
39	Buinerholt/Buinerbosch	2	State	Woodland of very outgrown coppice/high trees of Quercus robur	3
40	N.N. near Westeresch	1	Private	Woodland of high trees of Quercus robur and some Fagus sylvatica	3
41	Odoornerzand	25	State	Outgrown coppice of Q. robur and P. sylvestris on drift sand dunes	1,4
42	Valtherspaan/Wildgraven	7	State	Outgrown coppice of Quercus robur on sandy soil	3
43	Meerbos	5	State	High forest of Quercus robur, Fagus sylvatica and Ilex aquifolium	3
44	Oevermansbosje	10	Municipality	High forest of Q. robur and Ilex aquifolium, affected by storm in 1972	3
45	Tellingerbosch/Bosch Ma	4	State	High forest and outgrown coppice of Quercus robur	2,3
46	Oosterholten	5	State	High forest of Quercus robur and outgrown coppice of Betula sp.	3,2
47	Bruntingerbosch	4	State	Outgrown coppice of A. glutinosa/Betula sp. and high forest of Q. robur	2,3
48	Thijnsbosje	3	State	High forest and outgrown coppice of Q. robur and A. glutinosa; Ilex	3,2
49	Mantingerbosch	10	State	High forest and outgrown coppice of Q. robur and A. glutinosa; Ilex	3,2
50	Brunstinger Esch	2	NatMon	Outgrown coppice of Quercus robur and Prunus serotina; few P. abies	2
51	Nuilerbosch	10	State/Private	Outgrown coppice of Quercus robur and Betula sp. on moist soil	2
52	Lheederzand near Oosteresch	4	State	Outgrown coppice of Quercus robur along old arable fields	1
53	Oldengaarde (partly)	2	Private	Outgrown coppice of Quercus robur (part of an estate)	2
54	Boschoord	3	Private	High mixed deciduous and coniferous forest	4
55	Vledderhof/Vledderesch	3	DrLa	Outgrown coppice Quercus robur	2
56	De Delle	5	Private	Outgrown coppice Quercus robur	2
57	Moerhoven	3	Private	High forest of Quercus robur	3
58	Schieres	10	State	Outgrown coppice of Quercus robur along old arable field	1
59	Smeenholten	15	Private	Outgrown coppice of Q. robur along old arable field; partly P. serotina	1
60	Rheebruggen	15	DrLa	Outgrown coppice of high trees of Quercus robur and Betula sp.	2,3
61	Anseresch/Plantage	4	Private	Coppice and outgrown coppice of Quercus robur	2
62	Kinholt	4	Municipality	High forest of Quercus robur; Ilex aquifolium in sub-tree layer	3
63	Struikberg	5	Private	(Outgrown) coppice of Q. robur and Betula sp. on sandy bog-peat	2
64	De Veldkamp	1	Private	Remnant of outgrown coppice of Quercus robur and Betula sp.	2
65	Kloeterij	4	DrLa	Outgrown coppice of Quercus robur and Betula sp. on sandy soil	2
66	Schiphorst	4	DrLa/Private	Outgrown coppice of Quercus robur and Betula sp. on sandy soil	2
67	Lindenhorst	6	Private	Outgrown coppice of Q. robur and Betula sp.; standards of Q. robur	2,4
68	Balgenbosch (two parts)	4	Private	Degraded coppice of Q. robur and Betula sp.; a lot of Prunus serotina	1,2
69	Dunningen (partly)	4	Private/DrLa	Outgrown coppice of Quercus robur on sandy soil	2
70	Stapelerveld	3	Private	(Outgrown) coppice of Q. robur and Betula sp.; few coniferous trees	2
71	Bazuiner esbosje	4	Private	Outgrown coppice of Quercus robur and Betula sp. on sandy soil	2
72	Kloosterbosje in Westeindsche Stukken	2	Private	Open woodland of high trees of Q. robur and Betula sp. on moist soil	3

* Size is sometimes not exactly known and should be regarded as an indication.
† DrLa: Stichting Het Drentse Landschap; NatMon: Vereniging Natuurmonumenten; State: State property managed by the National Forest Service and in some cases by the Ministry of Defence.

Topographical Maps

Alberdingh, J. van. (1688) *Kaart van de moerassen en dijken gelegen tussen Coevorden en de Bellingwolder Schans.*

Epailly, d' (1811–1813) *French topographical map of Drenthe.* 1:20,000.

Huguenin (1820–1824) *Topographische Kaart der Provintiën Groningen en Vriesland.* 1:40,000.

References

Bieleman, J. (1987) *Boeren op het Drentse zand 1600–1910; Een nieuwe visie op de 'oude' landbouw.* Dissertation, Wageningen Agricultural University, Wageningen, 883 pp.

Blink, H. (1929) *Woeste gronden, ontginning en bebossching in Nederland voormaalsch en thans.* Mouton, 's-Gravenhage.

Buis, J. (1985) *Historia Forestis; Nederlandse bosgeschiedenis.* Dissertation, Landbouwhogeschool, Wageningen.

Elerie, J.N.H., Jager, S.W. and Spek, T. (1993) *Landschapsgeschiedenis van de Strubben-Kniphorstbos: archeologische en historisch-ecologische studies van een natuurgebied op de Hondsrug.* Stichting Historisch Onderzoek en Beleid/Regio- en landschapsstudies, Van Dijk and Foorthuis REGIO-PRoject, Groningen.

Kalb, J. (1984) *Van stuifzand naar woudreus; bosontwikkeling in Drenthe.* Staatsbosbeheer, Assen.

Kettner, A.J. (1997) Veranderingen in het bosareaal in Drenthe 1970–1993; Een GIS-studie naar veranderingen in het bodemgebruik met behulp van ARC-INFO. MSc thesis. Wageningen Agricultural University, Wageningen.

Koeman (1970) Een Franse topografische kaart van Drenthe uit de jaren 1811–1813. *Nieuwe Drentse Volksalmanak* 88, 89–101.

Laar, J.N. van and De Vries, I.G. (1985) *Heidebebossingen na 1900 in Drenthe en Overijssel.* MSc thesis, Landbouwhogeschool, Wageningen.

Lier, C.S., J. van and Tonkens, J. (eds) (1975) (Facsimile) *Hedendaagse Historie of Tegenwoordige Staat van het Landschap Drenthe 1792.* (first print 1792–1795), B.V. Foresta, Groningen.

Meeuwissen, T.W.M. (1991) Basisdocumenten *Bosvisie Noord-Nederland; mogelijkheden, ontwikkelingen, kansen.* Ministerie van Landbouw, Natuurbeheer en Visserij.

Ouden, J.B. den and Roosenschoon, O.R. (1994) *Van Meerbosch tot Oosterbos; historisch onderzoek in de Boswachterij Emmen.* MSc thesis. Wageningen Agricultural University, Wageningen.

Rackham, O. (1980) *Ancient woodland; its history, vegetation and uses in England.* Edward Arnold, London.

Rövekamp, C.J.A. and Maes, N.C.M. (1995) *Genetische kwaliteit inheemse bomen en struiken; deelproject: Inventarisatie inheems genenmateriaal in Drenthe.* IKC Natuurbeheer nr W77.

Scheper, M. (1989) *Drentse bossen: een selectiemethode gericht op natuurwaarden.* MSc thesis. Department of Forestry, Wageningen Agricultural University, Wageningen.

Vos, J.G. (1980) Ontstaan van de staatsbossen in de provincie Drenthe en de rol van de werkverschaffing hierbij. *Nieuwe Drentse Volksalmanak* 97, 50–66.

Waterbolk, H.T. (1984) Bossen. In: Abrahamse *et al.* (eds) *Het Drentse landschap.* De Walburg Pers, Zutphen, pp. 84–87.

Ecology and History of a Wooded Landscape in Southern Spain

10

T. Marañón[1] and J.F. Ojeda[2]

[1] IRNA, CSIC, PO Box 1052, 41080 Sevilla, Spain;
[2] Dept. Physical Geography and Regional Analysis, University of Sevilla, 41004 Sevilla, Spain

An extensive oak woodland, of about 1000 km^2, dominated by evergreen cork oak (*Quercus suber*) and semideciduous *Q. canariensis*, is found in southern Spain, near the Strait of Gibraltar, and contrasts with the paradigm of deforested Mediterranean mountains. Several factors, ecological, geographical and historical, have contributed to the origin and maintenance of this forested landscape. The rough relief and the acidic, nutrient-poor soils (derived from Oligo-Miocene sandstone) made this area unsuitable for cultivation. The oceanic influence favours the growth of oak trees. In particular, the cork oak is well suited to acidic soils and the humid Mediterranean climate. Three historical milestones seem relevant to the preservation of this woodland. Its location at a frontier during medieval times (thirteenth to fifteenth centuries) discouraged villages and reduced human pressure on the woodland resources. The rise of the value of cork helped to preserve the cork oak woodland during early nineteenth century industrial times. Contemporary consciousness about the conservation of woodland landscapes (somewhat unusual in the Mediterranean region) led to their designation as Los Alcornocales (meaning 'The cork oak woodlands') Natural Park, devoted to the eco-development of the region.

Introduction

When one thinks about a Mediterranean mountain landscape, the usual image that comes to mind is barren limestone with some pines killed by fire, or perhaps overgrazed shrubland on eroded soils. Deforestation has been extensive and repeated all around the Mediterranean Basin, and it is often cited as a classical example of non-sustainable use of natural resources (Thirgood, 1981) (Table 10.1).

Despite this paradigm of generally deforested mountain landscapes in the Mediterranean Basin (McNeill, 1992) there are some significant exceptions;

Table 10.1. Main agents of deforestation in the Mediterranean Basin (Thirgood, 1981).

Climatic change: Early twentieth century archaeologists (e.g. Huntington, 1915) believed that the decline of classical civilizations was due to a climatic deterioration and progressive desiccation.

Agricultural clearance: The increasing population growth and the improvement of agricultural technologies have caused an extensive transformation of forests into grasslands and croplands; this replacement has been almost complete in the lowlands.

Exploitation for timber and fuel: Especially important was the demand for ship building for the fleets of the Mediterranean naval powers (Phoenician, Greek, Roman, Venetian, Genovese, Spaniard and French) which was continuous until the end of the nineteenth century (e.g. Paola and Ciciliot, 1998). The need for fuel for the smelting of minerals has caused local significant impact in certain areas. More extensive has been the use of wood or charcoal for fuel, still important in rural areas of North Africa.

Wars and invasions: Woodland has suffered devastation by battling armies throughout history, and there are the present-day conflicts in the former Yugoslavia, the Near-East and Algeria.

Fire: Shepherds have traditionally used fires to improve grazing; nowadays, however, arsonists and negligent tourists start most of the summer forest fires.

Grazing: There is much literature about the pernicious action of the goat, the infamous 'black locust' of Mediterranean foresters, impeding woodland regeneration.

one is the extensive sylvo-pastoral *dehesa* covering about 55,000 km² in Western Iberia (Marañón, 1988; Rackham, Chapter 1, this volume). In this chapter we describe a lesser known oak woodland of about 1000 km² in the Strait of Gibraltar Region and propose a combination of ecological factors and favourable historical events to explain the exceptional survival of this wooded mountain landscape in the much deforested Mediterranean Basin. Why does this woodland exist and why has it not been transformed into a sylvo-pastoral savanna-like system?

Physical Factors

The southernmost tip of the Iberian Peninsula, the 'Gaditan Cape', lies between 36–37° N and 5–6° W, at the meeting point between the Atlantic Ocean and the Mediterranean Sea (Fig. 10.1). The climate is Mediterranean-type, with humid, cool winters and warm, dry summers. Most of the rain falls in winter, when temperature and evapotranspiration (ETP) are low, leading to an excess of water and runoff, but in summer the precipitation is almost nil, the temperature and thus ETP are high and, once the water stored in the soil is exhausted, there are two to three months of drought stress. The average annual precipitation is about 700 mm, but in the mountains it may increase locally up to 2000 mm, due to the geographical location (Fig. 10.1). The Guadalquivir valley and the gulf of Cádiz are open towards the Atlantic Ocean, facilitating the entrance of low-pressure, humid fronts coming from the southwest. When these prevailing, wet winds encounter the Aljibe and

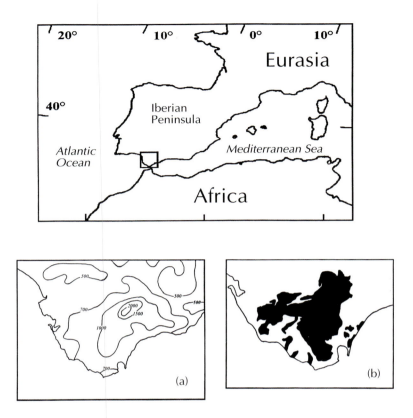

Fig. 10.1. Location of the Gaditan Cape, in the south of the Iberian Peninsula, and insert maps of annual precipitation isolines (a) (after JA, 1995) and of the Oligo-Miocene sandstone formation (b) (after Didon *et al.*, 1973).

Grazalema mountains, they discharge heavy, orographic rains. Another important weather peculiarity is the frequent formation of clouds and fogs during the summer in these mountains, which reduces the severity of the drought stress.

The Gaditan Cape is also peculiar in its geological features. Unlike the limestone mountains of the Alpine orogenesis that generally fringe the Mediterranean Basin, a sandstone rock of the Oligocene–Miocene period (Aljibe formation) forms a rough relief with heights of up to 1092 m (Aljibe peak), which is covered by thin, acidic, nutrient-poor soils (Fig. 10.1; CSIC, 1963; Didon *et al.*, 1973).

Oak Woodland

This physical scenario contains a woodland dominated by evergreen cork oak, *Quercus suber*, with an average density of 163 trees ha^{-1} and average trunk diameter of 23 cm (Ojeda *et al.*, 1994). *Q. suber* is a West-Mediterranean

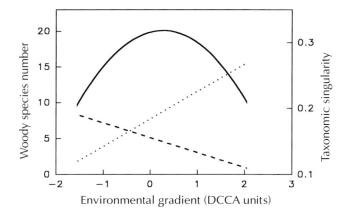

Fig. 10.2. Ecological trends of woody species richness (solid line), endemic species richness (dashed line) and taxonomic singularity (dotted line). Modified from Ojeda *et al.* (1994).

oak tree favoured by a less dry climate and acidic soils. A deciduous oak, *Q. pyrenaica*, also of West-Mediterranean distribution, but frost-tolerant, occupies a few thickets at altitudes higher than 900 m above sea level. In the more fertile and humid valley bottoms, the semi-deciduous, Iberian–North African oak, *Q. canariensis*, is dominant, probably displacing the cork oak by competition. Another two oak species are present in the area: the acidity-tolerant prostrate shrub, *Q. lusitanica*, found mostly in heathlands on ridges, and the calcicole shrub, *Q. coccifera*, at the fringes of the sandstone formation, on marl and clay soils (Ojeda *et al.*, 1994). Surprisingly, the otherwise common holm oak, *Q. rotundifolia*, is absent from the area although it grows on limestone mountains nearby.

The diversity of the tree species is low; one to four species were recorded on average in a set of 100 m transects. Exceptionally, up to seven species, including tall arborescent shrubs, were sampled in riparian forests (Ojeda *et al.*, 1994). European forests were decimated during the glaciations; only a few tree species persisted in refuges and were able to recolonize northwards (but see Adams and Woodward (1989) for an alternative explanation).

The shrub understorey is relatively species-rich. In a study measuring lineal cover in 100 m transects, up to 25 woody species per sample were recorded, and a trend towards higher species richness at intermediate levels of the environmental gradient, in the cork oak woodland, was found. Most of the endemic species were found at one of the gradient extremes, in the heathlands on ridges. At the other gradient extreme, in the understorey of *Q. canariensis* woodlands, shrubs tend to belong to less-diversified genera, that is they are more isolated from the taxonomic point of view (Fig. 10.2; Ojeda *et al.*, 1994, 1996; Arroyo, 1997).

The Gibraltar area is located at the western edge of the Mediterranean region, which borders with the Euro-Siberian (north), the Saharo-Arabian

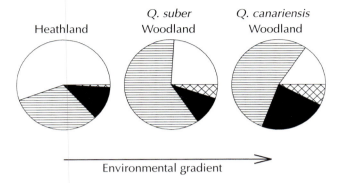

Fig. 10.3. Proportion of endemic (white), Mediterranean (horizontal lines), Mediterranean–Euro-Siberian (black) and Mediterranean–Macaronesian (crossed lines) species in three dominant types of woody plant communities. Modified from Ojeda *et al.* (1994).

(south), the Macaronesian (west) and the Irano-Turanian (east) regions. Most of the shrub species have a Mediterranean distribution. Endemic species are more numerous in heathlands on ridges than in woodland. Euro-Siberian species are important (but still less than 25%) only in the more humid, deciduous *Q. canariensis* woodland. There are few species with Macaronesian connections, but none with Saharo-Arabian or Irano-Turanian distributions. In summary, most of the woody species in these oak woodlands are truly Mediterranean, a considerable portion is endemic to the South Iberia/North Morocco region, and the influence from neighbouring floristic regions seems small (Fig. 10.3; Ojeda *et al.,* 1994, 1996).

Historical Milestones

The ecological history of the oak woodlands in the Strait of Gibraltar region from the last glaciation until the present is poorly known. There are few palaeoecological studies in the area (Gutiérrez *et al.,* 1996) and researchers have to deal with the added difficulties of finding suitable organic-rich sediments in this semi-arid environment, and distinguishing between several *Quercus* species of similar pollen types but contrasted habit and ecology (deciduous vs. evergreen, trees vs. shrubs). Precise documentation of woodland types and the extent of this area have been available only since the Cadastre of Marquis of Ensenada, which was produced as recently as 240 years ago (Bauer, 1991).

Given the general deforestation of the Mediterranean Basin, why does this extensive oak woodland of about 1000 km^2 persist? Why has it not been clear-cut, burnt, ploughed or overgrazed? We propose three 'favourable' historical events that have contributed to the preservation of this woodland.

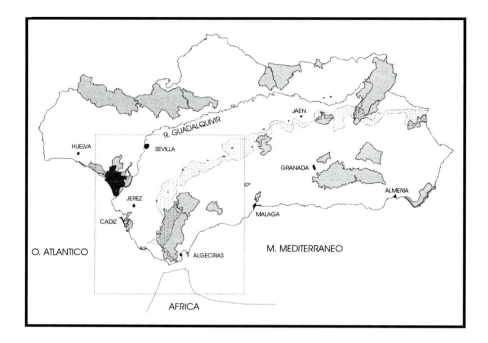

Fig. 10.4. Medieval frontier (after Torres, 1974) and location of Natural Parks (outlined areas). In the insert map, communication flows from Algeciras Bay are drawn. See also contrasted village patterns at both sides of the frontier.

Medieval frontier

The Arabs invaded North Africa and most of the Iberian Peninsula in the eighth century, and during the subsequent 700 years there were battles and moving frontiers between Arab and Christian territories. The last stronghold of the Arabs was the kingdom of Granada, which had a relatively stable territory for about 200 years, until it was conquered at the end of the fifteenth century. Between the Arab and Christian border lines (Fig. 10.4) there was a no-man's land. On both sides of that frontier we find villages today that still retain as part of the name a reference to their historical location (Spanish 'frontera' for frontier), for example Castellar de la Frontera and Jimena de la Frontera on the old Arab side, and Jerez de la Frontera and Arcos de la Frontera on the Christian side.

The between-frontiers situation impeded the settlement of villages, reduced the pressure of agriculture and grazing, and favoured the regeneration and persistence of wooded lands (López Ontiveros *et al.,* 1991). The present-day distribution of towns and villages is largely inherited from this medieval period. In fact, there are no human settlements inside the limits of the present 1700 km^2 of the Los Alcornocales Natural Park. Moreover, the fewer and larger villages in the old Christian territory can still be distinguished from the smaller and more dense villages on the old Arab side. Five present-day Natural Parks – Grazalema, Subbética, Mágina, Cazorla and Sierra María (Fig. 10.4) – are wooded areas following approximately that old frontier strip.

Historically, the Aljibe mountains have also been isolated from neighbouring towns. Although the nearby Bay of Algeciras harbours two very active ports, Algeciras and Gibraltar, it has very little relation with this abrupt, little inhabited area; instead it is more connected with the coastal towns of Cádiz and Málaga, and with the inland Ronda by the Guadiaro river (Fig. 10.4). However, a highway is currently being planned between Algeciras and Jerez crossing through this woodland.

Value of cork

A second historical milestone started with the invention in France, in about 1750, of the cork stopper for glass wine bottles. By the middle of the nineteenth century, the fabrication of cork stoppers, and thus the harvesting of cork oak bark, was industrialized on a large scale. In 1865 the harvesting of cork in public woodland in southern Spain was put out to tender for the first time (González *et al.,* 1996).

Cork is the outer layer of the bark of *Quercus suber,* and can be stripped off relatively easily. In this process the cambium must be left undamaged, allowing a new layer of cork to be regenerated. The extraction takes place every 9 years and, although it may shorten the life of the tree, it is not destructive for the woodland, that is, cork is a renewable resource.

The pricing of cork was crucial at that time because the period coincided in Spain with a political movement towards a general privatization of public lands, then owned by the Crown, the church or, in some cases, the village. The purpose of this 'desamortización' was to improve the productivity of the lands by taking them out of so-called 'dead hands'. Despite the laudable intentions the result was catastrophic for the woodland, because most of the newcomers buying the land immediately clear-cut the forest for profit, and cultivated the land – not always with success (Bauer, 1991; Blanco *et al.,* 1991).

In contrast, in the cork oak woodland of southern Spain, government foresters designed a woodland management plan, starting in 1890, to renew the population of oaks and to rationalize the harvesting of cork. That planning ended the chaotic traditional use of public woodland: cutting for fuel or for tannins (for the leather industry), or burning to improve the grazing. However, this 'production' management of woodland brought some negative effects, such as the elimination of ancient oaks, classified as non-productive after 160–200 years old. With this widespread practice we lost many witnesses of our woodland history. Another undesirable effect of the foresters' plan was the selective elimination of *Quercus canariensis*. This species had been used for pollarding and for acorn-feeding of pigs, but after about 1950 its economic value decreased and it was partly cleared out, despite its high ecological value (González *et al.*, 1996). Management plans need to take into account the effects of different silvicultural practices on biodiversity.

The historical interactions between humans and oak forests have resulted in two dominant wooded landscapes in southern Spain: the oak-savanna *dehesa* and the cork oak woodland (Ojeda, 1989). The holm oak, *Quercus rotundifolia*, produces valuable acorns, traditionally used to feed free-range pigs, so that the oak forest has been historically transformed into a savanna landscape of pruned trees and grassland understorey (Avila, 1988; Marañón, 1988). On the other hand, the acorns of cork oaks, *Quercus suber*, are less valuable as a foodstuff and the tree's main value resides in its cork, and as a result, a relatively dense woodland landscape is selected, which can be considered as a partly 'reconstructed' woodland (Avila, 1988).

Woodland conservation

The transition to a democratic regime in Spain coincided with an international sensitivity towards nature conservation which was recognized in the new Spanish Constitution (BOE 1976, article 45). Later, a national Law for the Conservation of Natural Areas and Flora and Fauna was passed (BOE, 1989). A model of autonomic regions is now being built in Spain and environmental policy is decided by regional governments. The Andalusian Region (in southern Spain) has a Law for the Inventory of Protected Natural Areas (BOJA, 1989), in which different types of protected landscape are recognized (e.g. Natural Park, Natural Reserve and Natural Monument), and protected areas now cover some 17% of the region. The regulation of environmental planning is established in the Law through the application of a set of plans (e.g. Plan for the Ordination of Natural Resources, and Plan for Use and Management).

Los Alcornocales Natural Park was declared in 1989, partly to protect the unusual wooded landscape, and partly because of its high biodiversity and richness in endemic species. About 23% of its area is publicly owned and 77% is private. More than 100,000 people live near the Park and use its natural resources. One of the objectives of the Park is to promote eco-development, by regulating traditional uses such as cork extraction, grazing by free-range cows, pigs and goats, hunting, harvesting of piñon-pine cones and heath burls (for pipe wood), and promoting new activities such as eco-tourism (Sánchez, 1994).

Table 10.2. Environmental and historical factors preserving oak woodlands in the Gaditan Cape (southern Spain).

Physical	Rough relief with peaks up to 1092 m
	Sandstone substrate covered by acid, nutrient-poor soils
	High annual precipitation (700–2000 mm) with summer fogs
Ecological	Dominance by *Quercus suber*
	High plant biodiversity and endemic richness
Historical	Location as medieval frontier (thirteenth–fifteenth century)
	Value of cork (nineteenth century)
	Woodland conservation (twentieth century)

Conclusion

The existence and preservation of this 1000 km^2 oak woodland in the much-deforested Mediterranean is the result of a combination of physical, ecological and historical factors (Table 10.2). The rough relief, sandstone substrate and high precipitation set the conditions for both a woodland dominated by *Quercus suber* and a high biodiversity and richness in endemic plants. The location as a medieval frontier limited the establishment of villages in this wooded area; the pricing of cork protected these cork oak woodlands in a period of woodland privatization and deforestation; and finally the future of the woodland seems optimistic with its legal declaration as a Natural Park in 1989.

Acknowledgements

The ecological study was carried out with Juan Arroyo and Fernando Ojeda, supported by the Spanish DGICYT (project 91/894), and Javier Sánchez, Director of Los Alcornocales Natural Park provided facilities. Vicente Jurado helped in tracing historical documents, Eulogio Parrilla Alcalá, Joaquín Mendoza Pérez and Juan Cara helped with the figures.

References

Adams, J.M. and Woodward, F.I. (1989) Patterns in tree species richness as a test of the glacial extinction hypothesis. *Nature* 339, 699–701.

Arroyo, J. (1997) Plant diversity in the region of the Strait of Gibraltar: a multilevel approach. *Lagascalia* 19, 393–404.

Avila, D. (1988) *Explotaciones agropecuarias en Sierra Morena Occidental.* IDR, Universidad de Sevilla, Sevilla.

Bauer, E. (1991) *Los montes de España en la Historia,* 2nd edn. MAPA, Fundación Conde del Valle de Salazar, Madrid.

Blanco, R., Clavero, J., Cuello, A., Marañón, T. and Seisdedos, J.A. (1991) *Sierras del Aljibe y del Campo de Gibraltar.* Diputación de Cádiz, Cádiz.

BOE (1989) *Ley 4/89, de 27 de marzo, de conservación de los espacios naturales y de la flora y fauna silvestre.* BOE, Madrid.

BOJA (1989) *Ley 2/89, de 18 de julio, por la que sea prueba el inventario de Espacios Naturales Protegidos de Andalucía y se establecen medidas adicionales para su protección.* BOJA, Sevilla.

CSIC (1963) *Estudio agrobiológico de la provincia de Cádiz.* CSIC, Sevilla.

Didon, J., Durand-Delga, M. and Kornprobst, J. (1973) Homologies geologiques entre les deux rives du détroit de Gibraltar. *Bulletin Societé Geologique Francaise* 15, 77–105.

González, A., Torres, E., Montero, G. and Vázquez, S. (1996) Resultados de cien años de aplicación de la selvicultura y la ordenación en los montes alcornocales de Cortes de la Frontera (Málaga), 1890–1990. *Montes* 43, 12–22.

Gutiérrez, A., Nebot, M. and Díez, M.V. (1996) Introducción al estudio polínico de sedimentos del Parque Natural de 'Los Alcornocales'. *Almoraima* 15, 87–92.

Huntington, E. (1915) *Civilisation and climate.* Yale University Press.

JA (1995) *Plan de Medio Ambiente de Andalucía.* Junta de Andalucía, Sevilla.

López Ontiveros, A., Valle Buenestado, B. and García Verdugo, F.R. (1991) *Caza y paisaje geográfico en las tierras béticas según el Libro de la Montería.* Junta de Andalucía, Córdoba.

Marañón, T. (1988) Agro–sylvo–pastoral systems in the Iberian Peninsula: *Dehesas* and *Montados. Rangelands* 10, 255–258.

McNeill, J.R. (1992) *The mountains of the Mediterranean World. An environmental history.* Cambridge University Press, New York.

Ojeda, J.F. (1989) El bosque andaluz y su gestión a través de la Historia. In: García, G.C. (ed.) *Geografía de Andalucía.* Tartessos, Sevilla, pp. 315–355.

Ojeda, F., Arroyo, J. and Marañón, T. (1994) Biodiversity components and conservation of Mediterranean heathlands in Southern Spain. *Biological Conservation* 72, 61–72.

Ojeda, F., Marañón, T. and Arroyo, J. (1996) Patterns of ecological, chorological and taxonomic diversity at both sides of the Strait of Gibraltar. *Journal of Vegetation Science* 7, 63–72.

Paola, G. and Ciciliot, F. (1998) Examples of woodland management related to timber supply for shipbuilding in the 18th century in western Liguria (NW Italy). In: Watkins, C. (ed.) *European woods and forests: studies in cultural history.* CAB International, Wallingford, pp. 157–163.

Sánchez, F.J. (1994) La gestión del Parque Natural de Los Alcornocales (Cádiz-Málaga): influencia en la población rural. In: *Simposio mediterráneo sobre regeneración del monte alcornocal.* IPROCOR, Mérida, pp. 23–26.

Thirgood, J.V. (1981) *Man and the Mediterranean Forest.* Academic Press, London.

Torres, C. (1974) *El antiguo reino nazarí de Granada (1232–1340).* Anel, Granada.

Landscape Evolution on a Central Tuscan Estate between the Eighteenth and Twentieth Centuries

11

M. Agnoletti[1] and M. Paci[2]

[1]Istituto di Assestamento e Tecnologia Forestale, Università di
Firenze, Italy; [2]Istituto di Selvicoltura, Università di Firenze, Italy

**The last two centuries of landscape evolution on a farm in central
Tuscany were investigated. The authors report information concerning
the history of the forest vegetation of the area from the Middle Ages. A
large-scale study defines the distribution of land use types from 1823
and analyses the direction of landscape changes. A small-scale study
investigates the factors influencing secondary successions. The evolution
of farming activities and population changes had influenced the
territory of the farm in the following ways: (i) growth of the forest and
shrubland cover, due to colonization of old fields and pastures, and
reduced landscape diversity; (ii) increased floristic diversity, due to
secondary successions in chestnut stands, pinewoods, oakwoods and old
fields; (iii) growth of structural diversity inside single landscape patches;
(iv) increased fire risk because of dense understorey; and (v) loss of
cultural heritage, in the form of local practices and traditions and loss of
the skill to use small-scale resources.**

Introduction

The object of this study is to define the evolution of the landscape in a part
of the territory of the Gargonza estate between 1823 and 1995. The changes
were closely related to social and economic factors which had a strong
influence on the structure and dynamics of forest and agrarian landscapes.
Large-scale and small-scale changes were studied in order to define the
general variations in land use and secondary succession. Another purpose of
this study is to suggest guidelines for the management of an area now mostly
devoted to recreation and tourism, an interesting application of historical
investigation to the management of the forest resources.

The Environment

The estate of Gargonza is named after the castle to which it belongs, placed on the hills of western Valdichiana Valley, near the town of Monte San Savino (Arezzo Province). It covers an area of approximately 700 ha, going from the plain to 600 m above sea level. The main geological formation of the territory is Macigno sandstone, with fluvial deposits and marsh in the valley.

The soils (sandy loam, pH 6.5, profile type A–C) belong to acid-brown soils of the old classification and to different classes of Inceptisols under modern soil taxonomy. The climate is sub-Mediterranean: mean annual temperature is 12.5°C, annual temperature variation 19°C, annual rainfall 925 mm, with a minimum during the summer when there may be two months of drought. Most of the area is included in the phytoclimatic hot under zone of *Castanetum,* according to the Pavari classification. The vegetation belt is the 'lower sub-Mediterranean' (Mediterranean series of *Quercus pubescens*) according to Ozenda.

Sources and Methods

We carried out a detailed analysis of approximately one-third of the farm (267 ha), situated on the upper side of the hills. This area had a small amount of cultivated land, but many pastures and much woodland in the past. The sources considered were a 'cabreo' of the second half of eighteenth century, the accounting books of the estate, the cadastre of Tuscany made in 1823, two maps of the estate made by the land agent around 1920–1930 and in 1958, and some aerial photo surveys made in 1954 and 1985. Other information was collected with the help of oral history and the analyses of material evidence. The actual situation on the ground was studied through field survey.

The large-scale investigation concerned five different land use types: cultivated land; pasture and wood-pasture; shrubland; woodland; and urban (the castle). The woodland was further divided into: pinewood; coppice; mixed woods (pine/oak); and chestnut orchards.

For statistical analyses we used: Hill's diversity number (Ludwig and Reynolds, 1988) related to effective number of land use types contributing to landscape diversity (N1 = e^H, where H is the diversity index of Shannon, calculated on the proportion of land use types); and the dominance index (DI) of Shannon and Weaver (1962) calculated on the proportions of different land uses (O'Neill *et al.,* 1988).

We also carried out small-scale investigations, using sample plots 30–80 m long and 10 m wide, divided into squares of 25 m². For each individual tree in the plots, species, height, age and cover were determined. The spatial distribution of woody species was studied through: Clapham dispersion index (variance/average ratio); and tests of association among woody species based on 2 × 2 contingency table (presence–absence) and χ^2 test, the null hypothesis being that species distributions are independent.

From Roman Age to the Birth of the Estate

Valdichiana was considered the granary of the Etruscan country in the first century AD; Plinius the Young describes it as a pleasant place, with a nice climate, quite populated, with flourishing agricultural activity and many forests. The fall of the Roman Empire and the barbarian invasions caused deep changes. In much of the valley was a swamp. People lived in villages surrounded by walls on the top of the hills, and used to cultivate vines, wheat and olive trees, along the hill slopes (Guidoni and Marino, 1972). The swamp extended for all the length of the valley, almost 40 km, and much of its width (25 km). Roads ran high on the hills with bridges to cross the swamp, each town had a port since the valley included a big lake linked with the Tevere River and Rome to the south, and the Arno River and Florence to the north.

The first information on the castle of Gargonza goes back to the thirteenth century, as it seems that Dante Alighieri stopped there one night on his way to Rome (Repetti, 1855). Between the thirteenth and fifteenth centuries the valley was divided between the rule of Florence, supporting the Pope, and the rule of Siena, supporting the Emperor. The situation was very unstable and several times hired armies conquered the towns of the valley, as they passed frequently from one lord to another. The forest, mostly oaks and chestnuts, was present in many areas, and also some timber trade was carried out in the eastern hills. At the end of the fifteenth century the Republic of Florence decided to start to dry the swamp, entrusting Leonardo da Vinci with the project. In 1503 he drew a detailed map of the valley, now collected in the Royal Library at Windsor, showing an archipelago of hills emerging from the swamp, which at that time had probably reached its greatest extent. However, only a small amount of the swamp was drained at that time. Halfway through the sixteenth century Florence gave Gargonza to the family of Della Stufa; in this period we have the first reliable information on its forests. The statute of Monte San Savino, the town nearby, reports the prohibition of cutting oak and chestnut trees, especially the big ones, in this area.[1] It is probable that the growth of population in this century had already had some effects on the woodland. A description of the seventeenth century divides the territory of the valley into cultivated area, pastures and swamp vegetation, but no forest is mentioned (Guidoni and Marino, 1972). In this period, works to dry the swamp were increasing and they would last until the final drainage of the valley in the nineteenth century. New arable land was created and farmhouses built in the plains.

In 1727, after the coming of the Corsi family, Gargonza was turned into an estate, divided into many smallholdings, each one with its farmhouse. In the second half of the eighteenth century the owner decided to make a topographic survey, all the fields were measured and the maps, together with the pictures of the farmhouses, collected in a cabreo. The drawings show very clearly the way the land was cultivated: the hills around each farmhouse

---

[1] Bilbioteca Comunale di Monte San Savino, Statuti di Gargonza, anno 1509.

were terraced, and each terrace was held up by a dry wall. Olives and vines were cultivated together, and maple, *Acer campestre*, used to train vines. Large areas were devoted to pastures and wood-pastures, while woodland covered most of the rest of the territory (Agnoletti, 1989).[2]

The Organization of the Estate between the End of the Eighteenth and the Beginning of the Nineteenth Century

Unfortunately not all the maps of the cabreo were preserved, but the organization of the estate can be understood from the accounting books made at the end of the eighteenth century and still present in the castle. Gargonza had 23 holdings: eight of them were placed on the higher side of the hills, between 450 and 600 m, six between 300 and 400 m, nine on the lower side of the hills and the plains. Cultivation was carried out with a 3 year rotation, wheat and rye were growing on one-third of the land, minor cereals were sown in the second third, while the rest was fallow land. Wheat was the most important crop, besides wine and olive oil; other crops were corn, barley, oats, vetch, lupins, great millet and millet. In the plains, flax and hemp were cultivated by the farmers, and the woven linen sold to merchants. However, the most important income of the farm came from livestock breeding, representing 52% of the total income; 450 sheep, 183 pigs, 25 mules and 85 cows were kept on the estate and were free to graze in the woodland. Between the eighteenth and nineteenth centuries important changes occurred. The owners decided to extend the cultivated areas, terracing more hill slopes and carrying out works for the management of watersheds, since the reduction of woodland increased erosion. In 25 years, the growth in wheat production was very high, almost 40%, but vine production also increased by 30% (Agnoletti, 1989). This caused the reduction in woodland cover, a common feature of Tuscany in this period. In fact the Grand Duke Pietro Leopoldo allowed wood cutting with no restrictions, and deforestation occurred all around the country.

The cadastre made by the Duke of Tuscany in 1823 is the first reliable document on the estate. The landscape described is quite complex showing many land uses. Cultivated land was divided into arable land, arable fields with olives, with vines, with poplars; pastures were also of different types: pastures, pastures with oaks, with olive trees, with mulberry trees. The forest had different features: high forests, chestnut orchards (for nut production), coppice, trees along the hedges, bushes. Cultivation played a different role according to the geographical position of the holdings. On the higher slopes of the hills, chestnuts and pastures were the most important element. The harvest of wheat was not sufficient to feed the farm families, so the harvest of chestnuts was very important to feed the peasants; in fact the castle had a drying room for chestnuts, as well as an olive mill. In the holding called 'Casali', at 530 m above sea level, cultivated land (2 ha) was surrounded by a stone wall, to protect it against grazing, while the rest of the land was chestnut orchard (14 ha), with

2 State Archive of Florence, Guicciardini Corsi Salviati, Cabreo di Gargonza.

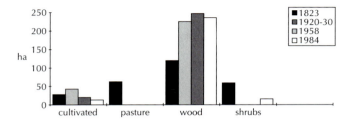

Fig. 11.1. Extension of cultivated fields, pasture, woodland and shrubland in Gargonza, from 1823 to 1984. Urban area was not considered because it remained unchanged.

some pasture inside. The lower slopes had a small amount of pastures and more cultivated areas, while the plains were entirely cultivated.

Landscape Changes between 1823 and 1984

Between 1823 and 1920 both cultivated land and woodland increased (Fig. 11.1). Over part of this time, 1814–1864, the population almost doubled, from 488 to 858 people.[3] The distribution of population in Gargonza followed two patterns: one was related to the life of the group living inside the castle, the second was more linked to farming, with many families distributed in the territory according to the location of the farmhouses. In 1931 most of the population lived in the farmhouses (80%), the total number of families was 133, with an average of 5.6 people in each family.

Pasture to woodland

Woodland and cultivated land spread on to pastures and shrublands. From 1823 to 1920–1930, pasture and wood-pasture disappeared to be replaced by cultivated land, mixed oak/chestnut forest, pine woods and chestnut coppice. Sheep were no longer kept; pigs and cows were reduced in number as well, and were kept in stall. Most of the woodland existing in 1920–1930 was coppice. More people meant more need for fuelwood and charcoal (much lighter to transport) and thus coppice was managed on the short rotation best suited for charcoal. Coppice was also used for farming activities (poles and tools) and to feed livestock. The rise in population is a general feature of the period and the nearby town of Monte San Savino showed a more than 50% increase. This meant more buildings and one of the most important economic activities, around 1870, was the production of bricks. In the area there were ten kilns, one of them in Gargonza, which needed enormous amounts of

[3] Sources for the analyses of population changes are found in the Archive of Monte San Savino (Arezzo): Censimento del 1814, Filza 1862; Note dei parroci e tassa familiare del 1836, Filza 3241; Registro della popolazione a norma del decreto 31/12/1864, volume quinto; VII° Censimento generale della popolazione del Regno d'Italia, 21/4/1931.

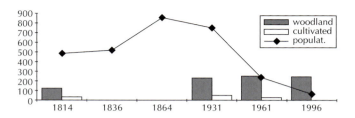

Fig. 11.2. Population, woodlands and cultivated land between 1814 and 1996 in Gargonza (y axis shows ha and number of people).

fuelwood collected from coppice. Some chestnut coppice was still utilized in 1885 using mules for logging since the coppice was far from main roads.

After the Second World War, there was a dramatic fall in the number of people living in Gargonza: peasants left the countryside and went to work in the cities (Fig. 11.2). This is the age of the so-called 'Italian Miracle' and the rise of Italian industry. Together with the strong reduction in population, there was a dramatic fall in fuelwood and charcoal production since people changed to fuels derived from petroleum for heating and cooking. Today fuelwood production is increasing again, but farmers seem to cut only those coppices close to the roads; more remote coppices are now ready to be turned into high forest stands.

Cultivated land

Between 1823 and 1920, cultivated land increased (+6%), but later there was a decrease of almost 11%. The decades after the First World War saw the greatest extension of cultivated areas, but also the start of the reduction of population. At Gargonza the arable land was mostly concentrated in the plains. In 1958 most of the cultivated land, in the area of our study, was turned into a plantation of *Pinus pinaster*, while in the remaining land we found natural pine forest and coppice. Only the farms of the lower hills were still cultivated, olive trees and vineyards were still present on the medium hills, while the upper hills are completely abandoned.

In 1984 there was still some cultivated land, but most of it had been afforested with pine stands or affected by early stages of secondary successions (Fig. 11.3). The pine (*Pinus pinaster* and some *Pinus nigra*) was planted for economic purposes but, since no thinning was done, the timber quality of these stands is poor. The packing industry did use these trees regularly until 1986, especially where pine trees were growing in mixed oak forest which produce better quality timber.

Shrublands

Shrublands played an important ecological and economic role. They are the result of degradation of the woodland or fire (the last big fire destroyed

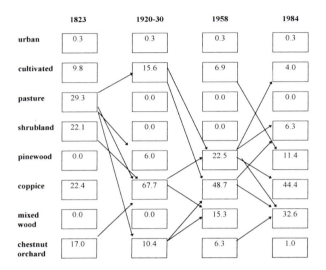

Fig. 11.3. Land use changes from 1823 to 1984, expressed as total surface percentages (267 ha). The arrows indicate the direction of changes.

272 ha of shrublands and forests in 1972). They covered 22% of the land in 1823, but seemed to disappear in 1920–1930 and 1958. However by 1984, 6% of the land was again covered with shrubs. The cadastre of 1823 did not use the word shrubs but 'scope' (brooms): this term did not mean degraded forest, but areas where *Erica scoparia* and *Erica arborea* were cultivated. The interest in these species was due to the production of brooms, then quite common in the area, and still being made today. Shrublands changed quite often over time, as it was easy to shift from coppice to shrubs and vice versa. According to some oral sources, in the sixties, woods were sometimes burnt to allow the growth of heather since its economic value was higher than fuelwood. In this period, cultivation of heather became perhaps more convenient than coppice, since many small businesses were producing brooms. Even in 1985 some lumbermen preferred to cut shrublands than to cut coppice.

Chestnut orchards

Until the end of the nineteenth century, stands of chestnut were very important for the population. In 1920 they decreased by 3%, by 1958 there was only one-third of the previous chestnut forest area, while by 1984 they had almost disappeared. The reduction in farmers' families in the upper slope of the hills had two effects: the first was that chestnut flour was not necessary anymore. Improved agricultural techniques and the extension of cultivation in the plains were giving more crops to feed the population. Secondly, the abandonment of the land meant that no one was left to care for this kind of forest. So the chestnut high forest degraded, due in part to diseases like *Cryphonectria parasitica*. Chestnut trees were cut and coppice and pine

Table 11.1. Dominance index D1 and diversity number N1, applied to land use type analysis.

	1823	1920–1930	1958	1984
D1	0.34	1.10	1.27	1.18
N1	3.54	1.66	1.41	1.52
D1*	0.48	0.93	1.12	1.20
N1*	2.48	1.57	1.31	1.21
D1†	0.50	0.82	0.73	0.78
N1†	4.83	3.52	3.85	3.67

* This analysis considers woods and shrubs together, urban, cultivated and pasture.
† Analysis considers cultivated land, pasture, shrubs and all wood categories separately.

replaced the old high forest; in 1984, there were 87% fewer chestnut stands compared to the beginning of the nineteenth century. Perhaps the most important species since the sixteenth century had nearly disappeared.

The detailed changes in area for each land use type are given in Fig. 11.3. In Table 11.1 the values for the landscape index are reported. The dominance index (D1) was lowest in 1823, when the highest landscape diversity was found (3.54 land use types). In 1920–1930 the dominance index increased, as woodland reached its greatest extent with the maximum value being reached in 1958. By 1984, shrubland had increased, mainly through colonization of fields by woody species, and maximum landscape uniformity was reached. Considering each woodland class, the highest dominance value was reached in 1920–1930 because of coppice extension.

Secondary Succession in Old Fields and Woodland: Small-Scale Analyses

Abandoned fields

In old fields, once cultivated with vines and olive trees, the invasion by tree species (*Acer campestre* and *Quercus pubescens*), shrubs (*Prunus spinosa, Cornus mas, Cytisus scoparius, Rubus* spp., *Rosa* spp.) and herbs, mainly *Graminaceae* and *Leguminosae*, is observed. Field maple (50%), olive tree (25%) and pubescent oak (25%) show an aggregated distribution (Id = 9.26**), with thick strips next to the lower edge of dry terrace walls. Cover is around 40%, with a multilayer structure showing an average height of the dominant layer of 6 m. Maple and oak are significantly associated ($\chi^2 = 4.38$, $P<0.05$). Regeneration near the base of dry walls is favoured by the spread of seeds by birds, and also by ecological factors such as moderate shading, reduced competition from herbs and protection against harvesting and browsing (Salbitano, 1987; Gandolfo, 1994).

On old olive cultivations invaded by shrubs, the resulting structure is characterized by a dense layer, 2 m high, of *Prunus spinosa* (70%) and *Crataegus monogyna* (5%), with scattered *Quercus pubescens* (12%) and *Olea europea* (13%), 6–8 m high. Cover is around 80–100%, and age analyses showed that most species invaded the fields within 20 years after the abandonment of cultivation. *Prunus spinosa* and *Crataegus monogyna*, both having a strong colonizing capacity, tend to be uniformly distributed (Id = 0.30). *Q. pubescens* and shrubs are not significantly associated in space. The thorny shrubs probably favoured the coming of oaks protecting them against browsing, in the first stages of colonization. During the colonization of the old fields, the first stages of succession are quite fast, but they can be slowed down in early stages dominated by shrubs (Puerto and Rico, 1988).

Chestnut orchards

Chestnut stands are present on the east–northeast slopes between 400 and 500 m above sea level. These woods have developed a structure showing two layers: the dominant one consists of old chestnut trees (80 trees ha^{-1}, average height 15 m, G = 4.16 m^2), mainly scarred by chestnut blight. The lower layer is made up of shoots sprouting from old trees, seedlings of *Castanea sativa* (maximum height 6–8 m), *Fraxinus ornus, Ostrya carpinifolia, Acer campestre, Quercus pubescens* and *Corylus avellana. Pinus pinaster* colonizes these stands in the openings. In the undergrowth, *Cytisus scoparius* is abundant, with *Brachypodium rupestre* and *Festuca ovina* common in the field layer. Chestnut stands are thus changing into a mixed hardwood forest.

Coppice-woods

Overgrown oak coppices prevail on the south and southwest slopes, between 450 and 650 m. Most of them are coppice with standards (*Quercus pubescens* and *Quercus cerris*), mixed with *Pinus pinaster* and *Pinus nigra*. There is selection amongst the shoots and some of them eventually overtop the others: thus the coppice is evolving into a mixed high stand of vegetative origin. In the regeneration layer (40–160 cm), under canopy cover, in addition to deciduous oaks, there are also *Quercus ilex, Sorbus torminalis* and *Robinia pseudoacacia*, while *Pinus pinaster* regenerates only in the openings. The most common shrubs in the undergrowth are *Cytisus scoparius, Juniperus communis, Erica scoparia* and *Rubus* spp.

Pinewoods

The *Pinus pinaster* stands were never thinned, thus the trees are quite drawn up, with crooked stems. The stands are dense (800 trees ha^{-1}), with a single layer structure (dominant height between 14 and 20 m, G = 12–25m^2). Mechanical stability is very low with many wind breakages, especially in the even-topped trees, and this is compounded by diseases such as *Dioryctria* sp. *Pinus nigra* stands are also very dense (750 trees ha^{-1}, dominant height 17 m, G = 20–22 m^2) but the stem shapes are better. Regeneration of pine occurs in the openings, formed by wind damage, where some extraction of timber has been carried out. Under less dense canopy cover, *Quercus pubescens,*

Fraxinus ornus, Robinia pseudoacacia, Acer campestre and *Castanea sativa* regenerate. On the other hand, under dense canopy cover, *Quercus ilex* prevails. This secondary succession of broadleaves in artificial conifer stands, is a common ecological process in Tuscany (Bernetti, 1987).

Conclusions

The abandonment of farming activities and population changes have had several effects on the territory of Gargonza.

- There has been growth of forest cover (woodland and shrublands) due to the colonization of old fields and pastures. Woodland is playing the role of a connective element among the patches of the ancient landscape mosaic. With its spread, landscape diversity has been reduced.
- There is increased biodiversity, due to secondary successions, in chestnut stands, pinewoods, oakwoods and old fields. Succession seems to go in the direction of mixed broadleaved woods.
- Structural diversity within individual landscape patches has increased, since colonization occurring in old fields is quite irregular. Also the structure of the old coppice, pinewoods and old chestnut stands is evolving towards a more complex pattern.
- Fire risks are increased because both in woodland and in old fields there is now a dense understorey.
- There has been a loss of cultural heritage, in the form of local practices and traditions, and loss of the skill to use small-scale resources. This is quite evident in the abandonment of many farmhouses and decay of the material evidence of human presence such as dry walls, dams, tracks and old roads. The cultural landscape survives in some land uses such as heather cultivation and small areas with chestnut, both as coppice and high forest.

The actual economy of the area is no longer devoted to agriculture but mostly to tourism and recreation, as the forest of Gargonza represents an important environmental resource for the nearby towns. The conservation of some evidence of the old chestnut high forest and coppice, as well as gradual landscape restoration, could be considered. Secondary succession occurring in artificial pinewoods could be accelerated by thinning, favouring the invasion of broadleaves. In abandoned terraces invaded with flammable bushes, the risk of fire would suggest some management oriented to the replacement of shrubland by plantations of local broadleaved species for timber production such as sessile oak, or reviving olive tree cultivation. It is also important to preserve some of the material evidence of the ancient organization of the territory.

References

Agnoletti, M. (1989) La fattoria di Gargonza fra '700 e '800. Elementi per una storia del bosco in Valdichiana. *L'Italia Forestale e Montana* XLIV, 67.
Bernetti, G. (1987) *I boschi della Toscana*. Edagricole, Bologna.

Gandolfo, G. (1994) *Indagine storico-biologica sull'insediamento della vegetazione forestale in consequenza della cessata coltivazione delle pendici terrazzate nel Ponente ligure (Valle Arroscia, Imperia).* Tesi di laurea, Università degli studi di Firenze.

Guidoni, E. and Marino, A. (1972) *Territorio e città della Valdichiana.* Roma.

Ludwig, J.A. and Reynolds, J.F. (1988) *Statistical ecology.* John Wiley and Sons, New York.

O'Neill, R.V., Krummel, J.R., Gardiner, R.H., Sigihara, G., Jackson, B., De Angelis, D.L., Milne, B.T., Turner, M.G., Zygmunt, B., Christensen, S.W., Dale, V.H. and Graham, R.L. (1988) Indices of landscape patterns. *Landscape Ecology* 1, 153–162.

Puerto, A. and Rico, M. (1988) Influence of tree canopy (*Quercus rotundiflora* and *Quercus pyrenaica*) in old fields succession in marginal areas of central-western Spain. *Acta Oecologica, Oecologia Plantarum* 9, 337–338.

Repetti, E. (1855) *Dizionario geografico, storico della Toscana.* Cinelli, Milano.

Salbitano, F. (1987) Vegetazione forestale ed insediamento del bosco in campi abbandonati in un settore delle Prealpi Giulie (Taipana-Udine). *Gortania, Atti del Museo Friulano di Storia Naturale* 9, 83–143.

Shannon, C.E. and Weaver, W. (1962) *The mathematical theory of communication.* University of Illinois Press, Urbana, Illinois, 125 pp.

An Insight into Past Climate via a Fossil Tree

12

Mesut Inan
University of Istanbul, Department of Forest Botany, 80895 Bahcekoy, Istanbul, Turkey

Macroscopic and microscopic examinations were carried out on the diagnostic characteristics of petrified trees found at the Big Gala Lake near Enez in the European part of Turkey. The trees belong to the White Oak group. The formation in which these petrified trees were found indicates that the material belongs to the Upper Miocene. *Quercus* taxa of similar morphological characteristics are presently growing in the area where the petrified material was found, which would indicate that the climate of the era was similar to that of today.

Introduction

Information obtained from a petrified tree found near Enez allowed estimation of the tree's age, and the climate of the area in the period when it was alive. Background information for the study came from Eroskay and Aytuğ (1982) who have previously described petrified trees which they had found in the East Ergene area. They also studied the climate of that period. The geological context was provided by Saner (1985) who defines five main geological structural features in the Saros Gulf and environs: (i) Hisarlidağ High; (ii) Enez Graben; (iii) Semadirek High; (iv) Saros Graben; and (v) Gelibolu Block. The two Highs and Gelibolu Block are pre-Miocene anticline structures. Enez and Saros Grabens developed in synclines. The Mid-Pliocene sequence makes transgressive overlaps on the flanks of the structural Highs. Enez Graben was filled by sedimentation from the Miocene and more recent periods. However, Saros Graben is still receiving sedimentation.

The research was carried out at Enez (40° 40' 20" north latitude, 26° 0' 30" east longitude), a village which has an altitude of 25 m and is located where the river Meriç (the ancient name is Hebros) meets the Aegean Sea. The present topography has changed dramatically since the Upper Miocene period. The river Meriç has brought a great amount of alluvium to the

Fig. 12.1. A view of the petrified tree.

connection point with the sea, filled the port with the soil and formed a delta. The port lost its importance and now the village Enez is 3.5 km from the sea (Basaran, 1988).

Material and Methods

According to Fengel (1991), two conditions under which ageing processes take place can be distinguished: (i) aerobic conditions, as prevailing in wooden

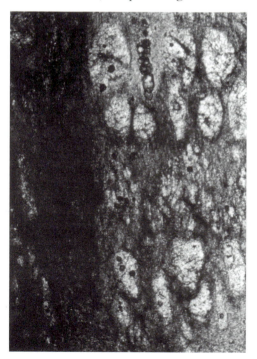

Fig. 12.2. The transversal section of the petrified tree × 400.

Fig. 12.3. The tangential section of the petrified tree × 400.

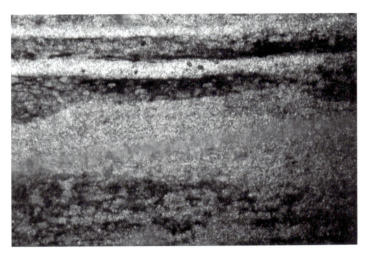

buildings and sculptures etc.; (ii) anaerobic conditions which apply to wooden items buried in the ground or submerged in water, such as foundation pillars, ships etc. Submersion and underground embedding initiate the very slow process of fossilization in which the cell wall substance is transformed into highly condensed compounds (coalification) or is substituted by minerals (silicification).

The petrified tree in this study (Fig. 12.1) which had followed the silicification route was taken to University of Istanbul, Faculty of Forestry, Department of Forest Botany. Small parts of the samples were kept immersed in hydrofluoric acid for 24 h. Then microscopic and macroscopic studies were carried out on those samples. For microscopic examination, transversal and tangential sections at about 30 μm thickness were prepared at the Laboratory of Geological Engineering, University of Istanbul.

For comparison, recent white oak (*Quercus hartwissiana*) samples were examined from our Wood Anatomy Laboratory (see Figs 12.4–12.6). To compare the climatic conditions of the past with the present, current climatic

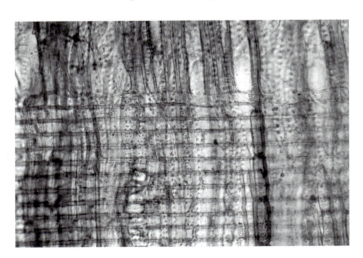

Fig. 12.4. The radial section of a recently grown tree × 250.

Fig. 12.5. The transversal section of a recently grown tree × 400.

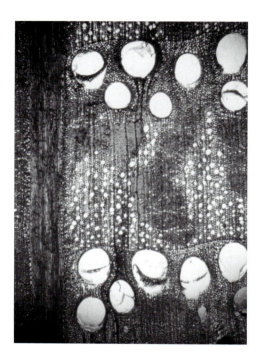

data of the region were obtained from Keşan Meteorological station. According to this data, collected over 36 years by the Thornthwaite Method, the prevailing climate type of the region is dry, mesothermal with a strong water deficiency in the summer period (C_1 B'_2 b'_3) (Anonim, 1974; Çepel, 1988).

Fig. 12.6. The tangential section of a recently grown tree × 100.

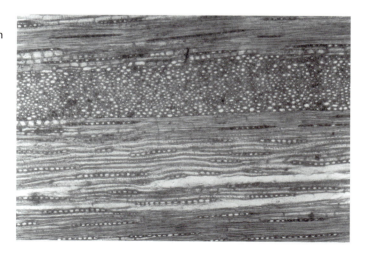

Results

The anatomical characters of petrified trees

The wood is heterogeneous; rays, homogeneous, uniseriate and multiseriate (±20 cells high); the tangential diameter of early wood vessels is 200–250 µm; in late wood their diameter is 20–25 µm (Figs 12.2, 12.3).

The anatomical characters of recently grown material (*Quercus hartwissiana*)

The wood is heterogeneous; rays, homogeneous, uniseriate and multiseriate (±24 cells high); the tangential diameter of early wood vessels is 250–300 µm; in late wood their diameter is 20–30 µm (Jacquiot *et al.*, 1973; Schweingruber, 1990; Bozkurt, 1992) (Figs 12.4–12.6).

These results show that the samples belong to a taxon of White Oak *Quercus* sp. group. The age of the sample is Upper Miocene based on the geomorphological studies carried out in the area (Göçmen, 1976; Sümengen *et al.*, 1983). There are enough similarities between the petrified *Quercus* and the present-day *Quercus* to suggest that the climate at that time must have been very similar.

Conclusion

These trees can be the source for climatic information of past periods. Based on the dendroclimatological studies, it is possible to estimate which periods were rainy, and which periods were dry. By the use of xylological research, we can ascertain from fossilized wood (coal, petrified trees etc.) past climatic conditions for various geographical localities. Turkey is very rich in such sources and further studies on these materials will yield significant information on the climate of Turkey and the geological ages.

Acknowledgement

This work was supported by The Research Fund of The University of Istanbul. Project number: Ö-III/14/120896.

References

Anonymous (1974) *Devlet Meteoroloji İsleri Genel Müdürlüğü Ortalama ve Ekstrem Kiymetler Bülteni.* The Ministry of Sources of Energy and Nature Press, Ankara.

Basaran, S. (1988) *Enez'in Tarihi Önemi, Arkeolojisi ve Turizm Bakimindan Oynayacagi Rol.* Ulusal Enez Kongresi.

Bozkurt, Y. (1992) Odun Anatomisi. *Üniversite yayin no. 3652, Fakülte yayin no. 415.*

Çepel, N. (1988) Orman Ekolojisi, İ.Ü. *Orman Fakültesi Yay. İ.Ü. Yay. No. 3518, Or. Fak. Yay. No: 399.*

Eroskay, O. and Aytuğ, B. (1982) Doğu Ergene Çanağinin Petrifiye Ağaçlari, İ.Ü. *Orman Fakültesi Dergisi Seri A., Cilt 32, Sayi 2.*

Fengel, D. (1991) Aging and fossilization of wood and its component. *Wood Science Technology* 25, 153–177.

Göçmen, K. (1976) Asagi Meriç Vadisi Taskin Ovasi ve Deltasinin Alüviyal Jeomorfolojisi, İst. Ün. Yay. 1990, Cografya Enstitüsü Yayini 80.

Jaquiot, C., Trenord, Y. and Dýrol, D. (1973) *Atlas d'Anatomie des Bois des Angiospermes*. Centre Technique du Bois.

Saner, S. (1985) Saros Körfezi dolayinin çökelme istifleri ve tektonik yerlesimi, Kuzeydogu Ege Denizi, Türkiye. Türkiye Jeoloji Kurumu Bülteni C. 28.

Schweingruber, F.H. (1990) *Anatomie Europäischer Hölzer*. Eidgenössische Forchungsanstalt für wald, Schnee und Landschaft, Birmensdorf (Hrsg), Haupt, Bern and Stutgart.

Sümengen, M., Terlemez, İ., Sentürk, K. and Karaköse, C. (1983) Gelibolu ve Güneybati Trakya'nin Jeoloji Haritasi.

13

What were Woods Like in the Seventeenth Century? Examples from the Helmsley Estate, Northeast Yorkshire, UK

Richard Gulliver

Carraig Mhor, Imeravale, Port Ellen, Isle of Islay PA42 7AL, UK

The nature and value of coppice-woodland on farms in Bilsdale in northeast Yorkshire in 1642 is contrasted with similar attributes of woods elsewhere on the Helmsley Estate. Details of the number and value of oak and ash trees in woods and on the Bilsdale farms are provided. Whilst the farm woods are moderately uniform, the non-Bilsdale woods show a great variety of structure, from woodland containing only standards to virtually all coppice-woodland. This variability is important for our understanding of the seventeenth century landscape. Furthermore it reminds us of the need to retain/recreate woodlands of many differing types of management and canopy structure in the twentieth century and beyond.

Introduction

The studies of Rackham (1971, 1980, 1986, 1990), Jones and Warburton (1993), and Peterken (1993) have produced the following generalized picture of change in woodland structure over the centuries since the Norman Conquest. In the early medieval period, coppice with standard woods were widespread in the lowlands of Britain and on the upland fringe. The coppice might be harvested on a fairly short rotation (e.g. between 4 and 8 years), and the standards felled from 25 to 70 years old (sometimes up to 100 years) to provide small to medium-sized timber (Rackham, 1974, 1990). Through the succeeding centuries the age at harvest of the standards increased and the length of coppice rotation also increased. From the eighteenth century onwards there was a tendency for woods to be managed chiefly for their standards (exclusively so in some places) with the underwood becoming less important. About this overall trend there was an enormous degree of spatial and temporal variation. Indeed it is this very variation which demands that long-term tendencies are only described in the most general of terms.

The state of our knowledge is intimately bound up with the nature of the records from the various historic periods and their subsequent survival. This account describes woodland structure and its variation on a very large estate in northeast Yorkshire in 1642; at a time just before the focus of interest moved away from the coppice element of woodland to the standards. The extent of medieval (*c.*1300) woodland in the area has been identified by Wightman (1968) and the management of oak woods in the eighteenth century described by Marshall (1788).

The Helmsley Estate

The Helmsley Estate of the Right Honourable Francis Lord Villiers ran from the town of Helmsley in northeast Yorkshire, northward up the valleys of Ryedale and Bilsdale, and included the southern edge of the North York Moors eastwards from Helmsley to Kirkdale.

In 1642, the annual value and area of each field on every farm in Bilsdale was estimated by Richard and Thomas Bankes in the Bilsdale Survey (NYCRO ZEW IV 1/5;[1] Ashcroft and Hill, 1980). For many fields, land use data and field names are also provided. Woodland on the estate was surveyed and recorded in the Wood Book (also NYCRO ZEW IV 1/5). Entries fall into three sections: (i) those relating to Ryedale and the land east of Helmsley; (ii) those relating to Bilsdale, which include all the trees and every block of woodland on each farm; and (iii) an addendum giving supplementary information for the whole estate.

The value of the herbage per acre in the woods in Bilsdale is given in the Bilsdale Survey; with the area of the wood and total value of the herbage being listed. All the Bilsdale woods were in the hands of tenants; for the non-Bilsdale woods, some were let and some were managed by the estate. Where the underwood is stated to belong to Lord Villiers, the values are estimated as if the coppice were to be cut and sold. Where the underwood is stated to belong to a tenant, the values appear to be estimated as if the tenancy were to be terminated and taken over by the landlord. The timber in all the woods and on all the farms appears to belong to the estate.

Values in this account are given in pounds sterling with the subsequent divisions of shillings and pence converted to two decimal places for ease of comparison. As the sterling system uses a base of 240, there will be small rounding errors if values are reconverted into shillings and pence. Areas are given in acres (2.4711 acres to the hectare), as this is the form in which the information was originally collected.

The Nature of the Information

The 42 non-Bilsdale entries in the main part of the Wood Book can be divided into six groups. Detailed analysis of the data was carried out on those 14 entries with a value of the underwood in the main part of the Wood Book (group 1, $n = 9$); or its addendum (group 2, $n = 5$). These entries refer

---

[1] NYCRO ZEW IV 1/5 Manuscript documents lodged at the North Yorkshire County Record Office, Northallerton, North Yorkshire.

unambiguously to woods and not to mixtures of woods and non-woodland trees. Twelve entries were in the East Park (group 3). Their location may have made their management atypical. They have a single, pooled value of underwood of £100. For these two reasons they were excluded from the main analysis. One entry refers to two named woods (Rivalkes cum Griff and First Abbott Hagg) in the Wood Book, but only one (Abbott Hagg [the word 'first' being absent]) is named in the addendum, which does however give an underwood value (group 4). Two entries in the addendum have underwood in two parts, one part only having a positive value, the other part being worth nothing (group 5). These entries also have either two woods named in the Wood Book and only one in the addendum (Hocliffe and Crinkle Carr); or multiple entries in the Wood Book and one in the addendum (Greencliff Hagg). Groups 4 and 5 were therefore not used in the main analysis. Seven entries are listed with no underwood value (group 6), and these are assumed to contain only standards. Lastly, six entries refer to trees in and around tenements, garths and closes, i.e. non-woodland trees (group 7). Overall the ratio of entries (some of which refer to several woods) of coppice with standards woods/all-standard woods is 17/6, i.e. excluding group 7 and the 12 entries in the East Park which are incapable of classification.

Only two types of tree appear in the records, oak and ash (*Fraxinus excelsior*). The species of oak (*Quercus robur, petraea* or hybrid) cannot be determined. That other tree species are not mentioned probably demonstrates their absence in the form of standards on the ground; but it may indicate that these other species, though present as standards, were of little value. Twelve of the 42 entries are oak or oak coppice-woods with no ash trees being recorded. There are no examples of ash-only woods, and the coppice-woods for which the most information exists (described subsequently) contain no ash standards.

Pollarded trees are described throughout the Wood Book as dodderells and this word has been retained as their low value suggests old trees. However, the word sapling has been used in the account in place of sampler in the original. Runt oaks are mentioned once; these are taken to be small, misshapen trees that were not pollarded.

Overall Relationships

The balance between coppice and timber

For the 14 woods in groups 1 and 2, coppice varied from above 90% of the total value to below 10% (Fig. 13.1a). The overall trend for woods to be managed less for coppice and more for timber over the period would result in any one wood passing through a series of phases where the value of the underwood declined progressively. In Fig. 13.1a, the woods have been arranged in descending order of percentage value of coppice. This pattern can be considered to be an analogue of the consequences of changes in management over time.

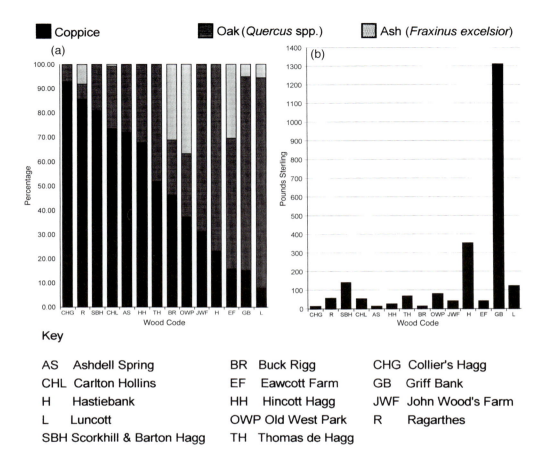

Key

AS	Ashdell Spring	BR	Buck Rigg	CHG	Collier's Hagg
CHL	Carlton Hollins	EF	Eawcott Farm	GB	Griff Bank
H	Hastiebank	HH	Hincott Hagg	JWF	John Wood's Farm
L	Luncott	OWP	Old West Park	R	Ragarthes
SBH	Scorkhill & Barton Hagg	TH	Thomas de Hagg		

Fig. 13.1. (a) Percentage value of the elements of 14 woods which have a value for the underwood; (b) total value of the 14 woods.

Oak is more important in value than ash as the non-coppice component of most of the woods, but ash is more valuable than oak at Buck Rigg and Old West Park. There is no particular association between the total value of the wood (Fig. 13.1b) and the percentage of the coppice element in the total value. Eleven entries have a total value of less than £150; and a value of the tree component of less than £60 (range £1.15 to £50.65; median £14.43). Griff Bank and Hastiebank have total values of over £300 (i.e. £313.09 and £354.33 respectively). Luncott has a value for the trees of £115.28, 92% of the total value of £125.28, i.e. it is radically different in character from the main group of 11 woods. To facilitate comparisons between woods, these latter three woods have been excluded from Figs 13.2 to 13.4. However, the sequence of the remaining 11 woods (high percentage coppice value on the left, low percentage coppice value on the right) has been retained.

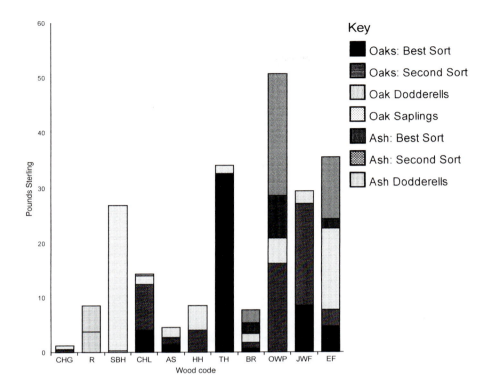

Fig. 13.2. The values of the various categories of oak (*Quercus* spp.) and ash (*Fraxinus excelsior*) trees in the 11 woods that had a total value for the trees of less than £60.

The 11 woods included in Fig. 13.2 show an enormous variation in the value of the non-coppice components. Scorkhill and Barton Hagg consist mainly of sapling oaks, possibly as a result of recent fellings: Thomas de Hagg is largely oaks of the 'best sort'. Ragarthes and Ragarth Hagg consist entirely of dodderell oaks and ashes, this may indicate a former wood-pasture which has been subsequently planted or has experienced natural regeneration. Carlton Hollins, Ashdell Spring and John Wood's Farm all show a balance between the three categories of non-pollarded oak. Buck Rigg (Sproxton), Old West Park and Eawcott Farm show a good balance between oak and ash components, and also between the different categories of oak and ash.

The differences in tree composition would be due partly to previous management and partly to soil conditions, particularly in relation to the degree to which ash is favoured or disfavoured. Around 1642 each wood was being managed in an individual way. This could have been an attempt to optimize return on existing structure as a result of carefully planned management, or the result of a *laissez-faire* situation whereby trees were harvested on an *ad hoc* basis as and when the need for timber or revenue arose.

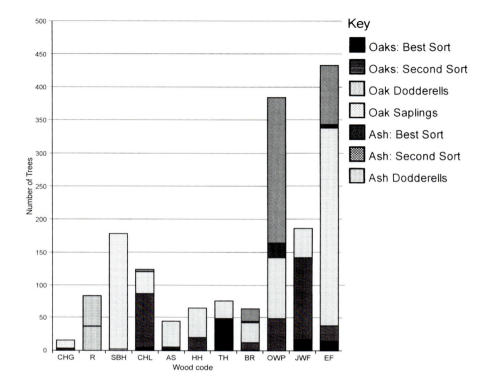

Fig. 13.3. The number of trees in the various categories of oak (*Quercus* spp.) and ash (*Fraxinus excelsior*) in the 11 woods that had a total value for the trees of less than £60.

The equivalent data for the 11 woods is presented as the number of trees in each category in Fig. 13.3. The impression of woodland structure gained from Figs 13.2 and 13.3 is very similar, especially where large numbers of low value oak saplings are present, as in Scorkhill and Barton Hagg and Hincott Hagg. The major differences relate to those woods with a number of the highly valued oaks of the 'best sort', such as Thomas de Hagg and John Wood's Farm.

For each entry in the Wood Book, a single value was used for each category of tree (Fig. 13.4). Oaks of the 'best sort' vary in value from £0.33 at Eawcott Farm to £1.00 at Carlton Hollins and up to £2.00 at Hastiebank (shown in Fig. 13.1a but not in Fig. 13.4), presumably due to differences in size and quality. There is a similar variation in the values of ashes of the 'best sort' from £0.33 at Eawcott Farm to £1.00 at Buck Rigg. In this set of 11 entries, oak pollards are valued at £0.10 to £0.167 and ash pollards at £0.10. Oak saplings range in value from £0.05 to £0.10.

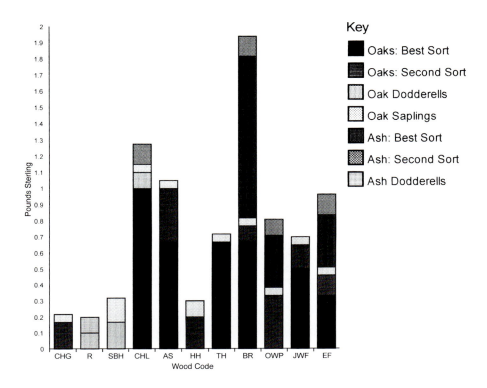

Fig. 13.4. The value per category of tree of oaks (*Quercus* spp.) and ashes (*Fraxinus excelsior*) in the 11 woods that had a total value for the trees of less than £60.

Non-woodland trees

The number of non-woodland trees on the 82 farms in Bilsdale (Table 13.1) can be compared with the 14 woods in groups 1 and 2. Ash saplings are not recorded in any of these 14 woods and are only listed for one of the total of 36 non-Bilsdale entries (groups 1–6), at Sleightholmdale. By contrast, 1193 ash saplings are recorded for the 82 Bilsdale farms. One interpretation of this difference might be a decision not to count the relatively numerous saplings in the woods but to enumerate ash saplings on the farms where they were scarcer and where the resultant trees would be more valued. Indeed, Sleightholmdale is one of the three 'out farms' on the estate, so the entry might include both woodland (coppice is definitely present on the other two 'out farms') and non-woodland trees. The number of oak and ash saplings on the farms is roughly similar, yet there are fewer adult ash trees compared with oak trees. Greater susceptibility to climatic factors, unauthorized removal by tenants and the sale of semi-mature ashes but not semi-mature oaks by the estate may all be contributory factors. The average number of trees per farm of all categories is 95.9.

Table 13.1. The number and percentage of the various categories of oak (*Quercus* spp.) and ash (*Fraxinus excelsior*) in the 14 woods that had a value for underwood and on the 82 farms included in the Bilsdale Survey.

Category of tree	Woods Number	Woods Percentage	Farms Number	Farms Percentage
Oaks: Best sort	971	17.83	777	9.89
Oaks: Second sort	1601	29.40	2223	28.28
Oaks: Third sort	0	0.00	141	1.79
Oak Dodderells	94	1.73	1781	22.66
Oak Saplings	1564	28.72	1126	14.33
Ash: Best sort	394	7.24	97	1.23
Ash: Second sort	678	12.45	420	5.34
Ash: Third sort	0	0.00	0	0.00
Ash Dodderells	143	2.63	102	1.30
Ash Saplings	0	0.00	1193	15.18
Total	5445	100.00	7860	100.00

Value of woodland and non-woodland trees

A detailed picture of the variety of values within each tree category for woods is provided in Table 13.2 which shows the number of records of each value class for the 36 non-Bilsdale wood entries (groups 1–6). Oaks of the 'best sort' range in value from £2 to 6s 8d (£0.33) and oaks of the second sort from £1 to 2s (£0.1). The range for ash is slightly narrower and shows no overlap in values; ashes of the best sort vary from £1 to 6s 8d (£0.33) and of the second sort from 5s to 8d (£0.25 to £0.033). Hence a much more accurate picture of the nature of the trees is provided by the values than is provided by the tree categories used by the surveyors. In Table 13.2 the values have been assigned to six valuation bands to allow comparisons to be made at an intermediate level of accuracy. Table 13.3 shows the actual total number of trees of each value for the 82 Bilsdale farms, with the same six point valuation band applied. The 'best sort' of oaks show a big range of values, but the other categories are relatively uniform. The greatest number of ash dodderells are valued at 1s 6d (£0.075), the same value as some ash saplings, suggesting that these dodderells contained very little useful timber.

The Nature of the Coppice

All the Bilsdale woods were in the hands of tenants. The value of the coppice is given either in the main part of the Wood Book or the addendum, or both. For three woods the value in the addendum is lower (Table 13.4). No indication of why this should be is presented in the original, although as Richard and Thomas Bankes gained more experience of the area, their valuations may have been progressively reduced.

Table 13.2. The number of records for each value class for trees in the 36 non-Bilsdale wood entries (groups 1–6 in the Wood Book) and a synthetic valuation band.

Tree category	Valuation band*																					Total
	1					2				3		4			5			6				
	Value in pounds sterling (pre-decimalization)																					
	£2	£1-10s	£1-6-8d	£1-5s	£1	18s	16s	15s	13-4d	10s	8s	6-8d	5s	4s	3-4d	3s	2-6d	2s	1-6d	1s	8d	
	Value in pounds sterling (post-decimalization)																					
	2	1.5	1.33	1.250	1	0.9	0.8	0.75	0.67	0.5	0.4	0.33	0.25	0.2	0.167	0.15	0.125	0.1	0.075	0.05	0.033	
Oaks: Best sort	3		1	1	4	1	7	1	5	1					3		1	1				29
Oaks: Second sort		3			3		1	1		11	9	1	2									31
Oaks: Third sort										3												3
Oak Dodderells															2			3				5
Oak Runts																				1		1
Oak Saplings																1		2	4	17		24
Ash: Best sort					2				1			5	8	2		1						19
Ash: Second sort									1		2	5		1	1	4	7	1			1	23
Ash: Third sort																		1				1
Ash Dodderells																		1				1
Ash Saplings																				1		1
Total	3	3	1	1	9	1	8	2	7	15	11	11	10	3	6	6	8	9	4	19	1	138
Total per valuation band	[17]					[18]				[26]		[24]			[20]			[33]				[138]

Note: * Applied by author.

Table 13.3. The number of records for each value class for trees in the Wood Book on the 82 farms in Bilsdale and a synthetic valuation band.

Tree category	Valuation band* 1			2			3				4					5		6				Total
	Value in pounds sterling (pre-decimalization)																					
	£2	£1-10s	£1	16s	15s	13-4d	12s	10s	8s	7s	6-8d	6s	5s	4s	3-4d	3s	2-6d	2s	1-6d	1s	8d	
	Value in pounds sterling (post-decimalization)																					
	2	1.5	1	0.8	0.75	0.67	0.6	0.5	0.4	0.35	0.33	0.3	0.25	0.2	0.167	0.15	0.125	0.1	0.075	0.05	0.033	
Oaks: Best sort	1	1	71	40	31	58	7	276	114	7	76	37	53		5							777
Oaks: Second sort									168		205		277	178	538	381	476					2223
Oaks: Third sort															88	53						141
Oak Dodderells															687			909	146	39		1781
Oak Runts																				0		0
Oak Saplings																				1126		1126
Oak total	1	1	71	40	31	58	7	276	282	7	281	37	330	178	1318	434	476	909	146	1165	0	Total oak 6048
Ash: Best sort						5		4			27	3	39	17	2							97
Ash: Second sort													6		15	84	113	195	7			420
Ash: Third sort																						0
Ash Dodderells																		23	75	4		102
Ash Saplings																			151	1042		1193
Ash total						5		4			27	3	45	17	17	84	113	218	233	1046	0	Total ash 1812
Grand total	1	1	71	40	31	63	7	280	282	7	308	40	375	195	1335	518	589	1127	379	2211	0	7860
Total per valuation band		[73]			[134]			[576]					[918]			[2442]			[3717]			7860

Note: * Applied by author.

Table 13.4. Coppice data from the Wood Book (1642) for the woods in Bilsdale and the unlocated woods.

Bilsdale woods	Value in pounds per acre from the main part of the Wood Book	Value in pounds per acre from the Wood Book Addendum	Age of coppice	Total value of coppice in pounds from the main part of the Wood Book	Total value of coppice in pounds from the Wood Book Addendum	Total area of coppice in acres	Number of trees	Value of the herbage in pounds per acre per year
Woods								
Benhill Banck Hagg and Little Close	–	–	–	–	–	33.33	–	0.05
Birk Hagg, Carrcoate	0.5	–	–	5	–	10	0	–
Byrk Wood	–	2	–	–	34.5	17.25	–	–
Crookley Farm	–	–	–	20	–	–	0	–
Eldermire Farm	–	–	–	3	–	–	0	–
Fangdale Hagg	–	–	–	–	–	3.5	0	0.075
Hawtherley Farm	–	–	–	30	–	–	0	–
Loebank Hagg, Newhouse Farm	–	–	–	–	–	2	–	0.05
Myrie Hagg, Helm House	1.33	0.67	6	21.33	10.67	16	0	0.1
Myrie Hagg, Wethercote	1.67	0.67	6	20	10	15	0	0.1
Newhouse Farm	–	–	–	2	–	–	–	0.1
North Hagg, Helm House	1	0.67	6	27	18	27	0	0.1
Spoute House Farm	–	–	–	5	–	–	0	–
Spring Hagg, Ewecott	–	–	–	14	–	–	0	–
The Hagg, Laurell Hall Farm	–	–	–	1.5	–	–	0	–
Wand Hagg, Carrcoat	0.5	–	6	28.75C	–	57.5	0	0.1
Unlocated woods								
Eawcott Hagg	–	1.5	7	–	30.75C	20.5	–	–
Footehead Spring	–	0.33	1	–	20.42	61.25	–	–
Footehead Spring, Part of	–	0.33	1	–	5.33	16	–	–
Middle Parte	–	1	–	–	8	8	–	–
Middle Straw Hagg	–	0.5	–	–	9	18	–	–
Will Hagg	–	–	–	–	19	–	–	–
Windle Straw Hagg	–	0.5	–	–	9	18	–	–
Median value	1	0.67	6	17	10.67	17.25	0	0.1

Note: C = Corrected value.

The age of coppice in the woods in Table 13.4 varied between 1 and 7 years, median 6 years. Even 1 year's growth had an attributed value of £0.33 per acre which suggests that the regrowth after cutting was rapid. The length of the leases is not stated, but it was normal to cut regularly within the term of a lease, for example, every 6 to 8 years for a 21 year lease. Most (possibly all) of the Bilsdale woods had no trees (Table 13.4), so the entire value of the wood was from the underwood. The double valuation for the three woods does suggest that the valuation process was not too reliable. Nevertheless the quality of the underwood must have varied enormously. At Wand Hagg, 6 years' growth is valued at £0.5 per acre, whereas at Eawcott Hagg, 7 years' growth is valued at £1.5 per acre. Taking the median value from the addendum of £0.67 per acre for 6 year old coppice of 17.25 acres, gives an annual income of £1.92 assuming an equal part is cut each year on a 6 year rotation, i.e. £0.11 per acre per year. This figure corresponds to harvesting a single ash of the second sort in Bilsdale of the most frequent value (£0.1) (Table 13.3). An annual valuation of £0.05 to £0.10 per acre has been attributed to the herbage. This valuation is presented without explanation in the survey, so it is not clear whether the herbage was normally cut; or whether it was expected that stock would be allowed into the more mature coppice, where the levels of damage to the underwood would be low. For Eawcott Hagg the tenant is described as having the herbage only, indicating that the herbage and coppice elements of woodland were sometimes treated separately.

The underwood at Middle Parte is described as 'Ingges [wetland] Alders' and the valuation for Hawtherlea Farm is for alder, byrks and hassel 'upon this farme'. This latter entry almost certainly refers to underwood and hence gives an indication of which were the important coppice species in Bilsdale.

For the non-Bilsdale *woods* only those mentioned in the addendum of the Wood Book have an acreage stated; a value per acre is also given (Table 13.5). The value of the coppice at Ragarthes and Ragarth Hagg from the main part of the Wood Book divided by the acreage gives a rate of £0.658 per acre, yet the value stated in the addendum is £1 per acre. The former value is used in Table 13.5 as this relates to the total value used in Fig. 13.1b, derived from the main Wood Book entry. The median value for non-Bilsdale woods is £1.17 per acre with a range of £0.17 to £2.00 per acre. The median value for age is 11 years, range 8 to 16 years. As with the Bilsdale entries, the figures, taken at face value, indicate a great variation in the quality of the underwood; 16 year old coppice at Collier's Hagg being worth £0.67 per acre but 10 year old coppice at Carlton Ragarthes being worth £1.50 per acre. The density of trees is extremely low, range 0.026–1.78 per acre, median 0.62 per acre, so differing degrees of competition between standards and coppice in different woods is unlikely to be a major factor.

Comparing Table 13.5 and Fig. 13.1a, the range of tree densities from Scorkhill and Barton Hagg (0.026 trees per acre) to Thomas de Hagg (1.78 trees per acre) takes us half way across the figure from 92.87% of value as coppice to 51.84%. Hence it seems likely that even those woods with a percentage coppice value of less than 50% still had a fairly low tree density.

Table 13.5. Coppice data for non-Bilsdale woods from the Wood Book (1642).

Woods	Value of coppice per acre* derived from total value and area	Age of coppice	Total value of coppice	Total area of coppice (and of wood)	Density of trees	Trees: oak or oak/ash: D = dodderell	Number of best and second sort trees
Abbott Hagg‡	1.00†	7	–	–	–	–	–
Ashdell Spring	0.17†	–	11.67	70.00	0.086	oak	6.00
Carlton Ragarthes	1.50†	10	26.63	17.75	–	–	–
Collier's Hagg	0.67†	16	15.00	22.50	0.133	oak	3.00
Greencliffe Hagg good part‡	2.00†	–	–	–	–	–	–
Greencliffe Hagg bad part‡	–	12	–	–	–	–	–
Greencliffe Hagg total‡	–	–	–	100.00	–	–	–
Hincott Hagg§	1.00†	–	18.00	18.00	1.11	oak	20.00
Hoecliff Hagg good part§	1.67†	14	–	–	–	–	–
Hoecliff Hagg bad part§	0.001†¶	NS	–	–	–	–	–
Hoecliff Hagg total§	–	–	–	99.50	–	–	–
Ragarthes and Ragarth Hagg	0.658*	–	58.50	76.00	1.12	D oak and ash	85.00
Scorkhill and Barton Hagg	1.5†	–	141.63	76.50	0.026	oak	2.00
Spring Wood	–	8	–	–	–	–	–
Thomas de Hag	1.33†	–	36.65	27.50	1.78	oak	49.00
Median value	1.17	11	26.63	70.00	0.62	–	13.00

* Rate given in the Wood Book Addendum.
† In the tenure of Kirckham.
‡ In the Lord's Hand.
§ In the Lord's Hand.
¶ Not included in the median.
NS, not supplied.
NB certain ambiguous entries in the Wood Book have been excluded, see text for details.

Table 13.6. Woody plants in a 200 × 4 m transect running obliquely downslope through closed canopy woodland at Sleightholme Dale in northeast Yorkshire, grid reference SE 667 885, in October 1982.

Scientific name	Common name	Seedlings	Standard tree	Standard bole (or 2–3 major poles which arise from a multi-stemmed base)	Under-storey/ small tree	Total
Acer campestre	Field maple	2			3	5
Betula pendula	Silver birch		1	1		2
Corylus avellana	Hazel	4			54	58
Crataegus monogyna	Hawthorn	9			5	14
Euonymus europaeus	Spindle				1	1
Fraxinus excelsior	Ash	6	19	1		26
Ilex aquifolium	Holly	7				7
Quercus petraea × *Q. robur*	Hybrid oak		1			1
Quercus robur	Pedunculate oak		1			1
Sorbus aucuparia	Rowan	1*				1
Ulmus glabra	Wych elm	2	1	8		11
Viburnum opulus	Guelder-rose				2	2
Total		31	23	10	65	129

* Dead.

Footehead Spring is described as being of one year's growth but 'cut confusedly'. Presumably the stools had been left in very irregular shapes. At Greencliffe Hagg (Table 13.5) the spring (underwood) had been much spoiled by keeping beasts in the wood; in this case the tenant paid no rent for the wood. The valueless section had 12 years' growth; 10 acres were unspoiled and were worth £2 per acre. Other indications of mismanagement of the underwood include Collier's Hagg 'ill preserved'; Windle' Straw Hagg 'very bad'; Hoecliff Hagg one-third 'very bad'; Middle Strawe Hagg 'very bad'.

In 1576 the hammersmithy at Rievaulx was rebuilt as a cold blast furnace; it worked until about 1647 and was associated with a bloom smithy 3.5 miles (5.6 km) up the River Rye (McDonnell, 1992). Both works required coppice-wood for making charcoal. McDonnell considers that the five entries for dodderell oaks (one for dodderell ashes) out of a total of 42, plus the entry for runt oaks, suggest a general decline in woodland management on parts of the estate as woods were cut over repeatedly to service the iron industry. The low density of standard trees in the six entries in Table 13.5 would support this view. He further considers that closure of iron working would aggravate the situation by removing a market which previously had been so important in determining woodland structure.

Taking the median value of the coppice (£1.17 per acre) and assuming a typical rotation of 11 years (this figure actually being the median age of the underwood that had been valued) with sections cut every year gives an annual revenue of £0.11 per acre, the same figure as for the Bilsdale and unlocated woods (Table 13.4).

Table 13.7. A comparison of the constancy and abundance of the main woody species in the British National Vegetation Classication, Community W8: *Fraxinus excelsior – Acer campestre – Mercurialis perennis* Woodland and W10: *Quercus robur – Pteridium aquilinum – Rubus fruticosus* Woodland.

Species		W8	W10
Trees			
Fraxinus excelsior*	Ash	IV(1–10)	II(1–8)
Acer campestre	Field maple	II(1–8)	–
Quercus hybrids	Hybrid oak	I(2–9)	I(1–10)
Ilex aquifolium	Holly	I(1–8)	I(1–7)
Betula pubescens	Downy birch	I(1–6)	I(1–9)
Sorbus aucuparia	Rowan	I(2–5)	I(1–5)
Quercus robur#	Pedunculate oak	III(1–10)	IV(1–10)
Betula pendula#	Silver birch	I(1–10)	II(1–10)
Acer pseudoplatanus	Sycamore	II(1–10)	II(1–9)
Ulmus glabra	Wych elm	II(1–10)	I(1–7)
Quercus petraea	Sessile oak	I(1–9)	II(3–10)
Shrubs			
Corylus avellana*	Hazel	V(1–10)	III(1–10)
Crataegus monogyna	Hawthorn	III(1–7)	II(1–7)
Acer campestre*	Field maple	III(1–7)	I(1–4)
Cornus sanguinea*	Dogwood	II(1–8)	–
Euonymus europaeus*	Spindle	I(1–5)	–
Ilex aquifolium	Holly	II(1–8)	II(1–9)
Viburnum opulus	Guelder rose	I(1–5)	I(1–4)
Sorbus aucuparia	Rowan	I(1–6)	I(1–5)
Mean shrub cover		48.00%	21.00%
Number of samples		429	379

Notes
The sequence of species follows Rodwell (1991). Roman numerals are constancy in 50 × 50 m plots. Bracketed numerals are domin values.
* Species somewhat preferential to W8.
# Species somewhat preferential to W10.

Modern Woodland Characteristics

The entries in the Wood Book include at least two woods on the west side of the Hodge Beck in Kirkdale/Sleightholmdale. The woodland sloping down to the east bank of the Hodge Beck has been classed as primary by Gulliver (1995). Table 13.6 shows the woody plants present in a 200 × 4 m downslope transect in this woodland in 1982. The shrub species give an indication of the likely composition of the underwood for some of the entries in the Wood Book occurring on limestone.

Traditionally, the management of coppice with standards involves the suppression of less valuable tree species so that they remain components of the underwood. Hence it is likely that several canopy species in modern woods (e.g. *Betula pendula* and *Ulmus glabra* in Table 13.6) would have been present as underwood species in seventeenth century woods in the area. Rackham (1990) states that *Betula* spp. appear to have increased in abundance in woods in the last 70 years; and *Ulmus* spp. seem to have been relatively more common as non-woodland trees and rarer in woods in the Middle Ages (Richens, 1983). It is possible that standards of species other than oak and ash were present in the 1642 woods but they were of so little value that they were not recorded. However, the fact that even oak saplings were counted and valued tends to mitigate against this interpretation. The transect (Table 13.6) lies beside a narrow, downslope track. This has permit-ted extra light penetration, allowing a better survival of the shrub species than in the body of the woodland. Hence, here we have a woodland struc-ture which still has the potential to be managed as coppice with standards, by the heavy thinning of the existing canopy. The more normal pattern in the area is for the shrub layer to be extremely sparse due to long periods of man-agement as high forest. In these cases conversion back to coppice with stan-dard woodland could only be achieved in the short term by extensive replanting of understorey species.

In a survey of the southern half of the North York Moors National Park (Anon, 1995), 236 stands were classified using the British National Vegetation Classification (NVC) (Rodwell, 1991). This survey area included the lands that constituted the Helmsley Estate in 1642. The most frequent stand type (26.7% of records) was W8, the *Fraxinus excelsior–Acer campestre–Mercurialis perennis* community; followed by W10, the *Quercus robur–Pteridium aquilinum–Rubus fruticosus* community (24.6% of records); and then W16, the *Quercus* spp.–*Betula* spp.–*Deschampsia flexuosa* community (22.9% of records). There was one occurrence of W1, the *Quercus petraea–Betula pubescens–Oxalis acetosella* community located on the banks of a side valley to Upper Ryedale.

The extensive data collection undertaken during the NVC ensured that the core characteristics of each community could be recognized throughout the UK, despite considerable geographic variation in species composition at individual sites. It is therefore proposed that this geographic robustness of the communities may also be associated with temporal stability. Table 13.7 shows the composition of woody species of W8 and W10. Three shrub species have a constancy of III or greater for W8, cf. one (*Corylus avellana*) for W10. The species present in many of the non-Bilsdale woods on the more basic soils on the Helmsley Estate in 1642 were probably broadly similar to those listed for W8; and on the less basic soils broadly similar to W10.

Discussion

The apparently short rotation coppice in Bilsdale is the natural response of a tenant optimizing the return on his lease. It is the very absence of standards in these woods that probably resulted in the desire to enumerate the trees that did occur elsewhere on the farms. The high percentage of saplings suggests that either young trees were planted or that some small seedlings growing in field boundaries were protected from grazing. Despite the absence of standard trees, the coppice-woods of Bilsdale represent a continuation of an ancient tradition of woodmanship.

The six group 6 wood entries that have no value for underwood in any part of the Wood Book must be assumed to be high forest. If so, they anticipate trends that were to develop over the next three centuries. The standards themselves may well have been quite small; and the remnants of the underwood were probably still evident; but the management almost certainly had a modern feel to it, especially as it embodied an emphasis on just two types of tree.

The non-Bilsdale coppice-woods stand between these two major systems. The low recorded incidence of standards (median value 0.62 trees per acre) suggests a closer affinity to pure coppice-woodland than to archetypal coppice with standards. For example, the densities of trees from Norfolk woods in the fifteenth century ranged from 5 to 40 per acre (Beevor, 1925); and 2 to 20 per acre for Suffolk woods in the seventeenth century (Rackham, 1974). However, it is just possible that the figures for the non-Bilsdale coppice-woods are misleading, and simply record a period after many standards had been felled. The fact that the survey was undertaken in the first place and that great care was taken to enumerate the non-woodland trees, may lend support to this interpretation. All the coppice-woods for which it is possible to work out a density of trees had either oak standards or oak and ash pollards.

The East Park Woods probably complete this range of woodland structures, with their many high value trees and only £100 value of underwood between 12 wood entries. The dual function of these woods as a landscape amenity as well as wood production units would probably favour the survival of mature and over-mature trees compared with woodlands outside the park.

The reports of the underwood being 'very bad' or 'cut confusedly' highlight the difficulties of woodland management in the period. Labour may have been abundant but not always available with the requisite skills; the situation may have been aggravated by a desire to cut much coppice quickly to service the iron works. Soil factors, species composition, inadequate boundaries and previous management all help to explain the variation in coppice values for underwood of similar ages (Table 13.4). However, coppicing was being undertaken successfully in several of the woods. How would these woods appear today if the market for coppice products had not collapsed? Factors which strongly influence broadleaved high forest in the twentieth century, such as rabbit and deer grazing of tree seedlings and saplings, vigorous growth of *Rubus fruticosus* agg. stands in lighter sections of woodland, progressive colonization of woods by *Acer pseudoplatanus* with its largely rabbit-proof seedlings, invasion by *Rhododendron ponticum* or other ornamental shrubs, would probably have had an even greater impact on coppice or coppice with standards woods.

The 1642 surveys cover all elements of the estate. Woodland and non-woodland timber were both very important. There is only a limited degree of comparison possible between the high forest of 1642 with its oak and ash canopies and the mixed species canopies in the area today. However, oak and ash are still present in good numbers as boundary trees in Bilsdale. It would therefore be possible to see how much this farmland landscape has changed in respect of tree complement since 1642. The various original values for oak and ash trees have been grouped into six valuation bands in Table 13.2 for woodland trees and Table 13.3 for farmland trees. Repeat surveys could be undertaken using the six point scale, and the balance of individuals in each band compared. The densities of trees on the farms in 1642 could be contrasted with those occurring today. Although both oak and ash are woodland trees, they are subject to intense competition when growing in woodland. Once the difficulties of establishment have been overcome, non-woodland areas may represent a more favourable habitat for these species. Despite a natural inclination to consider woods to have sharp boundaries and hence represent discrete landscape units, there is a strong interrelationship between a wood and its surroundings (Forman, 1995).

The woods on the Helmsley Estate in 1642 were extremely variable in the composition of their standards, in the balance of standards to underwood, in the quality of the underwood, and in the length of coppice rotation, (especially between farm woodland and woods elsewhere on the estate). Without the accurate data contained in the Wood Book, this variation would not be evident. Less detailed sources recording information from a small number of woods may give an artificially simplistic picture of the nature of woodland in the past. The Wood Book tells us not only how such woodland has changed on the estate over 350 years, but also that variety in structure and management has a long tradition. This is a message that twentieth century conservationists could heed with advantage.

Acknowledgements

The many people who have helped with this study are thanked most sincerely. A particular debt of gratitude is owed to Mr Michael Ashcroft, Mrs Jackie Chapman, Mrs Mavis Gulliver, Mr Paul Harris, Dr John McDonnell and Dr Charles Watkins.

References

Anonymous (1995) *North York Moors National Park: Phase II Woodland Survey: Year 2.* Ecological Advisory Service, Keighley.

Ashcroft, M.Y. and Hill, M. (1980) *Bilsdale Survey, 1637–1851.* North Yorkshire County Council, Northallerton.

Beevor, H.E. (1925) Norfolk woodlands from the evidence of contemporary chronicles. *Quarterly Journal of Forestry* 19, 87–110.

Forman, R.T.T. (1995) *Land mosaics: the ecology of landscapes and regions.* Cambridge University Press, Cambridge.

Gulliver, R. (1995) Woodland history and plant indicator species in North-East Yorkshire, England. In: Butlin, R.A. and Roberts, N. (eds) *Ecological relations in historic times: human impact and adaptation.* Blackwell, Oxford, pp. 169–189.

Jones, M. and Warburton, D.R. (1993) *Sheffield's woodland heritage*, 2nd edn. Green Tree Publications, Rotherham.

Marshall, W. (1788) *The rural economy of Yorkshire*. Cadgewell, London.

McDonnell, J. (1992) Pressures on Yorkshire woodland in the later middle ages. *Northern History* 28, 110–125.

Peterken, G.F. (1993) *Woodland conservation and management*, 2nd edn. Chapman & Hall, London.

Rackham, O. (1971) Historical studies and woodland conservation. In: Duffey, E. and Watt, A.S. (eds) *The scientific management of animal and plant communities for conservation*. Blackwell, Oxford, pp. 563–580.

Rackham, O. (1974) The oak tree in historic times. In: Morris, M.G. and Perring, F.H. (eds) *The British oak*. E.W. Classey, Faringdon, pp. 62–79.

Rackham, O. (1980) *Ancient woodland*. Edward Arnold, London.

Rackham, O. (1986) *The history of the countryside*. Dent, London.

Rackham, O. (1990). *Trees and woodland in the British landscape*, 2nd edn. Dent, London.

Richens, R.H. (1983) *Elm*. Cambridge University Press, Cambridge.

Rodwell, J.S. (1991) *British plant communities, volume 1, woodlands and scrub*. Cambridge University Press, Cambridge.

Wightman, W.R. (1968) The pattern of vegetation in the Vale of Pickering area *c* 1300 AD. *Transactions of the Institute of British Geographers* 45, 125–142.

Manx Woodland History and Vegetation

14

Rob Bohan

Department of Botany, Trinity College, University of Dublin,
Dublin 2, Ireland

A description of the status, history and management of woodland of the Isle of Man is presented. Origins of Manx woodland are discussed as are historic references to tree cover related to woodland still extant. A comprehensive examination of the cartographic record allowed ancient woodland to be identified for the first time on the island. Vegetation and tree demography of individual sites are also used to establish the antiquity of tree cover and species considered indicative of ancient woodland are discussed.

Introduction

The history and status of native woodlands on the Isle of Man has been generally overlooked in the relevant scientific literature of Britain and Ireland (Tansley, 1939; Peterken, 1996; Rackham, 1976, 1980, 1986). Garrad (1972a) summarizes the available knowledge. This chapter sets out to establish the past management and present status of the woodland vegetation on the island.

The Isle of Man (Fig. 14.1) covers 58,793 ha. It is about 56 km in length from the Point of Ayre in the north to the Calf of Man in the south and at its widest point, roughly Peel to Douglas, about 18 km wide. A slate massif covers the southern three-quarters of the island and is separated into three hill systems: the Mull, the Southern and the Northern Hills. The Northern Hills achieve the island's highest point at Snaefell Mountain, 621 m in height. The summits of these hills consist of narrowly formed peaks, linked by broad cols which separate extensive upland drainage basins. In their lower reaches these valleys become more deeply incised to form gorges and glens (Birch, 1964). The Central Valley lies between the Southern and Northern Hills and is drained by the rivers Neb and Dhoo. A southern lowland occurs to the southeast of the slate massif and is centred on the lower reaches of the Silverburn. Carboniferous limestone is found here (Dackombe, 1990). The island boasts

Key to sites

1. Dhoon Glen

2. Laxey Glen

3. Molly Quirke's Glen

4. The Nunnery

5. Cass-ny-Hawin

6. Colby Glen

7. Glen Maye

A. Glen Roy

B. River Dhoo

C. Santon Burn

D. Silverburn

E. Colby River

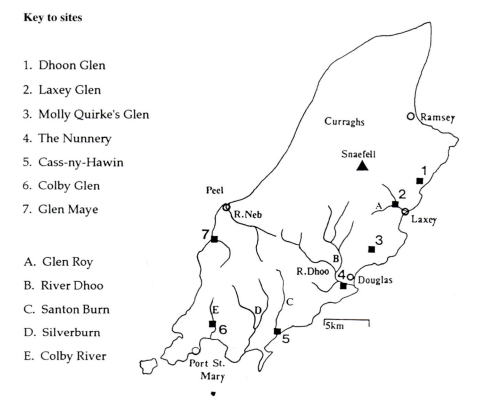

Fig. 14.1. The Isle of Man showing the principal topographic features including sites mentioned in text.

an Atlantic climate with mild winters and cool summers. Annual rainfall is generally high, ranging from 2286 mm in the Northern Hills to 676 mm at Port St Mary in the south (Allen, 1984). High winds from the west affect most of the island. Native woodland vegetation occurs on the sides of gorges, in glens and on cliffs; elsewhere it occurs in wet areas such as the Ballaugh Curraghs. Woodland cover is principally restricted to modern forestry plantations of exotic coniferous species (Davies, 1990). No area of native woodland occupies more than 10 ha.

Vegetation

Woodland vegetation from a number of site types including steep river gorges, glens and sea cliffs was examined. Floristic nomenclature follows Stace (1991) for vascular plants and Smith (1978) for bryophytes. The woodland canopy in the majority of sites is composed of exotic species such as *Fagus sylvatica, Acer pseudoplatanus, Larix* spp. and *Tilia* spp. Most of the glens have been planted up with hardwood species as well as with *Pinus sylvestris* which has allowed the development or survival of a woodland flora. The principal naturally occurring canopy tree in the glens is *Fraxinus excelsior*. In less fertile areas, as well as on damper soils, a mixture of *Betula* spp., *Salix cinerea* ssp. *oleifolia, Ilex aquifolium* and *Alnus glutinosa* dominate the canopy. On the cliff side at Cass-ny-Hawin, *Quercus robur* is most abundant. The understorey at most sites is dominated by native species with *Corylus avellana* being the main constituent. In several sites naturally regenerating *Ulmus glabra* and *Acer pseudoplatanus* occur. *Crataegus monogyna* was noted at several sites; those at Cass-ny-Hawin were both the oldest and most natural looking specimens. The field layers of the woods, given their limited extent, are remarkably diverse. In more acid areas *Vaccinium myrtillus, Dryopteris filix-mas, Rubus fruticosus* agg., *Deschampsia flexuosa, Blechnum spicant, Hyacinthoides non-scripta, Anthoxanthum odoratum* and *Pteridium aquilinum* form a patchy field layer. In base-rich and flushed areas (e.g. at riversides) species such as *Anemone nemorosa, Primula vulgaris, Allium ursinum, Brachypodium sylvaticum* and *Conopodium majus* form dense luxuriant stands. Several stands are affected by grazing, and as a result a suite of grazing-tolerant species such as *Anthoxanthum odoratum, Agrostis stolonifera* and *A. capillaris* occur. Ground layers within the stands are poorly developed and include such typical woodland bryophytes as *Polytrichum formosum, Mnium hornum, Pellia* spp., *Atrichum undulatum* and *Thuidium tamariscinum*.

Methods for the Study of Manx Woodland History

Three representative woodland sites supporting a broadleaved canopy were examined, each illustrating a different facet of woodland history. Historical literature was reviewed for descriptions of these sites and a cartographic survey commenced to establish their historical status. Biological surveys consisting of tree morphology and size class recording were undertaken. Topographical features were noted as were species composition and distribution. Using evidence gathered in the field, cartographic records and historical research, the histories of each woodland have been constructed. Cartographic sources are listed in Cubbon (1994).

Colby Glen (Grid Reference SC 232 708)

Vegetation

Colby Glen is a small wooded glen (Fig. 14.2), less than 2 ha in extent, to the southwest of the island. Non-native *Carpinus betulus, Acer pseudoplatanus, Fagus sylvatica* and *Pinus sylvestris* form a single-aged mixed canopy over the glen with native *Fraxinus excelsior* and *Ulmus* aff. *glabra* of similar age (their girths at breast height are 50–80 cm approximately). The understorey is principally composed of native species such as *Corylus avellana, Ulmus* and *Crataegus monogyna*. Naturally regenerating *Fagus* and *Acer* occur here as well. The field layer includes *Allium ursinum, Athyrium filix-femina, Brachypodium sylvaticum, Dryopteris dilatata, D. filix-mas, Galium aparine,*

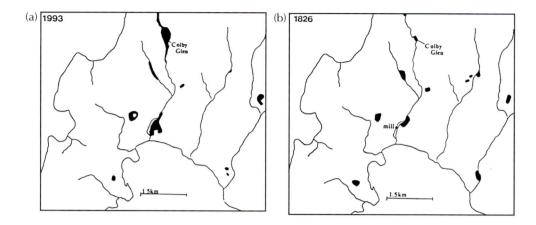

Fig. 14.2. Colby Glen (SC 232 708) to the southwest of the Isle of Man. Woodland cover is shown in black. Note that Drinkwater's map (b) of 1826 shows only a small area of the eastern side of the glen as wooded. Map (a) shows the distribution of woodland in 1993.

> *Hedera helix, Heracleum sphondylium, Hyacinthoides non-scripta, Luzula sylvatica, Phyllitis scolopendrium, Primula vulgaris, Ranunculus ficaria, Silene dioica* and *Veronica hederifolia.*

Ancient trees

One old *Fraxinus* in its second century of growth occurs on the south side of the glen. It appears to have been singled; the trunk being acutely angled at its base. The tree, unlike its comrades, is luxuriantly festooned with epiphytes such as *Polypodium* spp., *Isothecium myosuroides, Mnium hornum, Thuidium tamariscinum* and *Hypnum* cf. *cupressiforme.* Woody climbers include *Lonicera periclymenum* and *Hedera helix.*

Cartography

Colby Glen was not noted on any map of the island as wooded until Drinkwater's map (Fig. 14.2) of 1826 (Cubbon, 1994). This map illustrates trees present on the eastern side of the glen only.

Conclusion

Little evidence of antiquity, in the form of ancient trees or unusual woodland species, has survived at the site. The trees, being for the most part non-native and even-aged, are the result of planting. The cartographic evidence supports this conclusion. Present-day Colby Glen is a plantation probably no more than a century old with some relics from an early planting.

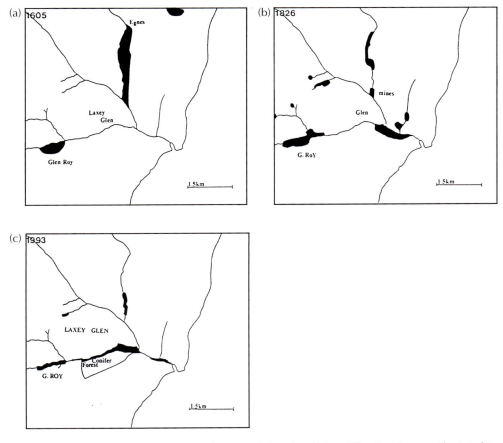

Fig. 14.3. Laxey Glen (SC 430 844) to the east of the Isle of Man. Woodland cover (depicted in black) in 1605 (a), 1826 (b) and in 1993 (c). Within the valley, woodland has been present since at least 1605. Only Glen Roy and one area to the north of the valley are shown as continuously wooded during this period.

Laxey Glen (SC 430 844)

Vegetation

Laxey glen is a wooded glen (Fig. 14.3) to the north of Douglas, amounting to some 4 ha in extent. The vegetation is a mosaic of wet woodland and scrub. Where drainage is free, the canopy is dominated by a mixture of *Acer pseudoplatanus* and *Fraxinus excelsior.* In wetter areas, old *Salix cinerea* ssp. *oleifolia* and rarely *Betula pubescens* dominate. *B. pendula* occurs lower down the valley (Allen, 1984). The canopy species are joined by *Ilex aquifolium* in the understorey. The field layer is composed of typical Manx woodland species, e.g. *Allium ursinum, Anemone nemorosa, Conopodium majus, Hyacinthoides non-scripta, Lonicera periclymenum, Luzula sylvatica,*

Ranunculus ficaria, Rubus fruticosus agg. and *Stellaria holostea*. Glen Roy, at the western end of the valley, supports two Manx rarities: *Phegopteris connectilis* and *Carex laevigata* (Allen, 1984).

Ancient trees

The largest trees in the glen are *Salix cinerea* spp. *oleifolia*. One tree, which appears to have been coppiced, has a stool *c*. 2.6 m in circumference and supports seven poles ranging in girth at breast height from 0.3 to 0.8 m. Given its considerable size, it is probably in its second century of growth. Many younger trees occur in the wood including a pure stand of *Corylus avellana*. These trees are even-aged and in an area that appears to have been cleared in the past. The glen has a long history of mining and these *Corylus* may have been planted and coppiced for pit props. They range in girth at base from 1.5 to 2.0 m.

Cartography

Speed's map (Fig. 14.3) of 1605 (see Cubbon (1994) for details of this and other maps listed below) clearly shows the glen as being wooded. Van den Keere (1627) also shows woodland in the glen as did Blaeu (1645) and Jannson (1646). Rocque's map a century later (1753) does not depict woodland at the site. Fannin's (1789) map also shows the glen as treeless. Fannin was, however, more concerned with depicting man-made features on the island such as roads and houses. Rocque's map was based on King's of 1656, itself based on Speed's of 1605. There is no evidence to suggest that Rocque visited Man, and given the derivative nature of his cartography, no conclusion as to the status of the woodland can be derived from his map. Drinkwater (1826) depicts the Valley in detail (Fig. 14.3). The upper reaches of the Glen (Glen Roy) are depicted as wooded; so too are the northwestern banks of the glen. The rest of the glen is, for the most part, shown as treeless.

History

Occasionally the glen has been described by visiting writers. Feltham (1798) described the glen as being covered in bleaching cloth, with the stream being used to clean it. He also commented on the mining of the valley, 'about one mile and a half up this woody valley you come to the new level, working in pursuit of a vein of lead'. Bullock (1816) describes the valley two decades later, 'Laxey is a village of little trade, composed of about thirty houses, the retreat of fishermen, but the glen is deserving of notice for the romantic beauty of its scenery. It is well planted with trees'.

Conclusion

The written and cartographic evidence suggests a long history of tree cover in the valley. The evidence suggests at least partial tree clearance of the glen in

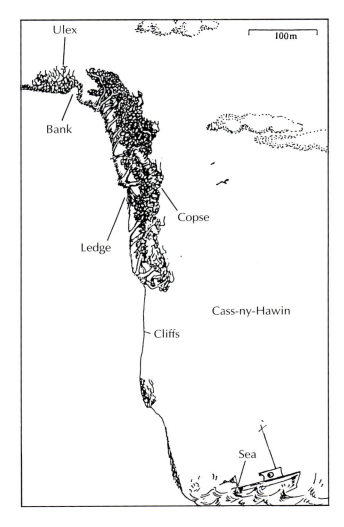

Fig. 14.4. Diagram illustrating the wooded cliffside at Cass-ny-Hawin.

the eighteenth century with replanting taking place soon thereafter which allowed the relict woodland flora to survive. Laxey Glen is, in conclusion, a disturbed ancient woodland.

Cass-ny-Hawin (SC 297 693)

Vegetation

Cass-ny-Hawin wood clings to a limestone cliff (Fig. 14.4) on the mouth of the Santan Burn. It is less than 2 ha in extent. Garrad (1972a, b) and Allen (1984) consider that *Quercus* x *rosacea* is the dominant canopy tree.

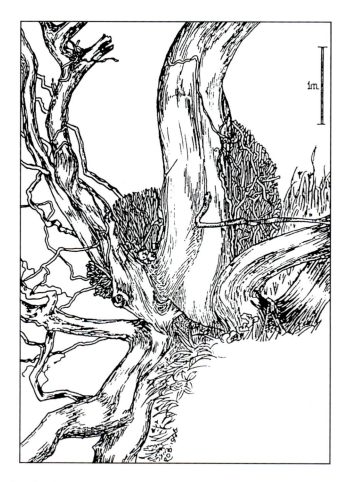

Fig. 14.5. Sketch of a coppiced *Quercus robur* from a cliff at Cass-ny-Hawin. The stool bears six poles ranging in girth from 1.5 to 0.5 m. The circumference of the stool is in excess of 3.5 m. The tree is probably in its fourth century of growth. Note the epicormic shoots surrounding the stool.

However, the field survey revealed *Q. robur.* to be the dominant tree (cf. Clapham *et al.*, 1989; Kelly, 1995). At the top of the cliff, an understorey of *Sambucus nigra* occurs. Elsewhere *Corylus avellana* and a single *Euonymus europaeus* (Dr L.S. Garrad, pers. comm.) occur in this layer. This is the only known site for the latter shrub on the island. The field layer is dominated by grasses including *Holcus lanatus* and *Brachypodium sylvaticum*. Other species present are either grazing tolerant or out of the reach of sheep. These include *Dryopteris dilatata, D. filix-mas, Heracleum sphondylium, Hyacinthoides non-scripta, Lonicera periclymenum, Rubus fruticosus* agg., *Silene dioica, Stellaria holostea, Umbilicus rupestris* and *Urtica dioica*.

Ancient trees

All the canopy trees are coppiced and multistemmed (Fig. 14.5). The tallest *Quercus* is less than 10 m tall due to the strong winds that blow against the cliff. It supports 12 poles ranging from *c.* 25–80 cm girth at breast height. Its stool is about 3.5 m in girth at base and its tallest branches reach *c.* 4.5 m in height from the ground. Despite the antiquity of these stools their epiphytic flora appears restricted. Another *Quercus* is some 10 m tall with five poles which range from 0.7 to 1.3 m girth at breast height, the stool being 3.2 m in circumference. Standing trunks of about 1.5 m in girth have been noted on some *Quercus* (Dr L.S. Garrad, pers. comm.) present in the wood. Several old and gnarled *Crataegus monogyna* are scattered through the wood. At the cliff top an earthen bank of about 0.3–2.0 m in height separates the wood from the pasture above.

Cartography

Woodland has not been mapped on the site even on the most recent 1 : 25,000 6th edition Isle of Man map (1993) because of the difficulty of depicting a vertical plane (the cliff) on a horizontal using modern cartographic formulae.

Conclusion

In the absence of cartographic and literary historical evidence, the morphology of the trees must be relied upon to establish the status of the wood. As the trees appear from their growth form, girth and their repeated coppicing to be in their fourth century of growth, the woodland is ancient *sensu* Peterken (1981).

Discussion

This chapter has demonstrated that although many of the woods are dominated by exotic canopy trees, their flora is for the most part native. Most of the woods examined had semi-natural characteristics. The long-term history of woodland is generally reconstructed through palynology, a relatively undeveloped discipline on the island (Russell, 1988; McCarroll, 1990; McCarroll *et al.,* 1990). Little is known of the Holocene vegetation of the island, although a late glacial pollen diagram exists (Mitchell, 1958). After the late glacial, *Betula* spp. formed woodland. In the Boreal period *Pinus sylvestris* and *Corylus avellana* joined this community. *Quercus* spp. arrived and dominated all areas except for the mountain tops and the wet valley bottoms where *Betula* spp. and *Alnus glutinosa* dominated, respectively. *Pinus* is believed to have later become extinct (Garrad, 1972a).

As the palynological evidence is poor, other disciplines have to be examined to establish the status of Manx woodland. Historical cartography has revealed the existence of woodland on the island throughout the modern

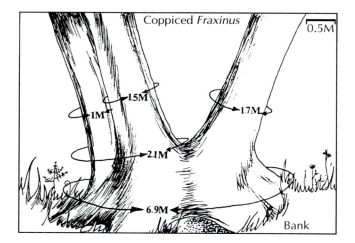

Fig. 14.6. Diagram of an ancient coppiced *Fraxinus excelsior* at Silverdale (SC 275 712). The tree possibly dates from before the dissolution of the Abbey at nearby Rushen in AD 1537.

historic period. Archaeobotanical examination of the woods for features indicative of antiquity has revealed a number of ancient trees. The finest example, a venerable *Fraxinus excelsior*, was found on a bank in the Silverdale area (SC 275 721). The coppiced *Fraxinus* was 6.9 m in girth at base and bore poles of between 1.0 and 1.7 m in girth at breast height (Fig. 14.6). The tree may have been an ancient boundary marker and might plausibly date from the Rushen Abbey period (dissolved in AD 1537) (Broderick and Stowell, 1973). A wood of old trees is described above for Cass-ny-Hawin.

The literary evidence of woodland survival on the island includes an eleventh century reference to a wood at Sky Hill. The late medieval chronicle *Recortys Reeaghyn Vannin as ny hellanyn,* notes for 1056 a *keyll er liargagh y clieau ta enmyssit Skyal* (Broderick and Stowell, 1973), i.e. a wood on the side of a hill called Sky Hill. A mid-thirteenth century reference in the same chronicle notes a thicket at Myroscough (at the Curraghs, SC 365 947). The chronicle also notes another thicket at Ardonan (SC 428 970), Broderick and Stowell (1973). Visitors to the island in the seventeenth and eighteenth centuries generally comment on a treeless landscape (Garrad, 1972a). Nonetheless, Robertson (1794) noted trees at the Nunnery (SC 370 754), a site still wooded today and with a continuous cartographic record for woodland from 1605 onwards. The wood has been heavily disturbed and now presents little evidence of naturalness. Robertson (1794) also noted trees near the Silverdale where the ancient *Fraxinus* (mentioned above) occurs. Bullock's (1816) observations of Laxey Glen's being 'well planted with trees' has been noted above. Biological evidence of antiquity other than old trees is harder to evaluate.

This chapter establishes the existence of three main historical categories: (i) Plantation woodland; (ii) Disturbed ancient woodland; and (iii) Ancient woodland. Plantations include such woods as the Dhoon Glen (SC 454 864), Colby Glen and Molly Quirke's Glen (SC 403 789), each of which is characterized by a poor cartographic and literary history, the presence of many non-native species and the absence of old trees or other archaeobotanical indicators of antiquity. Disturbed ancient woodlands include the Nunnery with a fine cartographic and literary history but poor archaeobotanical evidence and a vegetation dominated by exotics. Other disturbed ancient woods include Glen Maye, Laxey Glen and Silverdale. These woods have a variety of evidence to suggest their antiquity and are undoubtedly ancient woodland sites but have undergone a large degree of disturbance such as tree planting and, in two examples, mining (Garrad *et al.*, 1972). Ancient woodland is exceptionally rare on Man but does occur. Cass-ny-Hawin is a good example of a *Quercus* coppice with many features indicative of ancient woodland. The Isle of Man boasts some interesting woodland and would reward further research into both its history and vegetation.

Acknowledgements

I am very grateful to Dr L.S. Garrad for her help and advice in the study of Manx woodland history; Dr D.L. Kelly and Dr F.J.G. Mitchell for their advice and support; Mr Dave Boyce for introducing me to the Island's natural history; Dr George Peterken for comments on tree antiquity and also the Manx Museum Library staff for their help. The work was carried out through a grant from National Heritage Council of Ireland.

References

Allen, D.E. (1984) *Flora of the Isle of Man*. The Manx Museum and National Trust, Douglas.

Birch, J.W. (1964) *The Isle of Man: a study in economic geography*. Cambridge University Press, Cambridge.

Broderick, G. and Stowell, B. (1973) *Chronicle of the Kings of Mann and the Isles, Recortys Reeaghyn Vannin as ny h Ellanyn*. Learmonth Printers, Stirling.

Bullock, H.A. (1816) *History of the Isle of Man, with a comparative view of the past and present state of society and manners; containing also biographical anecdotes of eminent persons connected with that island*. Longman, Hurst, Rees, Orme and Brown, London.

Clapham, A.R., Tutin, T.G. and Moore, D.M. (1989) *Flora of the British Isles*. Cambridge University Press, Cambridge.

Cubbon, A.M. (1994) *Early maps of the Isle of Man, a guide to the collection in the Manx Museum*. Manx National Heritage, The Manx Museum and National Trust, Douglas.

Dackombe, R. (1990) Solid geology. In: *The Isle of Man, celebrating a sense of place*. Liverpool University Press, Liverpool.

Davies, G. (1990) Agriculture, forestry and fishing in the Isle of Man. In: *The Isle of Man, celebrating a sense of place*. Liverpool University Press, Liverpool.

Feltham, J. (1798) *A tour through the island of Man in 1797 and 1798*. Crutwell, Bath.

Garrad, L.S. (1972a) *The naturalist in the Isle of Man*. David & Charles, Newton Abbot.

Garrad, L.S. (1972b) Oak woodland in the Isle of Man. *Watsonia* 9, 59–60.

Garrad, L.S., Bawden, T.A., Qualtrough, J.K. and Scatchard, W.J. (1972) *The industrial archaeology of the Isle of Man*. David & Charles, Newton Abbot.

Kelly, D.L. (1995) A short guide to Irish oaks. *Moorea* 11, 24–28.

McCarroll, D. (1990) The Quaternary Ice Age in the Isle of Man, an historical perspective. In: *The Isle of Man, celebrating a sense of place*. Liverpool University Press, Liverpool.

McCarroll, D., Garrad, L.S. and Dackombe, R. (1990) Lateglacial and Postglacial environment history. In: *The Isle of Man, celebrating a sense of place*. Liverpool University Press, Liverpool.

Mitchell, G.F. (1958) A Lateglacial deposit near Ballaugh, Isle of Man. *New Phytologist* 57, 256–263.

Peterken, G.F. (1981) *Woodland conservation and management*. Chapman & Hall, London.

Peterken, G.F. (1996) *Natural woodland ecology and conservation in northern temperate regions*. Cambridge University Press, Cambridge.

Rackham, O. (1976) *Trees and woodlands in the British landscape*. Dent, London.

Rackham, O. (1980) *Ancient woodland, its history, vegetation and uses in England*. Edward Arnold, London.

Rackham, O. (1986) *The history of the countryside*. Dent, London.

Robertson, D. (1794) *A tour through the Isle of Man to which is subjoined a review of the Manks history*. Hodson, London.

Russell, G. (1988) The structure and vegetation history of the Manx hill peats. In: Davey, P. (ed.) *Man and the environment in the Isle of Man*. Bar. Brit. Series 54, pp. 39–49.

Smith, A.J.E. (1978) *The moss flora of Britain and Ireland*. Cambridge University Press, Cambridge.

Stace, C. (1991) *New flora of the British Isles*. St Edmundsbury Press, Suffolk.

Tansley, A.G. (1939) *The British Islands and their vegetation*. Cambridge University Press, Cambridge.

The History of the Coniston Woodlands, Cumbria, UK

15

Susan Barker

Environmental Sciences Research and Education Unit, Institute of Education, University of Warwick, Coventry CV4 7AL, UK

The Coniston Basin retains approximately 35% of its area as semi-natural pure and mixed sessile oakwood. This type of woodland is regarded as climax woodland for the English Lake District and its survival in an area there has been extensive woodland clearance over hundreds of years was ensured because of a local demand for coppice-wood. The woodlands on the interfluves are dominated by *Quercus petraea* and *Betula pubescens* whereas mixed deciduous stands containing *Ulmus glabra* and *Tilia cordata* are confined to the more inaccessible gorges. This chapter examines the extent to which silvicultural practices such as coppicing and charcoal burning have influenced the structure and composition of woodland communities. Evidence is provided to link the distribution of rare woodland species and communities to inaccessible areas which have a high degree of woodland continuity. Furthermore, the distribution of soil types also forms a series representing decreasing base-status in relation to interference by humans and topography.

Introduction

The Coniston Basin is located in the south of the Lake District National Park (Fig. 15.1), which is regarded as an area of outstanding natural beauty of national importance (Pearsall, 1969). Cumbria generally has a poor covering of deciduous woodland (about 5%) in common with the rest of the British Isles (Lake District Special Planning Board, 1978). The Coniston Basin, however, retains a large proportion (35%) of its area as woodland which makes an important contribution to the scenery, particularly the view as seen from the west-side of Coniston Water. The woods are particularly extensive on the steep valley slopes on the eastern side of the basin, where they often occur down to the shores of Coniston Water, while at higher altitudes they merge directly into the large conifer plantations of Grizedale Forest or open *Festuca/Agrostis* grassland and *Calluna vulgaris* heath. The woods are semi-natural in the sense that

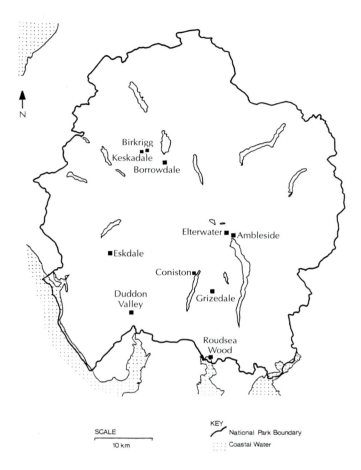

Fig. 15.1. The situation of the Coniston Basin in the Lake District National Park.

they are composed of native species: *Quercus petraea* is dominant with *Betula pubescens; Fraxinus excelsior, Alnus glutinosa, Ulmus glabra, Tilia cordata* and *Corylus avellana* occur frequently. This mixed and pure sessile oakwood has been regarded as the climax vegetation of the Lake District (Pearsall and Pennington, 1973) and is ecologically important because of an 'Atlantic' element in the flora (Steele, 1974; Ratcliffe, 1977).

These woods survived in an area where there has been extensive woodland clearance over the past hundreds of years, because of a local demand for coppice wood. However, because of extensive silvicultural practices such as coppicing over a long period of time, the woodland bears little resemblance to the structure of the primeval forest and this is one reason why they are considered of slightly less conservation value than some other sites in the area (National Trust, 1982). Although the Coniston Basin is one of the National Park areas designated to remain as broadleaved woodland (Lake District Special

Planning Board, 1978), there has been replanting with non-native hardwoods and conifers within it. Today, most of the woods are owned by the National Trust and Forestry Commission with a few privately owned. There is considerable variation in the distribution of the woodland vegetation communities, which form a series representing degeneration of mixed oak woodland to oak/birch woodland and ultimately heath. These communities are now considered against a background of human and vegetation history.

Human History

Mesolithic culture

The earliest evidence of humans in the Lake District is in the Mesolithic period when they probably had little impact on the vegetation (Iversen, 1949; Godwin, 1975). Smith (1970), however, considers that the vegetation at this time was composed in part of secondary communities. In the southern Lake District, Mesolithic communities lived a nomadic existence creating temporary clearings in the woodlands and had few tools apart from fire (Pearsall and Pennington, 1947). They were 'hunters and gatherers' and pollen analysis has shown that in this region, there was little effect on the vegetation (Millward and Robinson, 1974).

Neolithic culture (*c.* 5000 BP)

Incipient forest degeneration and the occupancy of the area by Neolithic humans are closely related. Clearance of woodland was largely for settlement, the cultivation of crops and for the feeding and pasturing of animals. Removal of the forest cover led to soil erosion and increasing base-deficiency, and consequently many areas within the original woodland zone became less suitable for forest growth (Pearsall and Pennington, 1947). Initially this was confined to upland areas but later spread into the lower zones. The grazing of stock was predominantly on the upland pastures but also took place within the woodland, particularly in the winter months. The effects of grazing within the woodland would have been: (i) an increase in Graminaceous species; and (ii) the prevention of natural regeneration of trees and shrubs. Nevertheless, some regeneration did take place, as light-demanding trees such as *Betula* regenerated at the expense of *Ulmus* (Walker, 1966). The decline of *Ulmus* in the area has been attributed to anthropogenic factors and in the Lake District the 'elm decline' is dated at *c.* 5000 BP (Pennington, 1965). It has been suggested that the main cause was the use of twigs and leaves of *Ulmus* to supplement the diet of stock and also for use as winter feed (Pennington, 1965). *Ilex* and *Hedera* are also thought to have been used in this way. Troels-Smith (1960) notes the importance of this browse material in maintaining animal populations greater than could be supported by the grass of Neolithic clearings. The only archaeological evidence of Neolithic cultures in the Coniston Basin is of stone-axe finds at Nibthwaite (Fig. 15.2) (Pearsall and Pennington, 1973) and the discovery of stone hammers in woodland near Kye Wood (Collingwood,

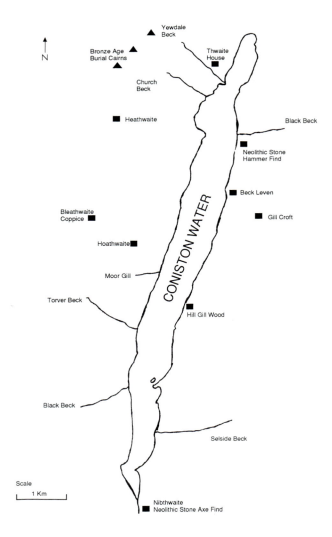

Fig. 15.2. Evidence of pre-historical cultures in the Coniston Basin showing stone axe finds, cairns and place-names with the Scandinavian element 'thwaite'.

1933). The hills around Coniston, however, have substantial evidence of Neolithic settlements, of particular renown is the Langdale stone-axe factory (Bunch and Fell, 1949) and there are also numerous burial and settlement sites (Pearsall and Pennington, 1973).

Bronze and Iron Age (c. 3800 to 2400 BP)

Although archaeological evidence of Bronze and Iron Age occupation in the Coniston Basin is small, indications from pollen analysis suggest that there was appreciable clearance of woodland at this time, particularly on the uplands (Pearsall and Pennington, 1973). The Bronze Age clearances were a result of an extension and intensification of the Neolithic way of life (Turner, 1970), but some of these clearances may only have been small and temporary (Pennington, 1970). In the British Isles in general, Iron Age clearances were more permanent and the clearances were accompanied by soil deterioration and erosion that prevented re-establishment of forest cover. Iron axes were particularly effective in clearing forest, and wood was required to make charcoal for use in smelting iron ore. Thus large areas of woodland were cleared (Turner, 1970). In the Lake District, because there is so little evidence of Iron Age occupation, it is thought that the Neolithic/Bronze Age culture remained largely undisturbed until the Roman invasion (Collingwood, 1933).

Romans (*c.* AD 200–300)

The Roman occupation had little impact in this area in comparison with the rest of the British Isles. They left their mark in the formation of roads and settlements but there is no evidence of appreciable clearance of the woodland except in the immediate neighbourhood of Roman stations, for example, at Ambleside (Pearsall and Pennington, 1973). The Romans, however, introduced to the area the art of smelting iron by the bloomery method, and so wood for the making of charcoal would have been locally in demand. A Roman Age rotary quern was discovered at Nibthwaite (Fig. 15.2), which indicates that cultivation of grain took place in this area (Pearsall and Pennington, 1973). The Anglo-Saxons and Danes who succeeded the Romans in the rest of the British Isles apparently never lived in Furness (Collingwood, 1896).

Vikings (*c.* AD 895)

Severe woodland exploitation followed the Norse land-takes which took place between AD 900 and 1000 (Pearsall and Pennington, 1947). These Norse invaders were spreading from bases in Ireland and the Isle of Man and were farmers whose farming system was based primarily on grazing. They also used wood for most purposes, cooperage work especially, and they smelted iron by the bloomery method using charcoal (Collingwood, 1896). This great demand for wood combined with the intense grazing of stock which was a consequence of the increased human population, led to a further deterioration of the woodland. There was an increase in cleared and settled areas. The tremendous influence of the Norse can be seen from the abundance of Scandinavian place-names in the area. The element 'thwaite' means woodland clearing (Pearsall and Pennington, 1973) and there are five 'thwaites' in the Coniston Basin (Fig. 15.2). Topographical names of frequent occurrence are fell (fjall), dale (dalr), beck (bekkr), gill (gil) and tarn (tjorn); for comparison the Norwegian form is bracketed (Pearsall and Pennington, 1973). The old name of Coniston Water is Thurston Water which is thought to be named after a Viking king, Thorstein, whose kingdom centred at Coniston (Kingston) (Collingwood, 1925).

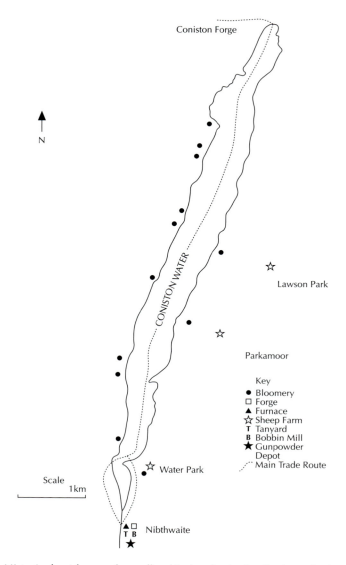

Coniston Forge

N

CONISTON WATER

Lawson Park

Parkamoor

Key
● Bloomery
□ Forge
▲ Furnace
☆ Sheep Farm
T Tanyard
B Bobbin Mill
★ Gunpowder
   Depot
⋯ Main Trade Route

Scale
1km

☆ Water Park

▲□
T B Nibthwaite
★

Fig. 15.3. Historical evidence of woodland industries in the Coniston Basin.

Cistercian Monks (AD 1127–1537)

In AD 1127, Furness Abbey was established and most of the east side of Coniston Water was under its ownership. In the original foundation charter the area was designated as forest (Collingwood, 1927). The monks were mainly sheep farmers but oxen and swine were also kept. Grazing of stock was predominantly on the upland pasture but by grazing the woodland as well, the number of sheep could be increased and consequently Furness Abbey became the largest wool producer in the region (Pearsall and

Pennington, 1973). In the middle of the fourteenth century, the Abbey established three large sheep farms in the Coniston Basin – Water Park, Parkamoor and Lawson Park (Fig. 15.3) (Collingwood, 1927) – which supported over 10,500 sheep (Cowper, 1898). In modern times, up to 16,500 sheep have been recorded (Collingwood, 1927), the increase being a consequence of the increased availability of other foodstuffs for winter feed. Currently, numbers of sheep are considerably lower because much of the land which previously formed these large sheep parks has been converted into plantations of conifers. The place-name element 'park' is of French origin indicating an enclosure for the purpose of sheep rearing or deer enclosure, introduced into England only after the eleventh century (Millward and Robinson, 1974). In addition there are several woodland place-names in the Coniston woods which indicate that extensive grazing took place:

- Grass Paddocks   GR SD 298914
- Cow Brown Coppice   GR SD 305932
- Dodgson Pasture   GR SD 300926

The monks also smelted iron and relied on local wood for the production of charcoal. Thus from 1150 onward there was the rise of the charcoal burning industry, with areas of woodland being enclosed and converted to coppice to ensure a regular supply of wood.

1537–present

In 1537, after the dissolution of Furness Abbey, most of the land came into the hands of the Crown and was then subsequently sold to tenants. Most of the original woodland had been converted into sheepwalk, enclosed as coppice or had degenerated to scrub. The woodland industries continued, and demand for wood began to exceed supply, especially when blast furnaces were introduced in 1711 (Marshall, 1958).

The woods of the Coniston Basin supplied timber for charcoal production from medieval times until around 1930 (Millward and Robinson, 1974) although the height of the industry was between 1711 and 1800, which correlates well with the height of iron manufacture in the area (Marshall, 1958). The first production in the region of iron using charcoal may have been in Neolithic times but the first documentary evidence is in the Coucher Book of Furness Abbey dated 1271 (Fell, 1908) where reference is made to charcoal pitsteads. Charcoal pitsteads are still in evidence today and appear as flat circular terraces on otherwise steep slopes and are linked by stone-covered packhorse ways. Virtually all the woods in the Coniston Basin have evidence of charcoal burning and a map of Dodgson and Bailiff Woods (Fig. 15.4) indicates the intensity of this industry. The making of charcoal in circular piles appears to have been the method adopted in Furness from the earliest times and remained unchanged until the industry ceased (Fell, 1908). The pitstead where the charcoal was made is a circular clearing 7 to 10 m in diameter, practically level and usually found in a conveniently sheltered position with a supply of water close at hand. A stake was placed at the

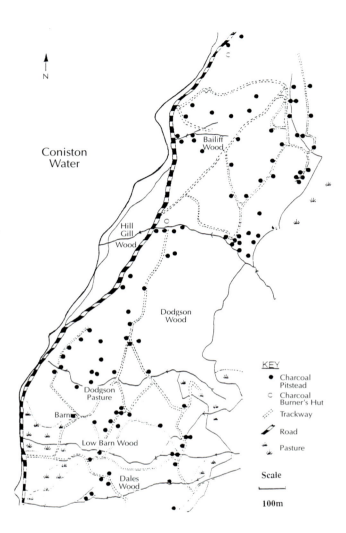

Fig. 15.4. The distribution of charcoal pitsteads in Bailiff Wood (GR SD 34/303930, from Marshall and Davies-Shiel, 1969) and Dodgson Wood (GR 34/304935).

centre and wood piled around concentrically, then the stack was covered with grass and on top was placed a layer of earth. The stake was removed and the pile fired (Fell, 1908). This method of production of charcoal had the advantage that the wood could be charred on site. The charcoal burners lived in the woods and up until 1770 they were accompanied by their families (Fell, 1908); remains of their dwellings can still be seen today.

The charcoal was used mainly for iron smelting and this took place in the near vicinity with many small bloomery hearths situated on the shore of Coniston Water (Fell, 1908). The remains of some bloomeries are visible today, but their age is uncertain. The descendants of the Norse may have been the first to smelt iron on the shores of Coniston Water, before the monks of Furness Abbey, but the date is put at no earlier than the tenth century (Collingwood, 1898). Large cinder accumulations at Beck Leven (GR SD 313950) and Parkamoor (GR SD 302932) indicate that smelting must have been carried on at both sites for a considerable period (Marshall and Davies-Shiel, 1969). Large amounts of slag also occur all along the western shore of Coniston Water in the shingle. However, a royal decree in 1564 prohibited the working of bloomeries for industrial purposes because of alleged denudations of the woodland and also to ensure an adequate supply of charcoal to the Mines Royal (Fell, 1908). The continued use of the bloomeries was allowed but only for the owners' own purposes. So important was the charcoal, that the ore which was mined in Low Furness was brought to the bloomeries on pack-horses or by boat along the lake. The bloomery method of producing iron existed until 1711 when blast furnaces were introduced (Marshall, 1958). The local sites producing charcoal were unable to meet the increased demand for the blast furnaces and charcoal was imported from Scotland (Marshall, 1958).

Heavy demands of the iron industry led to an almost complete exhaustion of local timber, and from as early as the sixteenth century the danger of depleting these woodlands was realized (Fell, 1908). Management policies were introduced which aimed at making maximum use of growing timber thus ensuring a lasting economic return. New woods were planted and existing timber supplies safeguarded. To this end the Coniston coppices were coppiced on a 16 year cycle (Marshall and Davies-Shiel, 1969). Coppice with standards was used where there was a need for larger timber. The standards also furnished seeds for natural regeneration and in some cases protected the coppice against frost. The coppiced wood was also sold to bark-peelers, charcoal burners, coopers, hoopers, bobbin turners and swill makers. There are five main types of coppice in the Coniston woodlands: coppice with standards included (i) *Corylus avellana* coppice with *Quercus petraea* standards; (ii) *Corylus avellana* coppice with *Fraxinus excelsior* standards; simple coppice types were (i) *Corylus avellana* coppice; (ii) *Quercus petraea/Betula pubescens* coppice; and (iii) *Alnus glutinosa* coppice.

The gunpowder industry also used charcoal made from *Juniperus communis* and *Alnus glutinosa* (Pearsall and Pennington, 1973). Gunpowder mills (established from 1764 onwards) were sited conveniently close to the Coniston woodlands and there was a gunpowder depot at Nibthwaite. The gunpowder was used for blasting in the Coniston coppermines, but with the introduction of new explosives such as dynamite, the industry declined and the last works to close were at Elterwater in 1930 (Pearsall and Pennington, 1973). Many other industries ensured the continuing wealth of the woodland: oak bark was used in the tanning of leather; oak timber was in demand for structural beams, and several types of wood, for example *Fraxinus excelsior* and *Acer pseudoplatanus*, were used in the production of bobbins. Swill making, which is an art of weaving split oak into a basket, has gone on in Furness for centuries and has still not entirely ceased. Such was the value of the woodland that most of the industries mentioned in this section were sited conveniently close to the source of supply (Fig. 15.3).

Fig. 15.5. The extent of the Coniston woodlands at three dates in history.

The value of the woodland may also have stimulated its expansion from the area shown on Yates' map dated 1786 (Yates, 1786), through the nineteenth century and to the present day (1974) (Greenwood 1818; Ordnance Survey 1974), as a result of amenity planting of hardwoods and conifers (Fig. 15.5).

Areas treated as coppice were enclosed to prevent sheep and deer grazing the shoots. Wood may have been used for fencing or primitive walls may have been constructed, but the drystone walls which are evident today originate from the period of land improvement from about 1750 and 1850 (Pearsall and Pennington, 1973). Today, the walls are in a degenerate state and sheep and deer are allowed to graze freely. Many of the coppices have not been cut for at least 50 years, and as a result have become dilapidated; in others the stools have been singled and the wood converted to high forest. The decline of coppicing is connected with the loss of former markets. Existing coppice is confined mainly to sections of more extensive high forest areas and, of the woods in the Coniston Basin named coppice, only Bleathwaite Coppice and Torver Coppice Wood show that structure now.

It is no longer economic to manage the woodlands as in the past because of cheaper alternatives in the fuel and construction industries, although a great deal of wealth lies in the amount of standing timber. Without the markets that ensured the survival of the woodlands in the past their future could be uncertain.

Table 15.1. The composition of the forest in Grass Paddocks (GR SD 298911) *c.* 5000 BP (after Pennington, unpublished).

Species	Pollen	Basal area (%)	
		*	+
Quercus	40	37	58
Tilia	10	54	35
Betula	10	5	1
Alnus	5	1	2
Corylus	15	(no correction factor available)	

Basal areas of *Ulmus* and *Fraxinus* below 3%.
NB Correction factors used to calculate basal area:
* Anderson's correction factor (Anderson, 1970).
+ Bradshaw's correction factor (Bradshaw, 1981).

Vegetation History

Pollen analysis of lake sediments in peat deposits in the Lake District are numerous (Franks and Pennington, 1961; Oldfield and Statham, 1963; Pennington, 1964) and outlines of the post-glacial vegetational history of the Lake District in general have been reconstructed (Pearsall and Pennington, 1947; Walker, 1966; Pennington, 1970). However, there are often differences between sites reflecting the individual history of each drainage basin (Pennington, 1965). Generally, pollen analysis shows that in the Lake District there has been a progressive decline in woodland cover since Neolithic times (Pennington, 1970).

A hollow located in Grass Paddocks (GR SD 298911) was found to be infilled with organic deposits, predominantly amorphous peat overlying a waterlain sediment (mud). Pollen analysis carried out by W. Pennington (unpublished) shows the composition of the woodland *c.* 5000 BP to be a mixed deciduous woodland dominated by oak and lime (Table 15.1). The pollen spectra then records three major clearance episodes of *Quercus* and *Tilia* which were correlated with a dated diagram from the sediments of Coniston Water (Pennington, 1983) to have occurred at: (i) AD 900–1000; (ii) AD 1200–1300; and (iii) AD 1600.

Declines of *Quercus* and *Tilia* allowed a corresponding increase in *Betula* which has been locally dominant from *c.* AD 100 to the present. *Ilex* also responded to the early clearances and was continuous from *c.* AD 600 with a maximum from AD 1000. There was temporary recovery of *Quercus* and *Tilia c.* AD 1400–1600 with a corresponding decrease of *Betula* and *Ilex*. However, the highest values for *Tilia* and *Fraxinus* occurred prior to 2000 BP. From AD 1600, *Tilia* has remained at its present density, because it is now virtually unable to regenerate by seed this far north (Pigott and Huntley, 1981), and hence declines at each clearance. *Quercus*, however, re-established after a minimum *c.* AD 1700, and expanded rapidly in the nineteenth century (as confirmed by the map evidence). The vegetation at the

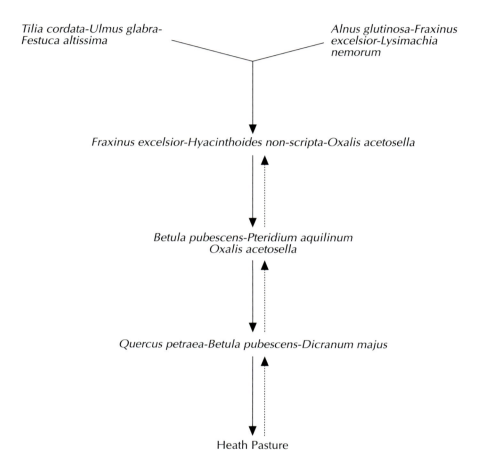

Mixed Deciduous Woodland

Tilia cordata-Ulmus glabra-Festuca altissima

Alnus glutinosa-Fraxinus excelsior-Lysimachia nemorum

Fraxinus excelsior-Hyacinthoides non-scripta-Oxalis acetosella

Betula pubescens-Pteridium aquilinum Oxalis acetosella

Quercus petraea-Betula pubescens-Dicranum majus

Heath Pasture

Fig. 15.6. The vegetation of the Coniston woodlands.

time of the clearance episodes appears to have been 'wood-pasture' with an open canopy of *Quercus* and *Betula*. *Calluna* remains below 5% so open heath was not present locally. Wood-pasture also agrees with the name of the wood, Grass Paddocks.

Present Vegetation

Five woodland communities were identified using National Vegetation Classification techniques after Rodwell (1991). These can be regarded as a series representing a change from mixed oak woodland to oak/birch woodland and ultimately heath (Fig. 15.6). The main slopes are occupied by

Quercus petraea–Betula pubescens–Oxalis acetosella (oak, birch and wood sorrel) woodland (W11) on brown earths and brown podzolic soils, with *Quercus petraea–Betula pubescens–Dicranum majus* community (W17) occupying the shallow podzolic soils on outcrops of grits and slates. *Fraxinus excelsior–Sorbus aucuparia–Mercurialis perennis* (ash, rowan, dog's mercury) woodland (W9) is more localized and associated with springs or streams where the soils are base-rich and usually weakly calcareous brown earths. The streams themselves are often entrenched in steep gullies with sequences of waterfalls called ghylls, characterized by *Tilia cordata* (lime), *Ulmus glabra* (elm) and *Festuca altissima* (wood fescue) on skeletal brown soils. At lower altitudes at the foot of the streams are small areas dominated by alder and ash.

The distribution of these communities is related to a complex of inter-related factors, namely, soil base-status, humus type, soil moisture, humidity and light intensity. However, there also appear to be further vegetational differences associated with the woodland industries and intensity of sheep grazing.

The mixed woodland stands in the ghylls approximate, in terms of composition, to woodland of about 5000 BP. These gorge sites are inaccessible and would have suffered less disturbance particularly from felling and coppicing; furthermore, leaching would be less pronounced as the gorges would be receiving runoff containing leached nutrients from the interfluvial slopes. As *Tilia cordata* is now unable to reproduce by seed in this area, these patches must be of long-standing woodland.

Several rare woodland species are confined to these areas because of the high woodland continuity but also because of their inaccessibility to sheep, e.g. *Hymenophyllum wilsonii* and *Festuca altissima*, respectively. *Hymenophyllum wilsonii* has a requirement for shaded conditions with high atmospheric humidity which in the Coniston woodland is conferred by an overhead canopy and deep gorges. However, it also has a requirement for woodland continuity since if the canopy and shrub layer is lost, light, dry conditions would prevail and *Hymenophyllum* would rapidly die out. *Festuca altissima*, a herbaceous species extremely susceptible to grazing, is confined to these gorge sites but for different reasons (Fig. 15.7). Experimental studies in the Coniston Basin (Barker, 1985) indicate that when the ghylls were fenced to exclude sheep, the fescue grew up to a distance to which sheep could reach through the stock netting. In small, fenced plots the fescue was transplanted successfully and spread, whilst in unfenced plots it was rapidly eliminated. The present restriction to ghyll woodland is imposed by a grazing regime and is preserved in the ghylls by the same factor which has conserved the former woodland, i.e. inaccessibility of the site. Other rare woodland species occur in the gorges possibly because of requirements for high soil base-status, e.g. *Hypericum androsaemum,* although historical factors may be important too. Thus, as a result of a high degree of woodland continuity and heterogeneous topography, these gorge areas are the most important sites in terms of ecology and conservation value because of the diverse flora containing rare species and their apparently irreplaceable nature.

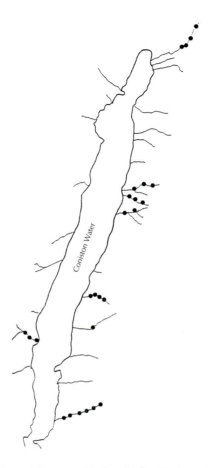

Fig. 15.7. The distribution of *Festuca altissima* in the Coniston Basin.

The woodland communities on the interfluves are of less value for nature conservation having been depleted of species and soil nutrients by centuries of intensive woodland industries and silvicultural practices, combined with naturally occurring soil and climatic changes. The specific impacts of the charcoal burning industry are several fold. The ash and charcoal removal from the woodlands has probably been the major contributing factor to the nutrient impoverishment of wide areas of the Coniston woodland. The charcoal pitsteads themselves physically influence the topography as level terraces on otherwise steep slopes and thus tend to be drier and have deep collections of leaf litter. There is no evidence that any enrichment of such areas which may have occurred in the past has any influence today; pitsteads generally support an impoverished flora typical of the interfluves, dominated by the *Quercus petraea–Betula pubescens–Dicranum majus* (W17) and *Betula pubescens–Pteridium aquilinum–Oxalis acetosella* (W11) communities. Some trees on the edge of the pitstead, facing inwards, still show evidence of scorching.

A more significant and localized impact on the woodland flora arises from the fact that up until 1770 woodland workers, their families and pack-horses lived in the woodland (Fell, 1908). Evidence of their dwellings can still be seen today. Human habitation increases the fertility of the soil, particularly in nitrates and phosphates (Pigott and Taylor, 1964) and in the Coniston woodland there are pockets of *Urtica dioica* adjacent to former dwellings, possibly where the pack-horses may have been tethered. Pigott (1971) regards the presence of *Urtica* in woodland as indicating areas where the soil has been disturbed and enriched by the excreta of animals or by wood ash. Although these sites have been long abandoned, phosphate ions are stable in the soil (Pigott and Taylor, 1964) and for this reason colonies of *Urtica dioica* flourish long after the habitation and pack-horses have gone.

Conclusion

Archaeological evidence and pollen analysis indicate that a more extensive mixed deciduous woodland once covered most of the Coniston Basin. Together with documentary evidence, the indications are that the deterioration of the Coniston woodlands was caused by: (i) centuries of sheep grazing; and (ii) removal of timber over many centuries particularly for charcoal burning. The evidence agrees on three major periods of exploitation: (i) Norse (AD 900–1000); (ii) Cistercian Monasteries (AD 1100–1500); and (iii) Industrial exploitation (AD 1607–1800).

In between these periods the woods were allowed to regenerate, although there is no indication as to whether regeneration was essentially natural or if planting took place. Communities in gorges, because of their inaccessibility, escaped grazing and the clearance episodes and are thus of considerable antiquity and conservation value whilst the communities on the interfluves are more recent, impoverished and thus of less value.

Acknowledgements

I thank Donald Pigott for his support and advice, Winifred Pennington for permission to use unpublished information and the National Trust for permission to work in woodlands at Coniston.

References

Anderson, S.Th. (1970) The relative pollen productivity and pollen representation of North European trees and correction factors for tree pollen spectra. *Danmarks Geologiske Undersogelse, Series II* 96, 1–99.

Barker, S. (1985) The woodlands and soils of the Coniston Basin, Cumbria. PhD thesis, University of Lancaster, Lancaster.

Bradshaw, R.H.W. (1981) Modern pollen representation factors for woods in southeast England. *Journal of Ecology* 69, 45–70.

Bunch, B. and Fell, C. (1949) A stone-axe factory at Pike of Stickle, Great Langdale, Westmorland. *Proceedings of Prehistorical Society* 15, 1–17.

Collingwood, R.G. (1933) An introduction to the prehistory of Cumberland, Westmorland and Lancashire, North-of-the-Sands. *Transactions of the Cumberland and Westmorland Antiquarian and Archaeological Society (N.S.)* 33, 163–200.

Collingwood, W.G. (1896) Furness, a thousand years ago. *Barrow Field Club Proceedings (N.S.)* 3, 36–44.

Collingwood, W.G. (1898) Reports on excavations at Springs Bloomery near Coniston Hall, Lancashire with notes on probable ages of Furness Bloomeries. *No 11 Transactions Cumberland and Westmorland Archaelogical and Antiquarian Society* 15, 223–228.

Collingwood, W.G. (1925) *Lake District history.* Wislon and Son, Kendal.

Collingwood, W.G. (1927) Ancient industries of the Crake Valley. *Barrow Field Club Proceedings Volume III, New Series,* 36–44.

Cowper, H.S. (1899) *Hawkshead: its history, archaeology, industries, folklore, dialect etc.* Bernrose and Sons Ltd, London.

Fell, A. (1908) *The early iron industry of Furness and district.* Hume Kitchen, Ulverston.

Franks, J.W. and Pennington, W. (1961) The late-glacial and post-glacial deposits of the Esthwaite Basin, North Lancashire. *New Phytologist* 60, 27–42.

Godwin, H. (1975) *History of the British flora,* 2nd edn. Cambridge University Press, Cambridge.

Greenwood, E. (1818) *Map of the county of Lancashire.* Lancashire Record Office, Preston.

Iversen, J. (1949) The influence of prehistoric man on the vegetation. *Danmarks Geologiske Undersogelse, Series (IV)* 3, 1–25.

Lake District Special Planning Board (1978) *Lake District National Park Plan,* Kendal.

Marshall, J.D. (1958) *Furness and the Industrial Revolution: an economic history of Furness (1711–1900) and the town of Barrow (1757–1897) with an epilogue.* James Milner, Barrow-in-Furness.

Marshall, J.D. and Davies-Shiel, M. (1969) *The industrial archaeology of the Lake Counties.* David & Charles, Newton Abbot.

Millward, R. and Robinson, A. (1974) *The Lake District,* 2nd edn. Eyre and Methuen, London.

National Trust (1982) *Biological survey, Lake District woodlands. Coniston (ii) summary and species list.* National Trust, Cirencester.

Oldfield, F. and Statham, D.C. (1963) Pollen analytical data from Urswick Tarn and Ellerside Moss, North Lancashire. *New Phytologist* 62, 53–66.

Ordnance Survey (1974) *1:25000 outdoor leisure map. The English Lakes, south-west sheet.* Ordnance Survey, Southampton.

Pearsall, W.H. (1969) *The Lake District. National Park Guide No. 6.* HMSO, London.

Pearsall, W.H. and Pennington, W. (1947) Ecological history of the English Lake District. *Journal of Ecology* 34, 137–148.

Pearsall, W.H. and Pennington, W. (1973) *The Lake District, a landscape history.* Collins, London.

Pennington, W. (1964) Pollen analyses from six upland tarns in the Lake District. *Philosophical Transactions of the Royal Society* B 248, 204–244.

Pennington, W. (1965) The interpretation of some post-glacial diversities at different Lake District sites. *Proceedings of the Royal Society* B 161, 310–323.

Pennington, W. (1970) Vegetation history in the north-west of England: a regional synthesis. In: Walker, D. and West, R.G. (eds) *Studies in the vegetational history of the British Isles.* Cambridge University Press, Cambridge, pp. 47–49.

Pennington, W. (1983) The vegetation history of the Coniston Basin. Unpublished, University of Leicester, Leicester.

Pigott, C.D. (1971) Analysis of the response of *Urtica dioica* to phosphate. *New Phytologist* 70, 953–966.

Pigott, C.D. and Huntley, J.P. (1981) Factors controlling the distribution of *Tilia cordata* at the northern limit of its geographical range. III. Nature and causes of seed sterility. *New Phytologist* 87, 817–839.

Pigott, C.D. and Taylor, K. (1964) The distribution of some woodland herbs in relation to the supply of nitrogen and phosphorus in the soil. *Journal of Ecology* 52 (suppl.), 175–185.

Ratcliffe, D.A. (1977) *A nature conservation review, Vol. I.* Cambridge University Press, Cambridge.

Rodwell, J. (ed.) (1991) *British plant communities, volume I: woodlands and scrub.* Cambridge University Press, Cambridge.

Smith, A.G. (1970) The influence of mesolithic and neolithic man on British Vegetation: discussion. In: Walker, D. and West, R.G. (eds) *Studies in the vegetational history of the British Isles.* Cambridge University Press, Cambridge, pp. 81–86.

Steele, R.C. (1974) Variation in oakwoods in Britain. In: Morris, M.G. and Perring, F.H. (eds) *The British oak. Its history and natural history.* E.W. Classey, Farringdon, pp. 130–140.

Troels-Smith, J. (1960) Ivy, mistletoe and elm. Climatic indicators – fodder plants. *Danmarks Geologiske Undersogelse Series IV* 4, 1–4.

Turner, J. (1970) Post-Neolithic disturbance of British vegetation. In: Walker, D. and West, R.G. (eds) *Studies in the vegetational history of the British Isles.* Cambridge University Press, Cambridge, pp. 97–116.

Walker, D. (1966) The late Quaternary history of the Cumberland lowland. *Philosophical Transactions of the Royal Society* B 251, 1–210.

Yates, W. (1786) *A map of the County of Lancashire.* Reprinted by J.B. Harley (1968) Historic Society of Lancashire and Cheshire. Messrs John Gardner, Liverpool.

16

Historical Ecology and Post-medieval Management Practices in Alderwoods (*Alnus incana* (L.) Moench) in the Northern Apennines, Italy

Diego Moreno, Roberta Cevasco, Sabrina Bertolotto and Giuseppina Poggi

Polo etnobotanica e storia, Dipartimento di Storia Moderna e Contemporanea, Universita' degli Studi di Genova, Via Balbi 6, 16126 Genova, Italy

The present structure and floristic composition of alderwoods (*Alnus incana* (L.) Moench) in the upper Aveto valley (Ligurian Apennines) are related to past land use practices, specifically a local cultural system ('alnocoltura') abandoned early in the present century and no longer remembered by local people. Cartographic and manuscript sources, from the seigneurial archives of the Doria Pamphilj family that ruled the valley from the sixteenth to the eighteenth centuries, show ancient sites of 'alnocoltura' in 1720. Here 'ronchi coperti', was carried out which involved clear felling, turf-stripping and burning, 1–4 years of crop cultivation, followed by subsequent grazing. There is archival evidence that the practice was still in use in 1822. These sites of 'alnocoltura' have been located, and vegetational and floristic data have been collected together with preliminary charcoal evidence from the soil.

Introduction and Research Methods

This research seeks to reconstruct the 'alnocoltura' system (and the associated practice called 'ronco coperto') found in the eighteenth century in the upper Aveto valley (eastern Ligurian Apennines), and to assess the effects of this practice on vegetation cover. Alderwoods are currently very widespread in the Aveto valley (Fig. 16.1). In the Apennines, *Alnus incana* is today normally sporadic. However, botanists in the nineteenth century documented a wider occurrence of *Alnus incana* in the Ligurian Apennines. For example, De Notaris (1844) marked the alder (*Verna gianca*, dialect) as 'in convallibus Apennini et alpium maritimarum vulgatissima [very common in the valleys of Apennines and maritime Alps]', and Chiappori (1876) describes its use as fodder in the upper Scrivia valley.

We have correlated the present distribution and ecology of alderwoods to

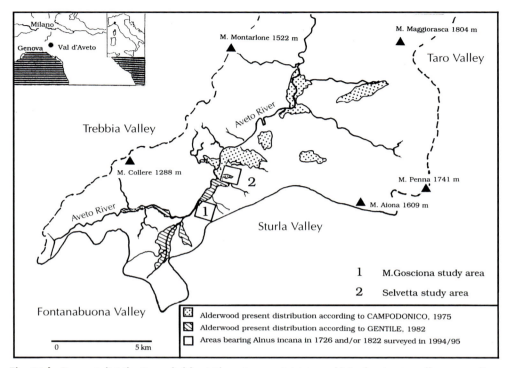

Fig. 16.l. Present distribution of alder (*Alnus incana* (L.) Moench) in the Aveto valley according to Gentile (1982) and Campodonico (1975).

the local history of agro-sylvo-pastoral practices in the Aveto valley. This reconstruction of 'alnocoltura' practice has been done through the use of documentary sources, historical cartography and field evidence. 'Alnocoltura' was classified by the Forestry Administration in 1896 as a multipurpose land use system involving land bearing trees being used for grain growing and wood-pasture. We have named the practice 'alnocoltura' to emphasize that it was a real historical agroforestry system: it is possible that further research on local archive material will provide us with a local name for it.

Oral sources have played an important role in understanding the present local conditions and recent dynamics of vegetation. The 'alnocoltura' system on the whole has been forgotten by local people, although some people remember the practice of 'ronco'. In the eighteenth century Ligurian mountains 'ronco' was one of the most widespread kinds of woodland agriculture: a form of shifting cultivation, involving burning, with different kinds of fire (Moreno, 1984).

Surveyed Area and Present Alderwood Distribution

The research area, the valley of the upper Aveto river (the main tributary of the larger Trebbia river), and particularly the area around Cabanne, a small

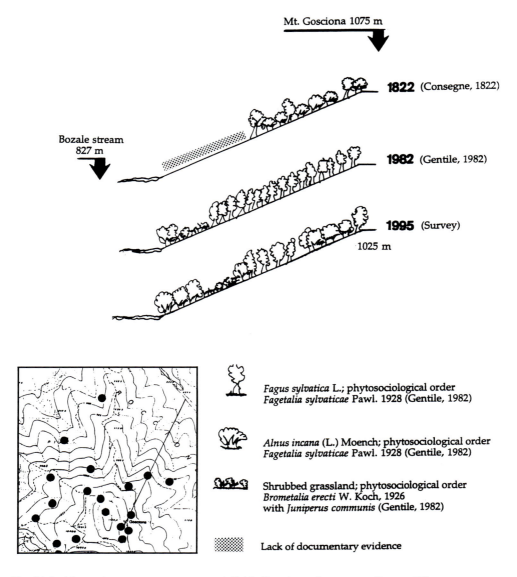

Fig. 16.2. Vegetation cover transect on a NE M. Gosciona slope according to different sources and fieldwork (1822–1995). (•) Alder woods.

village, was chosen because of the great extent of alderwoods and the continuity of 1720–1822–1982 cartographic and documentary sources for the area. The place is situated between 800 and 1100 m above sea level near the Po–Tyrrhenian watershed: the presence of mountains near the coast causes a high precipitation rate (over 2350 mm year^{-1} in Cabanne at 812 m) and much mist. The mean annual temperature is about 9°C. The climate on the whole is of 'Apennine' type with oceanic characteristics (Gentile, 1982). The underlying geology is mainly clay, limestones and sandstones of Cretaceous period, but there is no link between the geology and the alder distribution; *Alnus incana* is found on all soil types.

The present vegetation in the upper Aveto valley, as in most of the Ligurian Apennines, is characterized by the invasion of secondary woods and shrub species following the abandonment of fields, meadows and pasture from the 1960s. A few hay meadows (on the alluvial terraces), heathlands with *Cytisus scoparius* (or *Calluna vulgaris, Erica herbacea* and *Genista pilosa*) and grasslands with *Brachypodium* sp. and *Bromus erectus* or *Sesleria autumnalis* make up the more open vegetation types around Cabanne. The woodland vegetation is represented by ancient and secondary woodland composed of beechwood, oakwood with Turkey oak ('cerreta'), hazelwood, chestnutwood and strips of riparian wood (willows with *Salix purpurea, Salix elaeagnos,* alderwoods with *Alnus incana* and rarely *Alnus glutinosa*). However, *Alnus incana* also occurs mixed in with most of the other types.

Alderwood with *Alnus incana* is found both in narrow fringes along the Aveto river, the streams and other wet ground naturally favourable to *Alnus,* and on drier, level ground and slopes. Unusually, most of the alders are in the higher, dry sites (Fig. 16.2). This type of woodland has been studied by Campodonico (1975) and Gentile (1982) using the phytosociological method. They concluded that alderwoods in the upper Aveto valley are an unstable phase in the dynamic series of *Fagus sylvatica* or *Quercus cerris* and survive only because of pasture and coppices; but they do not consider the possible artificial origin of these alderwoods and the effects of the abandoned practice of 'alnocoltura'.

1994/1995 Surveys

Floristic surveys were made in 1994 and 1995 (Appendix 16.1) on selected sites following study of the eighteenth century topographical maps and the documentary sources ('Consegne' 1822), in order to understand the present woodland composition and dynamics as well as the dynamics of the alder population itself. The sites were chosen by locating the documented historical sites on the IGM sheets (1 : 25,000 scale) and then in the CTR sections (1 : 10,000 and 1 : 5,000 scale). Preliminary observations were as follows:

1. Riparian alderwood seems stable and contains the expected moist-nitrophilic species (*Aegopodium podagraria, Athyrium filix femina, Geum urbanum, Geranium robertianum, Salvia glutinosa, Urtica dioica, Sambucus nigra*).
2. Mountain alderwood, in contrast, shows a different dynamic and contains many beechwood species (*Asarum europaeum, Dryopteris filix-mas, Euphorbia dulcis, Geranium nodosum, Oxalis acetosella*); sometimes the field layer is dominated by *Senecio fuchsii* (Appendix 16.1, sites 1 and 4) or by *Rubus hirtus* W. et K. (sites 11–15) or *Rubus idaeus* L. according to the local

land use history or depending on the practices still at work in the woods. *Acer pseudoplatanus* ('piana' in dialect) and *Fraxinus excelsior* ('frascellanna' in dialect) are the common invading saplings in many abandoned alderwoods (sites 1 and 3–4).

3. Especially in the open ancient sites of pasture, we noted that *Alnus incana* is currently declining and being squeezed out by *Fagus sylvatica* (on the upper and cool slopes) or *Quercus cerris* (on the dryer ones) through a transitional phase with thorny shrubs (*Crataegus monogyna*, *Prunus spinosa*, *Rosa canina*) or *Cytisus scoparius*. Local oral sources confirm the tendencies described in **2** and **3** above.

Documentary and Cartographic Sources

For the historical reconstruction of 'alnocoltura' and the location of historical sites we have utilized a range of sources:

1. Six unpublished hand-drawn maps by M.A. Fossa, dated *c.*1720 (1 : 26,000–1 : 9000 scale) show ancient sites of 'alnocoltura' in the Aveto valley and are connected to a description of vegetation cover with the indications of the practices allowed in different sites ('ronchi coperti' or 'ronchi con fassine'). They date from between 1713 and 1726. In this text the maps are conventionally dated to 1720. Descriptions of soil uses are given in a bound manuscript bearing the title *Indice delle mappe o sia descrizioni di Territorii formate dal Dott. Marc'Antonio Fossa*.

2. The published cartographic sources we have used are: the *Gran Carta degli Stati Sardi in Terraferma* (1 : 50,000 scale) published in Turin, 1852; the 1 : 25,000 IGM Upper Aveto valley six sheets (1937 surveys); the *Catasto dei Terreni del Comune di Rezzoaglio* (1 : 2000 scale, 1951); the CTR (Carta Tecnica Regionale) 1 : 10,000 and 1 : 5000 scale (1979 survey).

3. The *Consegne dei Boschi* are 294 documents (describing over 1300 parcels) produced by landowners of four parishes in reply to the first Forestry Administration Act (1822) adopted in the area (Moreno, 1990). These documents give a detailed account of practices and soil utilization, especially the practice called 'ronchi', for each parcel of land.

4. The Provincial Commission occasionally describes the management of alderwoods in reply to an answer of 'svincolo forestale' (Forestry Law 1882) ('Verbali della seduta del Comitato Forestale di Genova', 28 April 1896).

5. Oral sources. A number of ethnobotanic interviews have been produced during fieldwork in 1994–1995. A more detailed account of these sources is provided in Cevasco (1995) and Bertolotto (1997).

The historical sites have been located by comparing the topographical features of the six maps by Fossa (streams, mountains, watershed, etc.) with the corresponding six Upper Aveto valley IGM sheets (1 : 25,000); abundant micro-scale place-names helped to locate sites. An example of a comparison between the historical descriptions of vegetation cover or soil uses from 1720 to 1982 is given in Table 16.1.

Table 16.1. Example of comparison between historical descriptions of vegetation cover or soil uses from 1720 to 1982.

No. in Fossa	Place-name	Fossa (1720)	'ronco' (1720)	'Consegne' (1822)	Gran Carta Stati Sardi (1852)	IGM (1979)	CTR (1979)	Gentile (1982)
9.11	La Selva	faggi	no	cespugli di ontani e faggi	Boschi	bosco	bosco di faggi	bosco di faggio
12	Selvetta	faggi et one	si	–	Gerbidi	bosco rado	bosco rado con sottobosco	praterie a bromo e paléo arbustate con ginepro
26	Cassine	–	si	faggi	Gerbidi	bosco rado	bosco molto rado con sottobosco	praterie/lande alte a *Cytisus scoparius* e *Pteridium aquilinum*
33	Costa secca (da metà in giù)	one	si	faggi e one	Prati	bosco rado	bosco rado con aree aperte	bosco ceduo di faggio/praterie a bromo arbustate con ginepro
20	Teccie (dove meno pendente)	one	si	–	Boschi	pascolo/ bosco rado	bosco di faggi	bosco ceduo di faggio
71	Poggio di Gossorna	–	si	faggi e one	Gerbidi	bosco rado	bosco rado	bosco ceduo a prevalenza di faggio

Translation:

Fossa (1720)
  faggi: beeches
  faggi et one: beeches and alders (local name)
  one: alders (local name)
'ronco' (1720)
  no: not allowed
  si: allowed
'Consegne' (1822)
  cespugli di ontani e faggi: bushes of alders and beeches
Gran Carta Stati Sardi (1852)
  Boschi: woodlands
  Gerbidi: (we can't translate Gerbidi; however we can bring it near to 'pasture' or 'wasteland')
  Prati: meadows

CTR (1979)
  bosco di faggi: beechwood
  bosco rado con sottobosco: thin woodland with shrubs
  bosco molto rado con sottobosco: very thin woodland with shrubs
  bosco rado con aree aperte: thin woodland with open areas
Gentile (1982)
  bosco di faggio: beechwood
  praterie a bromo e paléo arbustate con ginepro: grasslands with *Bromus erectus* and *Brachypodium* sp., shrubbed with *Juniperus communis*
  praterie/lande alte a *Cytisus scoparius* e *Pteridium aquilinum*: grasslands/heathlands with *Cytisus scoparius* and *Pteridium aquilinum*
  bosco ceduo di faggio/praterie a bromo arbustate con ginepro: coppice beechwood/ grasslands with *Bromus erectus* shrubbed with *Juniperus communis*
  bosco ceduo a prevalenza di faggio: coppice woodland *Fagus sylvatica* prevailing

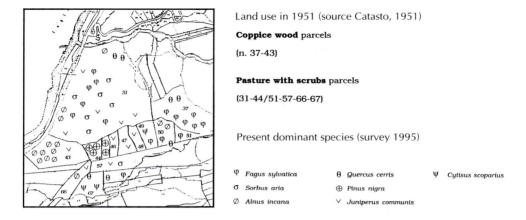

Fig. 16.3. Land use in 1951 (source Catasto, 1951) and present dominant species (survey 1995, 'Selvetta').

Historical Practices of 'Alnocoltura': an Example of the Long-term Effect on Alder Behaviour

The Gosciona Mountain area (Area 1 in Fig. 16.1) today shows relatively stable areas of alder and beech woodland. The present distribution of the alder in the upper part of the valley is due more to the long-term effect of the historical practice of 'alnocoltura' involving 'ronco' than natural soil features (geological composition and superficial water availability). This aspect is particularly evident when comparing (at different dates) a vegetation cover transect on a NE Gosciona slope (Fig. 16.2). An alderwood presently in contact with a beechwood on the upper part of the Gosciona slope (1025 m) is described in the 'Consegne' (1822), yet the vegetation map of the Aveto Valley, produced according to the phytosociological method (Gentile, 1982), does not predict any alderwood in the site other than near the Aveto river.

Another peculiar characteristic of this site is the well-defined geometrical shape of the parcel, possibly derived by the last 'ronco'. This was possibly about 1950, as the oldest standing alder trees in the parcels are around 45 years old. The floristic survey shows it to be poor in species with only *Alnus incana* in the canopy, covering 70% of the site surface. Perhaps as a consequence of the fire management, shrubs are mainly represented by *Rubus hirtus* and the only herb layer species is *Dryopteris filix-mas*. By contrast, the lower parts of the Gosciona Mountain are characterized by the presence of grasslands with *Juniperus communis*.

In other places with a good distribution of *Alnus incana* on the Gosciona Mountain, continuity of alder population is documented in our historical sources but nowadays they are characterized by different floristic features (Appendix 16.1).

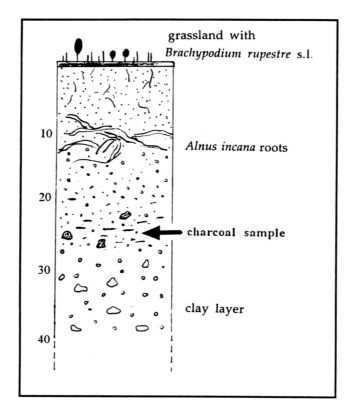

grassland with
Brachypodium rupestre s.l.

10

Alnus incana roots

20

charcoal sample

30

clay layer

40

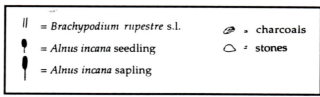

‖ = *Brachypodium rupestre* s.l.   ⌀ ᵔ charcoals

❗ = *Alnus incana* seedling   △ ᵔ stones

❗ = *Alnus incana* sapling

Fig. 16.4. Sketch of soil profile in one historical site of 'alnocoltura' ('Selvetta') showing the charcoal layers sampled.

Historical Practices of 'Alnocoltura': an Example of the Long-term Effect on the Soil

Area 2, 'Selvetta', was chosen for fieldwork because this is the only 'alnocoltura' historical site for whose vegetation cover we have a short description (Fossa 1720 and Table 16.1). In 1726, 'Selvetta' is described as covered by 'faggi et one' (beech and alder) and 'ronco' is allowed. In 1852 the site appears to be 'Gerbido' (neither woodland nor meadow nor cultivated land but wasteland or pasture). Today (Fig. 16.3), the pasture is abandoned (from about 1985) and we find many different and mixed

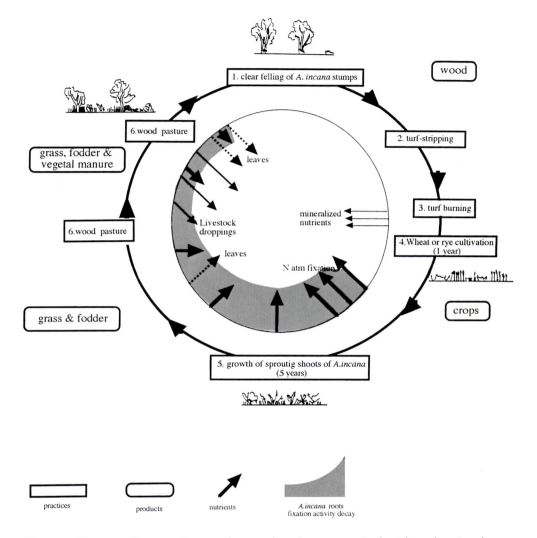

Fig. 16.5. The prevailing practices, products and nutrient sources in the 'alnocoltura' cycle according to the 1896 prescription. The cycle here lasts 5 years. At earlier periods the cycle could last from 5 to 10 years.

vegetation types depending on the previous land use (coppiced wood and pasture with shrubs according to Catasto, 1951).

On the steeper slope, a secondary beechwood is developing through a transitional and still open phase mainly with *Sorbus aria* (only a few of alders are present). On the flatter ground, *Brachypodium/Danthonia* grasslands are spreading, with many shrubs of *Juniperus communis, Erica carnea, Calluna vulgaris, Genista pilosa* and saplings of *Alnus incana*. These grasslands are becoming poorer and poorer in good fodder species. A coppice alderwood still survives on the lower slopes bearing a high *Rubus idaeus* and *Pteridium aquilinum* cover (see Appendix 16.1, site 6); moreover, the flatter sites show spreading clusters of young alder (sometimes in competition with *Quercus cerris, Cytisus scoparius* and *Pteridium aquilinum*). Soil samples (Fig. 16.4) suggest micro-charcoal layers at different depths which include alder charcoal. Further work by archaeologists and anthracologists is needed to confirm the date and historical significance of this charcoal which may be evidence of the burned alderwood in the 'alnocoltura' cycle.

Reconstructing Post-medieval 'Alnocoltura' Cycle

The 'alnocoltura' system adopted in the upper Aveto valley appears in the historical sources of the eighteenth and nineteenth centuries as a peculiar local type of the agricultural and woodmanship practice, broadly diffused in the mountains of northwestern Italy, called 'ronco'. The local cycle of 'alnocoltura' is reconstructed in Fig. 16.5 from a very fragmentary description in a prescriptive document of the Forestry Administration dated 1896. Descriptions in 1822 and 1720 give different details and this suggests that the practice varied from time to time.

The 'alnocoltura' cycle described in 1896 is a management system broadly similar to 'Hauberg-agriculture' (Pott, 1988). It started with the clear felling of alder stools (every 5 years), turf stripping and burning ('debbii a fuoco coperto') and 1 year of wheat or rye cultivation. Afterwards the woodland was grazed (the timings of pasture are not specified in the document). The rotation time, usually 5 years, probably reflected the time needed for the alderwood to reach a harvestable size. In 1896 the regime of pasture in the Aveto valley was regulated by the forest law and was very different from the eighteenth century common right regime regulated by the local statutes of Santo Stefano d'Aveto.

In the 1822 'Consegne' the management system is described with many variants (of which, up till now, we have made only preliminary analysis). The clear felling might occur at from 3 to 10 years, and the cycle period was related to when the wood could be utilized again for fire and fodder. 'Ronco coperto' (confined fire to manure the soil) and timber burning 'abbruciamento del legname' (unconfined fire) are quoted; it is probable that 'ronco coperto' helped to save useful fruit trees such as pear trees, cherry trees ('ceraso') and other leaf fodder trees such as Turkey oaks. The alder ('one' in dialect) in the words of 'Consegne' is classified as 'one' and as 'ontani selvatici'. This distinction is probably the same as the present-day local one between 'ontano negraio' (*Alnus glutinosa*) and 'ona selvatica' (*Alnus incana*). The rye cultivation (generally called 'terra seminativa') usually lasted much longer (from 1 to 4 years) than in 1896. Woodland sheep and goats grazing (or 'erbaggio') occurred after the crop and lasted until the trees had grown large enough to give burning wood again.

At different stages in the alderwood cycle there would have had to be prohibition of pasturage. Shepherds would have kept the sheep and goats from the sown land and the young trees. This may have been by shepherding rather than by physical restriction, but in some periods between 1720 and 1920 withered and plaited hawthorn hedges might have been used to protect growing trees from grazing animals. This could explain the interest in thorny shrubs shown in the 1822 'Denunce'.

Parcels affected by 'alnocoltura' were classified by locals as 'pasture' and not as 'coppice-wood' (even though alders were frequently cut as coppice-wood). Possibly the Forestry Prescriptions adopted in 1822, by forbidding pasture in the first 5 years after the crop, caused the legal death of this ancient practice. In the descriptions of 1822 is evident that alder had a peculiar shape and behaviour, being kept very low and coppiced.

In 1720, 'ronco coperto' was allowed in 'Domestici' sites and was prohibited in a lot of 'Forestri' sites where it was replaced by 'roncare con fassine' (literally, burning faggots). 'Roncare con ronco coperto' was allowed in a few of these sites and especially on the more level ground ('ne quali siti essendo meno pendenti si puo' permettere ii roncare'). To sow oats more than once after the 'ronco' was prohibited everywhere ('colla prohibitione di seminarvi doppo il ronco piu' di una volta l'avena o altre biade').

In the outer circle of Fig. 16.5 we summarize the economic products (wood, crops, grass and fodder, vegetable manure), in the middle circle the sequence of management practices and in the inner circle, the ecological basis of the system. The important nutrient sources during the whole cycle were: (i) the release of nitrogen by alder leaves and roots; (ii) mineralized nutrients from the turf burning which occurred in the first year; and (iii) live-stock droppings produced by grazing animals in the years after the last crop. Soil fertility was enhanced during the agricultural phase (1–4 years) by *Alnus* spp. having roots that fix atmospheric nitrogen through the action of colonies of the microorganism *Frankia* spp. The amount of nitrogen fixed is actually highest in the first 5 years of plant regrowth (Daniere *et al.*, 1986).

The regeneration of alder in the sites affected by the 'alnocoltura' cycle was based only on coppicing; today in the same places the regeneration is mainly from seedlings. In the area (and also in the nearby Fontanabuona and Vara valleys) the practice of *Alnus* co-plantation in farmland, meadows and sweet chestnut groves was in use until the 1950s. The increase in fertility brought about by this tree has been observed locally and is recognized in the agronomical–botanical literature since the end of the Middle Ages. This effect of the alder trees is not, however, mentioned in the agronomical literature of the classical age. If this gap is confirmed by further research, the adoption of 'alnocoltura' in the Apennines could be related to the important changes that occurred in local agro-silvo-pastoral systems during the Early Medieval Age.

Conclusions

In the upper Aveto valley, the present ecological behaviour of *Alnus incana* does not fit with simple phytosociological predictions. Its distribution on the upper slopes is related to the historical sites of 'alnocoltura'. In the late nineteenth century the 'alnocoltura' system fell into disuse either because it was not allowed or it was regarded as very inferior (as happened to all multipurpose local systems in the northern Apennines) following the adoption of the modern forestry regime in 1833. Many sites associated with *Alnus incana* and 'ronchi' were also common lands called 'Comunaglie'. The importance of commons has also declined during the nineteenth century and many have been converted into private property. In addition, during the first half of the present century, rural depopulation has led to the gradual abandonment of local coppicing and pasture practices. Topographic maps of the mid-nineteenth century (1852) help us to place the 'alnocoltura' system in a very different vegetation cover context, devoted to animal production, and relatively unaffected by changes in management since the late medieval and post-medieval age. According to our historical and topographical sources, the 'alnocoltura' cycle was practised in parcels of grazed land bearing pure populations of *Alnus incana,* or mixed with *Fagus sylvatica* and *Quercus cerris.*

Acknowledgements

The charcoal sample of *Alnus* was determined by G. Poggi at the Laboratoire de Paleobotanique et Paleoecologie of Montpellier University under the Galileo Programme scientific exchange. The authors acknowledge the help of Professor Jean Louis Vernet.

Cartographic and Documentary Sources

Cartographic sources

Fossa's maps

Six maps found in the Archivio Doria Pamphilj Landi in Rome and kept in file 10 of the Antico Archivio Genovese (S. Stefano d'Aveto) entitled 'Cartone che contiene sei mappe de' Territori delle Cabanne, Palazuolo, Ventarola, Mandirole, Calzagatta, Priosa, Scabbiamara, Gerba, Fossato, Mareto, Alpepiana, Monte, Vigosoprano e Vigomezzano fatte dal Dott. Marc'Antonio Fossa. Legate tutte dette mappe con cordonetto di seta cremisite'. The six maps have the following titles: *Esola, Ertola, Casalegio, Alpepiana; Territori di Scabiamara e Brignole e parte di quello delle Cabanne; Parte orientale de' Territorii di Cabbane, Palazolo e Ventarola; Parte occidentale de' Territori di Priosa e Annessi; Territorii di Vigosoprano e Vigomezzano;* and *Mandirole e Calzagatta.* They were made sometime between 1713 and 1726 but we do not know the exact date (these maps were mentioned in a *Nota de' Disegni* dated S. Stefano 19 Agosto 1726).

Corpo Reale di Stato Maggiore

Gran Carta degli Stati Sardi in Terraferma (1 : 50,000 scale) published in 1852 by the 'Corpo Reale di Stato Maggiore' sheet LXVIII Torriglia (Moreno, 1995).

IGM

The 1 : 25,000 IGM sheets (1937 survey): Borzonasca 83 II NE; Favale di Malvaro 83 II NO; S. Stefano d'Aveto 83 1 SE; Rovegno 83 I SO; and Ottone 83 I NE.

Catasto

'Catasto dei Terreni del Comune di Rezzoaglio' (1 : 2000 scale, 1951) sheet 108.

CTR

CTR (Carta Tecnica Regionale) 1 : 1000 and 1 : 5000 scale (1979) sections: 215090 M. degli Abeti; 215050 Rezzoaglio; 214120 Favale di Malvaro; 215010 M. Oramara; 215020 M. Maggiorasca; 215053 Brignole; and 215094 Cabanne.

Documentary sources

- Archivio di Stato di Genova (ASG) Prefettura Sarda, Boschi e Foreste – Consegne dei Boschi – busta 207.
- Archivio di Stato di Genova (ASG) Prefettura Italiana – Verbali della Seduta del Comitato Forestale di Genova (Vol. V 1894/1905, Seg. 28 VII 1896 – Archivio Stato Genova) (cf. Croce, 1987).

References

Bertolotto, S. (1997) Storia e copertura vegetale nell'Appennino: effetti delle coltivazioni temporanee in eta' storica in val d'Aveto. Tesi di laurea in Geografia storica dell'Europa, Universita' di Genova.

Campodonico, P.G. (1975) Ricerche fitosociologiche ed ecologiche sulla vegetazione ad *Alnus incana* (L.) Moench in val d'Aveto. Tesi di laurea in Scienze naturali, Universita' di Genova.

Cevasco, R. (1995) Storia e Geografia della copertura vegetale nell'Appennino: l'alnocoltura (*Alnus* sp.) nelle alte valli Aveto e Trebbia (secoli XVIII-XX). Tesi di laurea in Geografia storica dell'Europa, Universita' di Genova.

Chiappori, A. (1876) *La Silvicultura in Liguria.* Tip. Sorolo-Muti, Genova.

Croce, G.F. (1987) Effetti geografici della legislazione forestale in Liguria (XIX secolo). Tesi di laurea in Storia, Universita' di Genova.

Daniere, C., Capellano, A. and Moiroud, A. (1986) Dynamique de l'azote dans un peuplement naturel d'*Alnus incana* (L.) Moench. *Acta Oecologica* 7, 165–175.

De Notaris, J. (1844) *Repertorium Florae Ligusticae.* Reg. Typogr. Taurini, Torino.

Gentile, S. (1982) *Note illustrative della carta della vegetazione dell'alta Val d'Aveto (Appennino ligure).* CNR Collana del Programma finalizzato 'Promozione della Qualita' dell'ambiente'. Aq/1/123. Roma.

Moreno, D. (1984) The agricultural uses of tree-land in the North-Western Apennines since the middle ages. *Supplement du Journal Forestier Suisse,* 74, 77–88.

Moreno, D. (1990) *Dal documento al terreno. Storia e archeologia dei sistemi agro-silvo pastorali.* Il Mulino, Bologna.

Moreno, D. (1995) Une source pour l'histoire et l'archeologie des ressources vegetales, les cartes topographiques de la montagne ligure (Italie). In: Bousquet and Bressolier (eds) *L'oeil du cartographe et la representation geographique du Moyen Age a nos jours.* CTHS, Paris, pp. 175–198.

Pott, R. (1988) Impact of human influences by extensive woodland management and former land use in North-Western Europe. In: Salbitano, F. (ed.) *Human influences on forest ecosystems development in Europe.* ESF FERN-CNR, Pitagora Editrice, Bologna, pp. 263–278.

Appendix 16.1. Preliminary floristic data on 15 historical 'alnocoltura' sites ordered according to the present ecological features: species associated with *Fagus sylvatica, Alnus incana, Quercus cerris.*

	Site														
	1	2	3	4	5	6	7	8	9	10	11	12	13	14	15
Altitude (m)	1050	880	1090	1005	800	820	900	1000	880	974	810	1000	1025	1010	1100
Exposition	SW	E	N	NW	W	N-NW	E	–	S	N	W	N	NE	W	W
Tree canopy (a) %	85	80	80	85	60	75	70	80	85	65	70	65	70	60	65
Tree canopy (a) h	15	11	6.5	6.3	14	6	9	10	10	6	10	9	6	8	6
Low tree canopy (b) %					70										
Low tree canopy (b) h					8										
Scrub canopy (c) %	35	40	20	60	70	20	20	55	30	15	85	90	60	20	90
Scrub canopy (c) h	3.5	3.5	1	1	3	1.2	1.5	1.5	0.6	0.5	0.8	0.8	1	2	2
Grass canopy (d) %	95	70	40	90	70	80	80	70	30	60	40	40	2	90	35
Grass canopy (d) h	1	0.8	0.5	0.5	0.8	0.6	0.8	0.4	0.5	0.3	0.4	0.5	0.4	0.35	0.4
Species usually associated with *Fagus sylvatica* or *Carpinus betulus*															
Dryopteris filix-mas	+					+	+	+							
Acer pseudoplatanus (c)	3	2	1	+		1	+			+	1		2	+	
Euphorbia dulcis	2	+	2	3	+	1	1								
Prunus avium (a)		3	3	2	2		2		2			2			
Fragaria vesca	+				2	1	+		2	+	+				
Gentiana asclepiadea		1	+								2	+			
Stachys sylvatica	+			1	+				+						
Geranium nodosum	+			1	+		2	2							
Hepatica nobilis	+		+			1									
Prunus avium (c)			+		+				2						
Oxalis acetosella		+		1	+										
Asarum europaeum	2				+	1									
Pulmonaria officinalis		+		+			2								
Mycelis muralis	+					1									
Fagus sylvatica (a)								2	+(b)						
Fraxinus excelsior (a)				+		+					2				
Fraxinus excelsior (c)		2	2	2		+						2			
Corylus avellana (c)	2											+	1		
Luzula albida		1						3							

Species														
Avenella flexuosa	1													
Carpinus betulus (a)				3	2									
Brachypodium sylvaticum	+							+			+		+	
Anemone trifolia					2		1							
Prunus avium (b)			+		+									
Paris quadrifolia					1						+			
Lamiastrum galeobdolon	+				+									
Mercurialis perennis	1		+											
Primula vulgaris	+			+										
Pulmonaria saccharata	+			+										
Fraxinus excelsior (b)	1			2										
Aruncus dioicus			2					+						
Prunus avium (d)							1							
Laburnum anagyroides	+										+			
Acer pseudoplatanus (a)	1	+												
Euphorbia amygdaloides						+								
Lilium martagon														
Festuca heterophylla					+		+							
Platanthera bifolia				+										
Ranunculus lanuginosus					+		+							
Helleborus odorus	+													
Species usually associated with *Alnus incana*														
Alnus incana (a)	5	3	4	3	4	4	4	4	4	4	4	4	3	4
Alnus incana (c)	1			3	2	2	+/− 3	1	2	+	+	2		+
Rubus hirtus	2						1	1		4	4	4		5
Geum urbanum	1	+	2	3	+	2		1						
Aegopodium podagraria			2	3	1	3								
Senecio fuchsii	5		4	1	2		1	1						
Alnus incana (b)	1				2									
Alnus incana (d)					1		+ 1							
Rubus idaeus	+	1			3	2								

continued over

Appendix 16.1. *Continued.*

								Site							
	1	2	3	4	5	6	7	8	9	10	11	12	13	14	15
Geranium robertianum	3			1	2				1						
Sambucus nigra (c)	+	+	+	3					1						
Athyrium filix-foemina			+						1						
Salvia glutinosa				1		+									
Sambucus nigra (a)			2	2	2	+									
Sambucus nigra (b)															
Euonymus europaeus					2				1						
Circaea lutetiana							2								
Rubus ulmifolius									2						
Silene dioica									1						
Galium aparine									1						
Poa trivialis									1						
Plantago major					+										
Galium mollugo	+					+									
Symphytum tuberosum	+					+									
Solanum nigrum					+										
Stellaria media	+														
Species usually associated with *Quercus cerris*															
Crataegus monogyna (c)		+	+		3		+	3	+	1	1	3	1		+
Pteridium aquilinum					+	3			2			2			3
Quercus cerris	+	+			3				2	2	2	2		3	
Malus sylvestris		+								2		1			
Prunus spinosa			+		2	+						3		+	
Galeopsis speciosa										3	3				
Juniperus communis										3					
Crataegus monogyna (a)					+	+					1			+	
Knautia drymeia		1			2		1								2
Pyrus pyraster		+								2					
Rosa canina									+					2	
Clematis vitalba					+				+						
Ajuga reptans	+								+						

Species	Cover values (sites 1–15)
Acer campestre	2, 3
Teucrium scorodonia	
Sorbus aria	1
Cratagus monogyna (b)	1, 3
Quercus cerris (c)	+, 1
Fraxinus ornus (c)	+
Quercus pubescens (c)	+
Cornus sanguinea	+, +
Grassland species	
Brachypodium rupestre s.l.	2, 2, 3, 3, +, 3
Dactylis glomerata	3, 2, 2, +
Sesleria autumnalis	2, 4
Agrostis tenuis	1, +, +
Knautia arvensis	2
Carex pallescens	1
Trifolium medium	1
Helianthemum nummularium	+
Genista germanica	+
Lilium croceum	+
Festuca rubra	+, +

Sites:
1. Allegrezze
2. Calzagatta (case Bertè)
3. Allegrezze
4. Cappelletta di Allegrezze
5. Tacora
6. Selvetta
7. Case Bertè–Calzagatta
8. Sorgenti Aveto
9. Pianette (Selva)
10. M. Gosciona
11. Campetti–M. Gosciona
12. M. Gosciona
13. M. Gosciona
14. Villa–Tomarlo
15. M. Gosciona

Note: the numbers in the table for each species are the percentage cover according to the Braun-Blanquet scale:
5 = a percentage cover of 75–100; 4 = 50–75; 3 = 25–50; 2 = 5–25; 1 = 1–5; + = < 1%.
(a)–(d) refers to type of canopy under which the plant is found.

Habitat Alterations Caused by Long-term Changes in Forest Use in Northeastern Switzerland

17

Matthias Bürgi

Group for Nature and Landscape Protection, Swiss Federal Institute for Forest, Snow and Landscape Research, Zürcherstrasse 111, CH-8903 Birmensdorf, Switzerland

In the nineteenth and twentieth centuries, forest use and management changed in response to changing demands put on the forests, and altered the character of forest habitats. This development is reconstructed and analysed as a regional case study for the Swiss lowlands. About 590 forest management plans from 46 communities of the Unter- and Weinland (canton of Zürich, Swiss central plateau) were evaluated, taking account of reasons why the plans had been prepared. The plans span the period from 1820 to the present.

In 1820 more than two-thirds of the forest area was taken up by stands of simple coppice and coppice-with-standards. During the nineteenth and early twentieth centuries, the so-called minor forest uses as wood-pasture, litter-collecting, agricultural use and use of oak bark were given up. In the first half of the twentieth century, a trend to higher stand density and larger growing stock per area occurred. Overall, forests became more closed and darker because of the shift to high forest, larger growing stock and a move from pine to beech. Moreover, the decline of the minor forest uses brought a decline in related habitats. Some of these habitats, for example poor soils, are nowadays of prime interest for nature conservation. Today the amount of fallen, dead wood is very likely to be higher than ever in the period under study.

Introduction

Approximately 60–70% of all plant and animal species of Switzerland occur in forests, although only 28% of the land area is forested (Meyer and Debrot, 1989). Forests therefore play an important role for biodiversity protection and changes in the use and management of the forest bring about changes in habitats for plants and animals (Bürgi, 1997a).

Many publications that report biodiversity losses in Switzerland (e.g. red lists: Landolt, 1991; Duelli, 1994) mention the cessation of traditional forest management practices as one of the causes. Due to lack of knowledge about these forms of use and management and therefore the habitat structure in former times, the consequences of the neglect of these uses can often only be discussed in general terms. Because these uses and management practices differed from one region to another, they can only be reconstructed on a local scale. Thus a regional case-study approach was adopted. Changes in the use and management of the forests were investigated and these changes then interpreted in terms of habitat alterations. In this way, historical evidence can help to analyse the causes of the reductions in biodiversity and enable us to make efforts to preserve and restore it more efficiently.

Materials and Methods

The study area of *c.* 340 km² is situated in the northern part of the canton of Zürich, in the northeastern part of the central plateau of Switzerland. Today, in the study area, more than one-third of the land is covered by forests, of which about one-third is privately owned. The other two-thirds are divided up among about 70 public forests owned mainly by the 47 local communities. Without human influences, these forests nowadays would chiefly consist of beech. Along the rivers would be riparian forests, and in some more exposed places, pine and oak–hornbeam forests.

Since 1820, about 590 forest management plans have been issued for the public forests in the study area, covering planning periods of 10 to 20 years (Staatsarchiv Zürich ZAK III-IV 88/07 and 90/21, various plans in the archives of several forestry offices). These plans were meant to ensure a sustainable forest yield. Every plan referred to the forests of one owner, usually a community or the state. The plans were written by forest engineers, usually by those in charge of the forests, and some run to over 100 pages. The earliest plans have a description of the forest, a short report about its use and management and guidelines for future management. Normally, a map of the forest area was also included. After about 1850, the plans were made more comprehensive, and increasingly included tabulated information describing the stands or the planned felling quantity. Since about 1900, they have also contained inventories of standing timber. In this study, these different parameters were extracted from the plans and then interpolated for each decade, and summed up to assess change in the forests for the whole region.

Forest management plans also contain qualitative information, which is difficult to evaluate in an appropriate manner. One has to be aware of the fact that these plans are official papers, written for community authorities and higher forestry officers. Thus, they are to some extent biased by the interests of their authors and their superiors. Topics such as so-called minor forest uses were hardly ever given the importance that they had for the people who were depending on them. Nevertheless it was possible to extract some information about different aspects of habitats in the forests, if a critical approach to the sources was used.

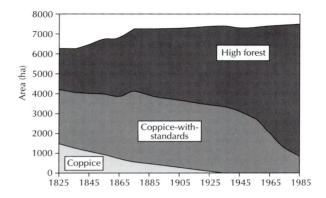

Fig. 17.1. Development of the forest types, 1825–1985.

Information from the Forest Management Plans

Development of forest types, 1825–1985

The forests in the study area can be classified according to three forest types: coppice-forests, coppice-with-standards-forest and high forest. In the forest management plans these terms were also used but often this did not correspond to the actual structure of the stands, rather to the intentions of the forestry officials regarding the future management of the stands. In terms of habitat alteration, these intentions have little relevance. Therefore, the reconstruction of the development of the forest types (Fig. 17.1) had to be based directly on the descriptions of the stands.

In 1825, more than two-thirds of the forest area was covered by coppice or coppice-with-standards. Simple coppice almost completely disappeared during the nineteenth century, whereas most of the decline of coppice-with-standards occurred after the middle of the twentieth century. Today, only about 10% of the forests consist of such stands. However, coppicing had already been given up in the 1920s, and the last regular coppice clearing was conducted in 1958. Thus, the stands identified as coppice-with-standards in the second half of the twentieth century were already turning into high forests.

Early plans show that during the 1820s, many coppice-stands were cut on a stem-by-stem basis. Therefore, the expression 'all-aged coppice-forest' would be a more appropriate term for them. High forests were also treated in a selection forest system before clear-cutting became more popular. In the study area, the clear-cutting system was common between 1850 and 1910. Today, group-selection management is generally preferred.

Development of growing stock and the number of stems, 1925–1985

Since about 1920, standing timber inventories by full calipering are part of the forest management plans. These inventories provide precise information on

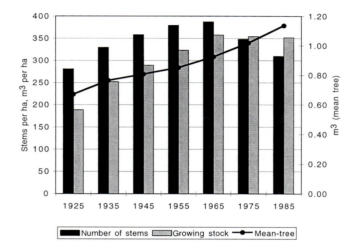

Fig. 17.2. Development of growing stock and the number of stems, 1925–1985.

the growing stock and the number of stems in every stand and forest. Only stems above a given diameter, normally 16 cm, are measured. Sometimes, changes in the methods, e.g. from full calipering to a sample survey, make results difficult to compare. Still, for the whole area and all three forest types, some tendencies are clear (Fig. 17.2).

From 1925 to 1965 a trend to higher stand density and larger growing stock per unit area occurs. The number of stems started to decline after 1965, but the growing stock remained constant. The mean tree size, expressing the ratio of growing stock per stem, increased steadily, and nearly doubled from 0.67 m³ per stem in 1925 to 1.14 m³ per stem in 1985. Thus the main change took place in mean tree size, whereas the number of trees per area in 1985 is not much higher than in 1925. The distribution of the stems in diameter classes reveals that the numbers of trees smaller than 24 cm declined, whereas the number of trees measuring more than 36 cm in diameter increased from only 11% of the total number of trees in 1925 to 32% in 1985. The general stand structure had completely changed within only 60 years.

Development of the four main tree species, 1925–1985

The main tree species are spruce, *Picea excelsa*, pine, *Pinus sylvestris*, beech, *Fagus sylvatica* and oak, *Quercus robur, Quercus petraea*. In 1925 and 1935 pine was the most important tree species (Fig. 17.3a), but since then has strongly decreased. In 1985 the density of pine was only 28% of the value in 1935. Since the middle of the century, spruce has been the most important tree species. Beech also became more common during the twentieth century. In 1985 there were four times as many beech standing in the forest as in 1925. No significant change took place as far as the number of spruce and

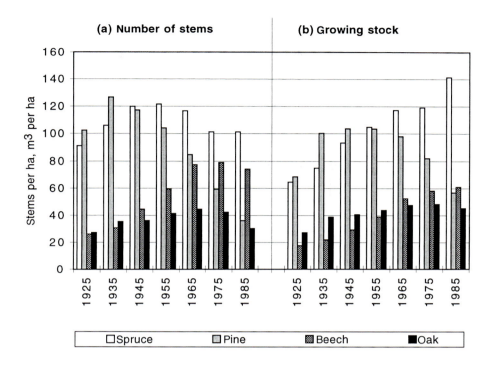

Fig. 17.3. Development of the distribution of stems (a) and growing stock (b) in diameter classes, 1925–1985.

oak are concerned. The development of the main tree species is slightly different when the growing stock is considered (Fig. 17.3b). For instance, the decline in pine was less strong, indicating an increase in the mean tree size. Furthermore, the increase of spruce was much stronger when expressed by growing stock than by numbers of stems. The relatively smaller increase of beech shows that the average beech is much smaller in 1985 than the average spruce, despite their similar frequency.

How can this shift from pine to beech be explained? In the nineteenth century, the increase in high forest was closely connected to an increase in coniferous tree species. Whereas pine-seeding was preferred in the first half of the century, spruce-plantation was preferred in the second half. Since the turn of the century, promotion of the ideas of 'natural forestry' marked a return to the pre-nineteenth century ideas that the proportion of deciduous trees should be increased. Thus, since the middle of the 1950s, generally more than half of the planted seedlings were deciduous species, beech being the most important. In the 1960s and 1970s, plantations of spruce regained importance, but soon the total amount of planting decreased again because natural regeneration became more popular. Today the proportion of artificial regeneration is comparatively low.

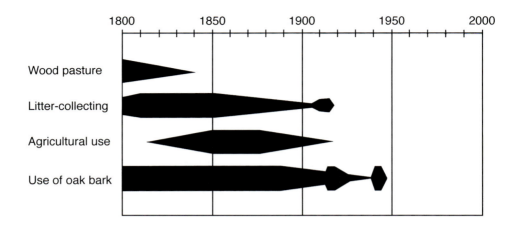

Fig. 17.4. Conceptual diagram depicting the impact of the main so-called minor forest uses.

Therefore, changing forestry practices seem to be the main cause for the change in species composition. There seems to be a temporal sequence of the species preferred by forestry, from pine to spruce and finally to beech (Fig. 17.3). This is reflected first in the number of stems in the stands, whereas the change in growing stock takes longer to occur. In the number of stems (Fig. 17.3a), peaks for pine, spruce and beech are visible. In the development of the growing stock, only for pine is there a distinct peak, developed about 25 years after the peak in the number of stems. Due to the overall tendency to a higher mean tree size, the other species have not yet shown a peak in growing stock.

Minor forest uses

In the study area, the most important minor forest uses were wood-pasture, litter-collecting, agricultural use and the commercial use of oak bark. Forest management plans are an unsatisfactory source for examining the development of minor forest uses. Nevertheless, together with information extracted from other sources, it is possible to get an impression of their importance (Fig. 17.4), although it remains difficult to quantify these uses.

Wood-pasture disappeared in the first half of the nineteenth century due to changes in agricultural practice. Indoor feeding of cattle during the summer and an increased productivity of the meadows were part of the innovations propagated in the 'agricultural revolution' towards the end of the eighteenth and in the early nineteenth century (Pfister and Messerli, 1990), and allowed the abandonment of wood-pasture. As indoor feeding became common practice, the demand for litter increased. Foresters tried to stop litter-collecting, because it led to an impoverishment of the soils, but it remained common throughout the nineteenth century (Fig. 17.4), especially in dry years, probably because of smaller yields from the hay meadows. During World War I, litter collecting was revived in some communities of the study area.

The agricultural use of the clearings in the high forests and of coppice-stands that were cleared for conversion into high forest was also quite common in the nineteenth century. After the outbreak of the potato disease in September 1845 (Fritzsche *et al.*, 1994), such practices increased, because soils in forest clearings were not infected. Moreover, the agricultural use of the clearings offered an additional income to the communities and reduced the cost of transforming the coppice-forests into high-forests. It thus made sense to clear-cut these areas and to rejuvenate them artificially after a farming interval of 2 or 3 years. Such temporary agricultural use was tolerated or even supported by the foresters because it allowed the propagation of the idea of modern forestry (Bürgi, 1997b). In the second half of the nineteenth century, foresters became aware that growing crops on clearings had unwanted side-effects, as trees planted or sown afterwards grew more slowly. Clearings in the study area were probably last farmed in 1912.

Unlike the uses mentioned so far, the commercial use of oak bark was apparently never opposed by the foresters. On the contrary, selling the bark for tanning purposes was very attractive because of the income it produced. Bark was partly obtained from oak-coppice stands but also taken from under-wood in the coppice-with-standards stands rich in oak. Unfortunately the market price fluctuated strongly, which was often deplored by the foresters. The introduction of quebracho wood and artificial tanning products set an end to the use of oak bark at the beginning of the twentieth century, although it was temporarily reinstated during both World Wars.

Habitat Alterations in the Mirror of Forest Management Plans

The decline of coppice and an increase of the area occupied by high forest corresponds to a decline observed in plants typical for coppice-with-standards forests (Landolt, 1991). Today, high forests are managed in a group-selection system that avoids clear-cutting and, therefore, clearings in the forest are relatively rare. Clearings are important for light-demanding vegetation communities (Barkham, 1992; Gilgen, 1994) and associated insects such as many butterfly species (Schiess and Schiess-Bühler, 1997). Furthermore, the ending of coppice management has led to a complete absence of brushwood stages. Such stages are also important for many butter-flies (SBN, 1991).

From the increase of growing stock and number of stems, one can conclude that the forests have become darker in general; a conclusion that is supported by Kuhn (1990) and Egloff (1991). Kuhn (1990) compared phyto-sociological surveys from the 1930s to the 1960s with surveys from 1984–1988. In the northern central plateau he found that, while nitrogen indicators had increased, light indicators had decreased.

The growing number of large trees per unit area may be advantageous for cavity-nesting birds. Yet few trees, if any, are not harvested and so few can reach biological maturity. Only a few snags are to be found in these forests, although this has probably been the case throughout the past 150 years. It cannot be estimated to what extent the old standards in former coppice-with-standard forests provided standing dead wood.

There has been an increase in the proportions of deciduous trees, through a shift from pine to beech. By analogy with the species list of insects in studies by Southwood (1961) and Kennedy and Southwood (1984) on insects associated with various trees in British forests, this shift may result in an impoverishment of species richness. Furthermore, the tendency towards darker forests does not only occur at the level of forest types and growing stock but also at the level of tree species, because beech provides more shade than pine. So-called 'natural forestry' is not necessarily best for nature conservation. While the increase of beech may be seen as a move towards more natural forests, since beech would be dominant without human influence, this tendency can also threaten species that depend on light forests, if beech replaces a tree species that transmits more light to the ground, as pine does.

Today the amount of standing and particularly lying dead wood is very likely to be higher than ever in the period under study, as a result of the declining demand for firewood after World War II and the low price of timber. The decline in mainly agricultural minor forest uses is also relevant for nature conservation. Their effects on the forest ecosystem were often deplored by the foresters. For example, litter-collecting made the soils depauperate in nutrients (Schenk, 1996). Today, however, poor soils are scarce. According to Plachter (1991), species depending on poor soils are particularly at risk and, therefore, poor forest soils are of prime conservation interest. The complete segregation of land use – here forestry, there agriculture – can be seen as a success of regulated land management. However, it has brought about a loss of biodiversity. Forestry alone is not responsible for this development: given the economic and social changes since the beginning of the nineteenth century, the forest officers' freedom to act should not be overestimated and their decisions and actions have to be judged by the knowledge and circumstances of the time.

Differing assessments of the shift from pine to beech show that there is a lack of discussion about the values behind the different aims of nature conservation: the desire for a more natural forest versus the conservation of light-demanding species. Ecological history can provide information that lets decisions in conservation management be more conscious of the underlying values.

References

Barkham, J.P. (1992) The effects of coppicing and neglect on the performance of the perennial ground flora. In: Buckley, G.P. (ed.) *Ecology and management of coppice woodlands*. Chapman & Hall, London, pp. 115–146.

Bürgi, M. (1997a) *Waldentwicklung im 19. und 20. Jahrhundert. Veränderungen in der Nutzung und Bewirtschaftung des Waldes und seiner Eigenschaften als Habitat am Beispiel der öffentlichen Waldungen im Zürcher Unter- und Weinland*. Diss. ETH Nr. 12'152.

Bürgi, M. (1997b) Benutzung und Bewirtschaftung der Wälder im 19. und 20. Jahrhundert – eine Fallstudie über das Zürcher Unter- und Weinland. *News of Forest History* 25/26, 119–130.

Duelli, P. (1994) *Rote Listen der gefährdeten Tierarten in der Schweiz*. BUWAL, Bern.

Egloff, F.G. (1991) Dauer und Wandel der Lägernflora. *Vierteljahrsschrift der Naturforschenden Gesellschaft in Zürich* 136, 207–270.

Fritzsche, B., Lemmenmeier, M., König, M., Kurz, D. and Sutter, E. (1994) *Geschichte des Kantons Zürich. Band 3, 19. und 20. Jahrhundert*. Werd Verlag, Zürich, pp. 519.

Gilgen, R. (1994) Pflanzensoziologisch-ökologische Untersuchungen an Schlagfluren im schweizerischen Mittelland über Würmmoränen. *Veröff. Geobot. Inst. Eidgenöss. Tech. Hochsch.* 116, 127.

Kennedy, C.E.J. and Southwood, T.R.E. (1984) The number of insects associated with British trees: a re-analysis. *Journal of Animal Ecology* 53, 455–478.

Kuhn, N. (1990) Veränderungen von Waldstandorten. *Ber. Eidgenöss. Forsch.anst. Wald Schnee Landsch.* 319, 47.

Landolt, E. (1991) *Rote Liste. Gefährdung der Farn- und Blütenpflanzen in der Schweiz*. BUWAL, Bern, pp. 185.

Meyer, D. and Debrot, S. (1989) Insel-Biogeographie und Artenschutz in Wäldern. *Schweiz. Z. Forstwes.* 140, 977–985.

Pfister, Ch. and Messerli, P. (1990) Switzerland. In: Turner, B.L. (ed.) *The earth as transformed by human action. Global and regional changes in biosphere over the past 300 years*. Cambridge University Press, Cambridge, pp. 641–652.

Plachter, H. (1991) *Naturschutz*. Gustav Fischer, Stuttgart/Jena, pp. 463.

SBN (1991) *Tagfalter und ihre Lebensräume*. SBN, Basel.

Schenk, W. (1996) Waldnutzung, Waldzustand und regionale Entwicklung in vorindustrieller Zeit im mittleren Deutschland. *Erdkundliches Wissen* 117, 325.

Schiess, H. and Schiess-Bühler, C. (1997) Dominanz-minderung als ökologisches Prinzip: eine Neubewertung der ursprünglichen Waldnutzungen für den Arten- und Biotopschutz am Beispiel der Tagfalterfauna eines Auenwaldes in der Nordschweiz. *Mitt. Eidgenöss. Forsch.anst. Wald Schnee Landsch.* 72, 3–127.

Southwood, T.R.E. (1961) The number of species of insects associated with various trees. *Journal of Animal Ecology* 30, 1–8.

18

The Investigation of Long-term Successions in Temperate Woodland Using Fine Spatial Resolution Pollen Analysis

Fraser J.G. Mitchell

Department of Botany, Trinity College, Dublin 2, Ireland

Traditional approaches to woodland succession provide data for one or two generations of the dominant canopy trees. The restricted temporal scale of such data limits their value in the elucidation of the long-term dynamics of woodland. Pollen–analysis from small sites within woodland has been demonstrated to reconstruct vegetation on a local spatial scale. The application of pollen–vegetation correction factors to independently dated pollen profiles generates reconstructions of dynamics at the woodland stand scale for thousands of years. Such data can be used to test existing successional models over long time periods. The impacts of climate change and human interaction can also be traced over long time-scales. These themes will be illustrated using data from the Killarney oak woods, SW Ireland; Bialowieza Forest, E Poland and the Ringarooma temperate rainforest, Tasmania.

Introduction

The most widespread technique for studying temperate forest succession is that adopted by Sernander (1936) in his study of Fiby Urskog, Sweden, where he recorded the cohort of mature trees present, the fallen trees uprooted and snapped by windthrow and the regeneration of saplings that would make up the next cohort. This approach has been integrated into the SILVI-STAR model where the field data of past, present and future forest canopy dominants are modelled to provide simulations of forest structure in the past and the future (Koop, 1989). An alternative technique is to describe forest stands of different ages but which have similar origins and follow similar dynamics. Miles (1981) has demonstrated the application of this technique by illustrating the succession of *Betula* woodland on acid heathland in Scotland.

Both of the above techniques are widely used and have provided valuable data on temperate woodland succession. The short timeframe is the main restriction of the former approach, at best it can provide data on three generations of canopy dominants but time-scales in excess of 300 years are rare in the literature. The latter technique may also suffer from a restricted timeframe and is also sensitive to the assumption that matching vegetation communities in different locations have had similar origins and follow similar dynamics. Olson (1958) and Henderson and Long (1984) have tested this assumption on the sand dune succession at Lake Michigan. The straight-forward succession from grass to poplar (*Populus*), then pine (*Pinus*) to black oak (*Quercus*) and finally a beech/maple (*Fagus–Acer*) climax as first described by Cowles (1899) was found to be far more complex and related to spatial variations in soil mineralization rates and fire history.

An alternative approach to those described above is pollen analysis. This technique has the advantage that long time records (thousands of years) of vegetation change can be reconstructed from a single site. The main restriction of this approach is poor spatial resolution. Traditional pollen analysis uses sediments from large lakes or bogs which collect pollen from a radius of tens of kilometres (Jacobson and Bradshaw, 1981). The pollen from numerous vegetation communities is mixed before deposition and so identifying dynamics in a single community is restricted by uncertainties. Surface sample measurements and modelling studies in northwest Europe and the United States have shown, however, that the pollen records contained within small hollows in forests are dominated by very local pollen rain (Jacobson and Bradshaw, 1981). Northwest European data suggest that, for such sites, the relevant pollen source area is of 20–30 m radius (Andersen, 1970; Bradshaw, 1981a; Mitchell, 1988). A recent, more sophisticated model suggests that in landscapes dominated by forest, the relevant pollen source area may be a 50–100 m radius (Sugita, 1994) and this has been supported to some degree by empirical data from the United States (Jackson and Wong, 1994; Calcote, 1995). Despite differences in proposed pollen source areas, small hollow pollen analysis from forested sites does provide data on forest dynamics at the forest stand scale. If such data are chronologically controlled they can be likened to analysis of vegetation patterns in forest stand quadrats through time. Pollen analysis from small hollows, used in conjunction with other data sources such as fossil charcoal analysis, can provide a suitable time perspective for describing temperate woodland succession or testing existing models of succession.

In this chapter, small hollow pollen analysis from three, distinctly different, temperate woodland types are presented to illustrate the technique and elucidate the succession at each site.

The Ringarooma Temperate Rainforest, Tasmania

In northeastern Tasmania, *Nothofagus cunninghamii*-dominated temperate rainforest is confined to gullies in the mountain plateau (Ellis, 1985). The vegetation outside the gullies is principally *Eucalyptus* sclerophyll woodland that is currently exploited as commercial forest. The two methods described in the introduction have been employed to model the dynamics of the rainforest. When fire is absent for up to 400 years, *Eucalyptus* becomes replaced

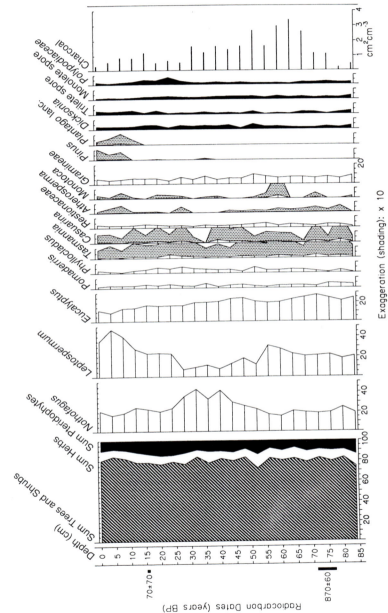

Fig. 18.1.   Summary percentage pollen diagram from Ringarooma Rainforest, Tasmania. Redrawn from Dodson *et al.* (1998).

by woodland dominated by *Nothofagus cunninghamii* (Gilbert, 1959; Jackson, 1968; Mount, 1979; Ogden and Powell, 1979; Ellis, 1985). In the continued absence of fire, *Atherosperma moschatum*, which has a greater shade tolerance than *Nothofagus cunninghamii*, is considered to be the potential forest dominant (Read, 1985; Read and Hill, 1985). Fine spatial resolution pollen analysis of four small hollow profiles was undertaken in the Ringarooma rainforest to test these models. Summarized data from one of the profiles are presented here, whereas the full data from all the profiles and detailed discussion appear in Dodson *et al.* (1998).

The pollen diagram from Ringarooma covers a 900 year period (Fig. 18.1). The European introductions of *Pinus* and *Plantago lanceolata* date to AD 1850. Investigation of contemporary pollen rain and the associated vegetation at this site indicate that *Eucalyptus* and *Leptospermum* have higher pollen representation than *Nothofagus*, and that *Atherosperma* is considerably under-represented by its pollen (Dodson *et al.*, 1998). The highest *Eucalyptus* pollen and charcoal values are found in the lower part of the diagram, indicating the importance of fire-maintained *Eucalyptus* populations. *Eucalyptus* declined in correlation with declining charcoal values and *Nothofagus* expanded in response to reduced fire. The associated decline in *Leptospermum* was most probably related to reduced flowering of this shrub under the shade burden of the *Nothofagus* canopy. The subsequent decline in *Nothofagus* pollen is not associated with increases in charcoal or *Eucalyptus* and is assumed to have been due to the death of old *Nothofagus* trees. *Leptospermum* in the understorey responded to the increased light conditions formed by these canopy gaps. The succession sequence presented here is consistent with that of the other three profiles investigated (Dodson *et al.*, 1998).

These data confirm the general model for Tasmanian *Nothofagus*-dominated rainforest, that in the absence of fire for up to 400 years, *Eucalyptus* is replaced by woodland dominated by *Nothofagus cunninghamii*. The pollen data also demonstrate that the *Nothofagus*-dominated rainforest can maintain itself in the continued absence of fire. Scattered *Atherosperma moschatum* trees can be found at the site today but the pollen data indicate that this taxon has not increased in abundance over the last 900 years despite the long absence of fire at this site. The lowest values of *Atherosperma* pollen coincide with the maximum *Nothofagus* cover so despite the former's greater shade tolerance it does not appear to have expanded at all at Ringarooma.

Bialowieza Forest, Poland

Bialoweiza Forest extends for over 1200 km² and straddles the Polish/Russian border. Most of the forest is now exploited for commercial timber, but in 1921, 47 km² of old growth forest in the Polish section was set aside as a strict forest reserve (Bialowieza National Park). The most comprehensive ecological description of the forest has been published by Falinski (1986). A mosaic of forest types make up the national park, principally stands of *Tilia cordata* and *Carpinus betulus* with *Quercus robur* (*Tilio–Carpinetum*) but

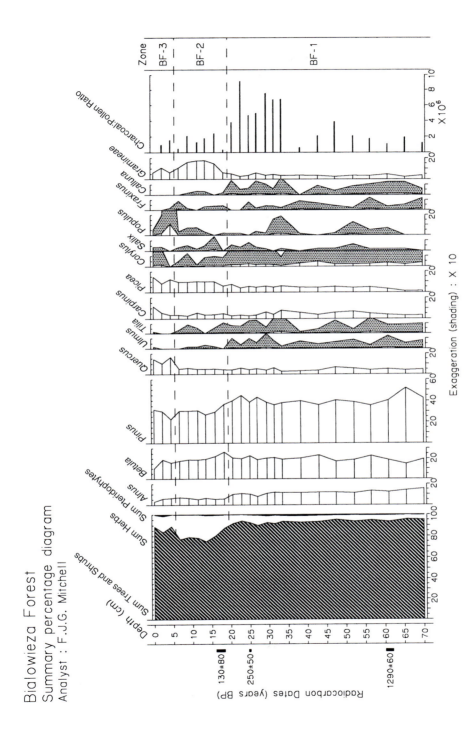

Fig. 18.2. Summary percentage pollen diagram from Bialowieza Forest, Poland. Redrawn from Mitchell and Cole (1998).

also conifer-dominated stands with *Picea abies* and *Pinus sylvestris*. The national park is popularly referred to as primeval although this conflicts with historical and field evidence of human exploitation (Peterken, 1996).

The scale and integrity of Bialowieza Forest has attracted numerous ecological studies from which a range of successional models have been developed. Fine spatial resolution pollen analysis offers the possibility to investigate the status of this forest and test some of the proposed successional models. Pollen analysis of two small hollow profiles from the northern sector of the national park have been completed (Mitchell and Cole, 1998). The summary diagram presented in Fig. 18.2 is from a small basin that measured 100 by 30 m. This site is larger than typical small hollows and so the pollen source area is considered to have a radius of up to 1 km (Jacobson and Bradshaw, 1981).

The record covers about the last 1500 years and has been divided into three zones to aid interpretation (Fig. 18.2). Forest structure in zone BF-1 appears to have been stable with tree pollen in excess of 90% throughout. This section of the forest was dominated by *Pinus* but both *Picea* and *Carpinus* increase during the zone. The high charcoal values at the top of this zone, especially at 22.8 cm, may represent the catastrophic fire of 1811 that swept through the forest. Zone BF-2 opens about 150 years ago and is marked by a substantial increase in *Gramineae* pollen; *Picea* expanded and there was an associated reduction in *Pinus*. Since the fifteenth century, Bialowieza forest has been managed principally as a hunting estate and consequently high game densities were maintained especially during the period 1892–1915 (Falinski, 1986, 1988). A major fire followed by a period of increased grazing pressure would have facilitated the expansion of *Picea* and *Gramineae* as seen in zone BF-2. Falinski (1988) also reports the widespread regeneration of *Picea* under broadleaved canopy trees during this period. The upper zone, BF-3, records the recovery of the forest from the high grazing regime following the establishment of the National Park. *Populus* is the first tree to respond to reduced grazing pressure and it was in turn replaced by expanding *Quercus* and *Carpinus* populations. Many observers have commented on the small size of many trees in the Park and this relates to the cohort of regeneration following the grazing reduction (Koop, 1989).

The pollen data from Bialowieza demonstrate the integrity of the forest. The section investigated has not been felled within the last 1500 years and is certainly ancient *sensu* Peterken (1981). However, the structure and composition of the wood has clearly been influenced by high grazing pressures earlier this century. The apparently stable woodland depicted in zone BF-1 has been transformed into one containing a higher proportion of broadleaved trees. All the successional models published from Bialowieza consider *Quercus* to be an ephemeral species of secondary successions in the forest, consequently the current composition of this section is unlikely to be sustained in the long term.

The pollen diagram from a smaller adjacent site, however, provides evidence of a continuous *Quercus*-dominated canopy for 600 years (Mitchell and Cole, 1998). This profile is located next to Koop's (1989) SILVI-STAR plot

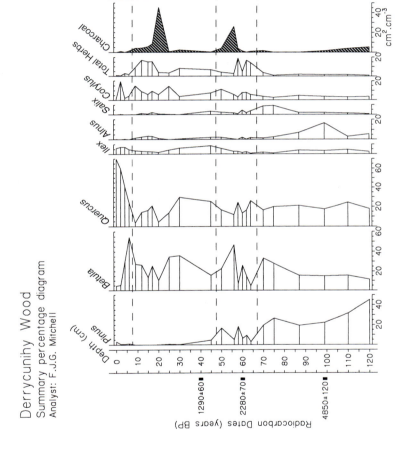

Fig. 18.3. Summary percentage pollen diagram from Derrycunihy Wood, Ireland. Redrawn from Mitchell (1988).

where the oldest *Quercus robur* individual was found to be 358 years old. The largest *Quercus* tree within 20 m of the profile was 52 cm DBH which is equivalent to trees about 150 years old in Koop's age/DBH relationship (Koop, 1989). It can thus be concluded that at this pollen site, *Quercus* has been able to maintain itself for more than one generation.

The contrast in the dynamics between the two sites indicates that climate change (e.g. the Little Ice Age) has not had an overriding influence on woodland composition. This contrast further emphasizes the dangers of the blanket application of successional models based on detailed studies at a limited number of sites to extensive vegetation communities.

Derrycunihy Wood, Ireland

In prehistoric times Ireland was totally covered by woodland (Mitchell *et al.*, 1996), yet today, this country has the lowest percentage cover (0.6%) of native woodland in Europe after Iceland. The greatest concentration of this native woodland is located in the southwest in the Killarney Valley. The vegetation of Derrycunihy Wood has been described by Kelly (1981) who classified it as a *Blechno–Quercetum* association (Braun-Blanquet and Tüxen, 1952). This wood contains some of the largest trees in the region and many believe that it represents the last vestiges of the primeval woodland that once covered the country. However, some historical records indicate that the wood was exploited in the past (Watts, 1984). Small hollow pollen analysis was used to reconstruct woodland composition and dynamics from prehistoric times to the present day (Mitchell, 1988). This enabled the status of the wood to be assessed and the impact of human exploitation to be quantified.

The pollen record from Derrycunihy Wood (Fig. 18.3) illustrates the lack of any long-term stability in woodland composition. Prior to 2000 BP, the wood was composed of a mixture of *Quercus* and *Pinus*. Larger-scale pollen studies on the west coast of Ireland demonstrate that *Pinus* declined to extinction around 4000 BP in association with renewed development of blanket bog (Bradshaw and Browne, 1987; O'Connell, 1990). However, at Derrycunihy Wood where *Pinus* was growing on mineral soil in association with *Quercus*, it survived for a further 2000 years and this has been confirmed from other small hollow sites in the southwest of Ireland (Little *et al.*, 1996; Cooney, 1996). The decline in *Pinus* around 2000 BP is associated with fluctuations in *Betula* and *Quercus* and substantial increases in herbaceous taxa and charcoal. These data are suggestive of human disturbance dating to the Iron Age. Other evidence for human exploitation in the region at this time is poor (Monk, 1993).

The subsequent recovery of the woodland following disturbance does not include *Pinus* which became extinct at this site (and at other sites investigated). The secondary woodland was more open as indicated by the higher representation of *Betula*, the understorey shrubs of *Ilex* and *Corylus* and herbs. Increases in charcoal and disturbances to the woodland canopy towards the top of the diagram relate to human activity in the eighteenth and early nineteenth centuries. This evidence for disturbance coincides with documentary records of charcoal production for iron smelting, timber extraction and grazing in Derrycunihy Wood (Mitchell, 1988). Assessment of the complete pollen records from Derrycunihy Wood and the other sites in Killarney indicate that following disturbance, the botanical and structural diversity of the woods were reduced. The almost monospecific *Quercus petraea* canopy at Derrycunihy Wood today reflects this and the early nineteenth century silvicultural preference for this species. The development of the *Quercus*-dominated canopy is clearly illustrated in the upper levels of the pollen diagram (Fig. 18.3). At no time over the last 5000 years has *Quercus* had such a dominance in Derrycunihy Wood. The reintroduction of *Pinus* to the region by eighteenth century foresters is also recorded in the upper levels of the diagram.

   The pollen data from Derrycunihy demonstrate that the present composition and structure of the wood bear little resemblance to its primeval counterpart. The complete dominance of *Quercus* in the canopy today has been brought about by repeated human exploitation and intervention at the site. The 5000 year pollen record demonstrates that this composition is unlikely to be sustainable in the long term without substantial intervention. Despite this history, woodland has been maintained on the site for at least the last 5000 years, and the wood is still exceptionally diverse in both floral and faunal communities. The wood must therefore be classified as ancient (*sensu* Peterken, 1981).

Conclusions

The case-studies presented above were selected from temperate woodlands that differed significantly in geographical location, species composition and history to demonstrate the wide applicability of the technique of fine spatial resolution pollen analysis to the study of woodland history and succession. The main limiting criterion is the availability of stratified organic deposits that contain well-preserved pollen. Such deposits appear to be widely available in the temperate zone and similar studies have been published from Europe, America, Australia and Japan.

   The close spatial definition of the pollen source area in these studies, described in the introduction, provides the basis for producing reliable pollen–vegetation correction factors. These factors account for variations in pollen production and dispersal characteristics between taxa so that pollen diagrams can be converted into vegetation diagrams. Such exercises have been attempted in the past (e.g. Bradshaw, 1981b; Andersen, 1984) but have been confined to correcting for principal tree taxa only. While this is a valid exercise, the development of more sophisticated models in the future will allow palynologists to make more comprehensive use of the quantitative data that small hollow pollen analysis generates in pollen–vegetation reconstructions. For the time being, the pollen data still provide a valuable source of *relative* data. Correction factors of the principal tree taxa have been published for each of the case-studies above and can be consulted for more in-depth interpretation of the pollen data.

   Review of the fine spatial resolution pollen data from small hollow and mor humus profiles in Europe confirms Peterken's (1996) view that European woodland has evolved through a long history of association with human activity. While this technique gives us the ability to qualify and quantify the impact of this association, it also demonstrates that the great diversity of woodland types in Europe is also a product of their remarkable history.

References

Andersen, S.T. (1970) The relative pollen productivity and pollen representation of north European trees, and correction factors for the tree pollen spectra. *Danmarks Geologiske Undersolgelse* Series II 96, 1–99.

Andersen, S.T. (1984) Forests at Løenholm, Djursland, Denmark at present and in the past. *Det Kongelige Danske Videnskabevnes Selskab. Biologiske Skrifter* 24, 1–210.

Bradshaw, R.H.W. (1981a) Modern pollen representation factors for woods in south-east England. *Journal of Ecology* 69, 45–70.

Bradshaw, R.H.W. (1981b) Quantitative reconstruction of woods in south-east England. *Journal of Ecology* 69, 941–955.

Bradshaw, R.H.W. and Browne, P. (1987) Changing patterns in the Post-glacial distribution of *Pinus sylvestris* in Ireland. *Journal of Biogeography* 14, 237–248.

Braun-Blanquet, J. and Tüxen, R. (1952) Irische Pflanzengesellschaften. In: Lüdi, W. (ed.) *Die Pflanzenwelt Irlands*. Verlag Hans Huber, Bern, pp. 224–420.

Calcote, R. (1995) Pollen source area and pollen productivity: evidence from forest hollows. *Journal of Ecology* 83, 591–602.

Cooney, T. (1996) Vegetation changes associated with Late-Neolithic copper mining in Killarney. In: Delaney, C. and Coxon, P. (eds) *Central Kerry. Field Guide No. 20.* Irish Association for Quaternary Studies, Dublin, pp. 28–32.

Cowles, H.C. (1899) Dune floras of Lake Michigan. *Botanical Gazette* 27, 95–117, 167–202, 281–308, 361–391.

Dodson, J.R., Mitchell, F.J.G., Bögeholz, H. and Julian, N. (1998) Dynamics of temperate rainforest from fine resolution pollen analysis, Upper Ringarooma River, north-eastern Tasmania. *Australian Journal of Ecology* (in press).

Ellis, R.C. (1985) The relationships among eucalyptus forest, grassland and rainforest in a highland area in north-eastern Tasmania. *Australian Journal of Ecology* 10, 291–314.

Falinski, J.B. (1986) *Vegetation dynamics in temperate lowland primeval forest. Ecological studies in Bialowieza forest.* Junk, Dordrecht.

Falinski, J.B. (1988) Succession, regeneration and fluctuations in the Bialowieza forest (NE Poland). *Vegetatio* 77, 115–128.

Gilbert, J.M. (1959) Forest succession in the Florentine Valley, Tasmania. *Papers and Proceedings of the Royal Society of Tasmania* 93, 129–205.

Henderson, N.R. and Long, J.N. (1984) A comparison of stand structure and fire history in two black oak woodlands in north-western Indiana. *Botanical Gazette* 145, 222–228.

Jackson, S.T. and Wong, A. (1994) Using forest patchiness to determine pollen source areas of closed-canopy assemblages. *Journal of Ecology* 82, 89–99.

Jackson, W.D. (1968) Fire, air, water and earth – an elemental ecology of Tasmania. *Proceedings of the Ecological Society of Australia* 3, 9–16.

Jacobson, G.L. and Bradshaw, R.H.W. (1981) The selection of sites for paleovegetational studies. *Quaternary Research* 16, 80–96.

Kelly, D.L. (1981) The native forest vegetation of Killarney, south-west Ireland: an ecological account. *Journal of Ecology* 69, 437–472.

Koop, H. (1989) *Forest dynamics. (SILVI-STAR: A comprehensive monitoring system).* Springer-Verlag, Berlin.

Little, D.J., Mitchell, F.J.G., von Engelbrechten, S.S. and Farrell, E.P. (1996) Assessment of the impact of past disturbance and prehistoric *Pinus sylvestris* on vegetation dynamics and soil development in Uragh Wood, S.W. Ireland. *The Holocene* 6, 90–99.

Miles, J. (1981) *Effect of birch on moorlands.* Institute of Terrestrial Ecology, Cambridge.

Mitchell, F.J.G. (1988) The vegetational history of the Killarney Oakwoods, SW Ireland: evidence from fine spatial resolution pollen analysis. *Journal of Ecology* 76, 415–436.

Mitchell, F.J.G. and Cole, E. (1998) Reconstruction of long-term successional dynamics in Bialowieza Forest, Poland. *Journal of Ecology* (in press).

Mitchell, F.J.G., Bradshaw, R.H.W., Hannon, G.E., O'Connell, M., Pilcher, J.R. and Watts, W.A. (1996) Ireland. In: Berglund, B.E., Birks, H.J.B., Ralska-Jasiewiczowa, M. and Wright, H.E. (eds) *Palaeoecological events during the last 15,000 years.* John Wiley, Chichester, pp. 1–13.

Monk, M.A. (1993) People and environment: in search of the farmers. In: Twohig, E.S. and Ronayne, M. (eds) *Past perceptions. The prehistoric archaeology of south west Ireland.* Cork University Press, Cork, pp. 35–52.

Mount, A.B. (1979) Natural regeneration processes in Tasmanian forests. *Search* 10, 180–186.

O'Connell, M. (1990) Origins of Irish lowland blanket bog. In: Doyle, G.J. (ed.) *Ecology and conservation of Irish peatlands.* Royal Irish Academy, Dublin, pp. 49–71.

Ogden, J. and Powell, J.A. (1979) A quantitative description of the forest vegetation on an altitudinal gradient in the Mt Field National Park, Tasmania, and a discussion of its history and dynamics. *Australian Journal of Ecology* 4, 293–325.

Olson, J.S. (1958) Rates of succession and soils changes on southern Lake Michigan sand dunes. *Botanical Gazette* 119, 125–169.

Peterken, G.F. (1981) *Woodland conservation and management.* Chapman & Hall, London.

Peterken, G.F. (1996) *Natural woodland. Ecology and conservation in northern temperate regions.* Cambridge University Press, Cambridge.

Read, J. (1985) Photosynthetic and growth responses to different light regimes of the major canopy species of Tasmanian cool temperate rainforest. *Australian Journal of Ecology* 10, 327–334.

Read, J. and Hill, R.S. (1985) Dynamics of *Nothofagus*-dominated rainforest on mainland Australia and lowland Tasmania. *Vegetatio* 63, 67–78.

Sernander, R. (1936) The primitive forests of Granskar and Fiby: a study of the part played by storm-gaps and dwarf trees in the regeneration of the Swedish spruce forest. *Acta Phytogeographica Suecica* 8, 1–232.

Sugita, S. (1994) Pollen representation of vegetation in Quaternary sediments: theory and method in patchy vegetation. *Journal of Ecology* 82, 881–897.

Watts, W.A. (1984) Contemporary accounts of the Killarney woods 1580–1870. *Irish Geography* 17, 1–13.

Ecological Changes in Bernwood Forest – Woodland Management during the Present Millennium

19

Rachel C. Thomas

English Nature, Northminster House, Peterborough PE1 1UA, UK

The ecological history of Bernwood Forest is traced from the Middle Ages to the twentieth century, examining the relative importance of high forest and coppice systems, the role of grazing and browsing, open space, scrub, wood edge and dead wood. The implications of these management changes on the abundance of several species of Lepidoptera is considered.

Introduction

This chapter draws together available information from a wide range of sources, direct and indirect, to reconstruct the woodland composition, structure and management of Bernwood Forest at different times during the present millennium, based, unless otherwise stated, on detailed work in Thomas (1987) and Broad and Hoyle (1997).

In common with other sites, there are few records for particular species before the nineteenth century except for some timber trees and game animals. However, for Bernwood, there are good records of woodland plants and animals, especially the Lepidoptera, throughout the twentieth century. The historical ecology of the Bernwood area can be reconstructed from historical documents and an understanding of the modern ecology. It is possible to assess how important different habitats were historically, and to speculate on the previous abundance of species known from Bernwood this century. The records show that Bernwood underwent major changes in management practice during the millennium. This is in apparent contrast to some other woods such as Hayley Wood (Rackham, 1975) and Bradfield Woods (Peterken, 1981) which have been reconstructed in similar detail. As Rackham (1975) warns 'woodland history is not simple, even at the regional level, and … we must be cautious in drawing generalisations and applying them to particular sites'.

© CAB INTERNATIONAL 1998. *The Ecological History of European Forests* (eds K.J. Kirby and C. Watkins)

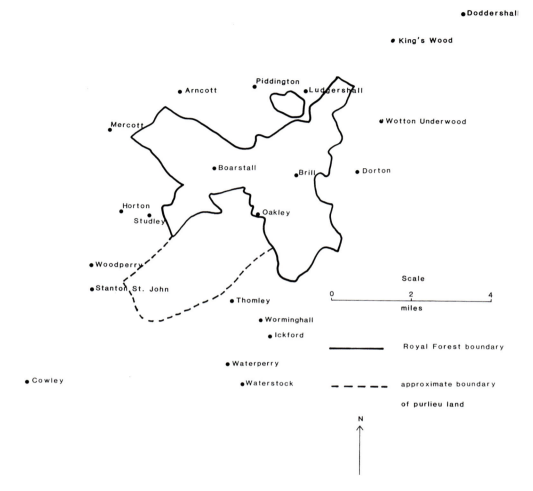

Fig. 19.1. Bernwood Royal Forest and townships, *c.* 1600.

Woodland History

The management of Bernwood Forest can be traced from the Middle Ages to the present day through approximately four different phases. The Forest straddled the Oxfordshire/Buckinghamshire border approximately 9 miles (15 km) northeast of Oxford. The underlying geology is mainly clay, with outcrops of sandstone and alluvium. At its greatest extent in the twelfth century, the Forest stretched for 10–15 miles from Stanton St John in the southwest to Steeple Claydon in the northeast, Figs 19.1 and 19.2. Individual woods, traced throughout the period, are listed in the Ancient Woodland Inventory (Spencer and Kirby, 1992; Kirby *et al.*, Chapter 26, this volume), reflecting continuity since at least AD 1600.

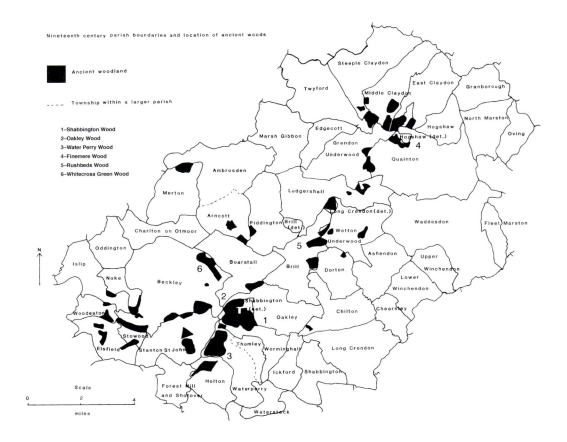

Fig. 19.2. Nineteenth century parish boundaries and location of ancient woods.

In the Middle Ages, the Forest landscape was a mosaic of woodland, arable, pasture, meadow, wood-pasture and settlement. With the exception of wood-pasture, these land uses continued throughout the succeeding centuries and are still present today, although in different proportions. Some medieval woods were enclosed and managed as coppice, or coppice-with-standards, while others (probably the majority) were treated as wood-pasture or a grazed form of coppice. Timber was felled for construction while pollard bollings and poles, as well as the underwood, were used largely as firewood. Certain management practices seem to have been more important in some woods than in others. The Forest supported three species of deer which were hunted for sport, for meat, and gifted by the King to stock deer parks.

	1200s	1300s	1400s	1500s	1600s	1700s	1800s	1900s
Woodland structure and species composition	Wood-pasture and some coppice					'Sustainable' coppice and coppice-with-standards		High forest
			Timber felled and extracted					
		Assarting			Woodland clearance		Woodland clearance	
		Native species only					Exotic trees introduced	
Grazing or browsing		Domestic stock and deer						Deer
				Sheep				
		Some woods fenced and gated		Common rights extinguished, overgrazing		Most woods fenced and gated		
Open space		Temporary open space in response to felling, cutting and extraction routes				Open space following coppicing		Insect glades
						Fixed ride system established		
Scrub/wood edge	Scrub development in response to grazing pressure				Scrub cleared			Managed scrub
	'Thorn' cut as fuel/hedging							
Dead wood	Dead wood in pollard bollings, collected as fuel							Dead wood gathers

Fig. 19.3. Main management changes in Bernwood Forest at various dates.

	1200s	1300s	1400s	1500s	1600s	1700s	1800s	1900s	Examples of species using habitat
High forest	*	*	*	*	*	**	**	***	Light crimson underwing moth *Catocala promissa*
Coppice followers		*	*	*	**	**	***	*	Lead-coloured pug *Eupithecia plumbeolata*, Speckled yellow *Pseudopanthera macularia*
Bare ground	*	*	*	**	**	*	**	**	Barred rivulet *Perizoma bifaciata*
Moist grassland	***	***	***	***	***	**	**	*	Narrow-bordered bee hawk-moth *Hemaris tityus*
Dead wood	*	*	*	*	*				Various Oecophorids
Conifer							*	**	Pine hawk-moth *Hyloicus pinastri*
Scrub/wood edge	*	*	*	**	**	**	**	*	Black hairstreak *Strymonidia pruni*

Fig. 19.4. Importance of different habitats in Bernwood at various dates.

Towards the end of the sixteenth century, changes occurred. The Crown showed less interest in the area and disafforestation occurred in 1632. At this time Bernwood probably still had a well-wooded appearance, although the management of the Crown woods was neglected and the timber exploited. Common rights were extinguished and commoners allocated parcels of land in lieu of their common. To achieve this, much of the wood-pasture in the north of the Forest was cleared.

During the seventeenth to nineteenth centuries, management changed from exploitation to careful husbandry. In common with many other woods (Collins, 1985), coppice-with-standards had developed over much of the Forest area by the mid-eighteenth century and continued largely unchanged for 150 years. As elsewhere, this system broke down in the late nineteenth century and was followed by periods of exploitation after the two World Wars. During the twentieth century, high forest management, largely with introduced commercial conifer species in mixture with oak, was introduced.

Ecological Changes through History

The main management changes are summarized in Fig. 19.3, and the relative importance of different habitats in Fig. 19.4.

Woodland structure and species composition

Bernwood retained its native flora until the mid-nineteenth century and some compartments still support semi-natural vegetation today. The modern semi-natural composition results from changing grazing and woodland management during recent centuries. With the exception of the impact of deer browsing in the last two decades, and timber extraction post-1945, the remaining semi-natural compartments have received very little management this century. These compartments may, therefore, be considered as the core historic types, upon which management practices in previous centuries operated.

These compartments are largely oak, *Quercus robur* – ash, *Fraxinus excelsior*, woods with a mixed underwood of hazel, *Corylus avellana*, thorn, *Crataegus monogyna, C. laevigata* and *Prunus spinosa*, sallow, *Salix* species and maple, *Acer campestre*. There are scattered stands of aspen, *Populus tremula*, and wild service, *Sorbus torminalis*. In modern terms, the woods can best be described as a mixture of *Quercus robur–Pteridium aquilinum–Rubus fruticosus* and *Fraxinus excelsior–Acer campestre–Mercurialis perennis* woodland (NVC W10 and W8) depending on the underlying soil.

European larch, *Larix decidua*, was introduced to Shabbington Wood around 1850 and other species including sweet chestnut, *Castanea sativa*, beech, *Fagus sylvatica,* Norway spruce, *Picea abies*, Scots pine, *Pinus sylvestris*, and Lawson's cypress, *Chamaecyparis lawsoniana*, were introduced in the first half of the twentieth century. The majority of the woods in the Forest were replanted in the 1960s.

Two main management practices, wood or timber production and grazing, have operated in the Forest. The balance between them has influenced the landscape and ecology of the area. Timber and wood production have been achieved through simple coppice, coppice-with-standards, selection systems and high forest–clear cut systems at different times. These have all been influenced by grazing stock or deer.

During the period of the Royal Forest, and especially towards the end of the sixteenth century, three systems operated in parallel. Timber was probably felled from many different parts of the Forest. Individual stems, usually oak, were selected according to need, but especially from Panshill Wood (Fig. 19.5) which may have been managed to favour timber, rather than coppice. Bernwood was only a moderate producer of timber compared to other forests. Most coppice compartments contained timber trees at stocking rates of between 2 and 30 per acre (*c.* 5–70 ha^{-1}), probably reflecting their size. This is a lot less than in modern high forest stands but equivalent to those advocated for coppice-with-standards systems where standards are widely spread to ensure vigorous underwood under the standards (Fuller and Warren, 1990). The semi-natural stands in Oakley Wood today contain approximately 100 trees per acre (250 ha^{-1}) or more and regrowth of underwood is poor (Waring, 1990). The lower density of standards was probably maintained as long as coppice management was viable. Many were extracted during the late 1940s, but the number of small standards has increased subsequently as broadleaf woodland management has declined.

In the later years of the Royal Forest, timber was exploited by the hereditary forester. New extraction routes were cut and saw pits dug. Standing trees were damaged through rough cutting of limbs to help extract other loads. These activities created disturbed woods with muddy tracks, and bare spoil heaps where the underwood had been flattened. Grazing animals may have exacerbated the problem by keeping the clearings open. These conditions probably favoured ruderal species – thistles, coarse grasses and birch – which invade the open areas today.

The crown coppices were, in theory, cut on a 21 year rotation, having been opened to commoners' stock for about the final 10 years. Latterly this custom too was exploited by the hereditary forester who put additional stock into some compartments and excluded commoners' stock from others. In 1534, an account of the underwood in 20 coppices records 11 as either mostly, only or chiefly thorn, seven of thorn and ash, maple or hazel, one of hazel and sallow, and one 'of very smalle value but only scrubbis of thorne' (Letters and Papers of Henry VIII, VI no. 406, PRO, SP1/75 fols. 187r–193v). These coppices, with underwood described as 30, 40 or even 60 years old, may have been rather open and scrubby in appearance; the result of heavy grazing. Aspen, present in the woods today and not browsed by the deer, may have been a significant element, although not a particularly valuable one. Hazel is only mentioned twice. It is interesting to note that ash was cut as part of the underwood and not considered as a timber species, in contrast to the situation today. The ground flora may have been impoverished as grazing-sensitive species were reduced and the woody regrowth stunted.

In modern woodland, grazing reduces regeneration of woody species and the abundance and height of herbaceous species including dog's

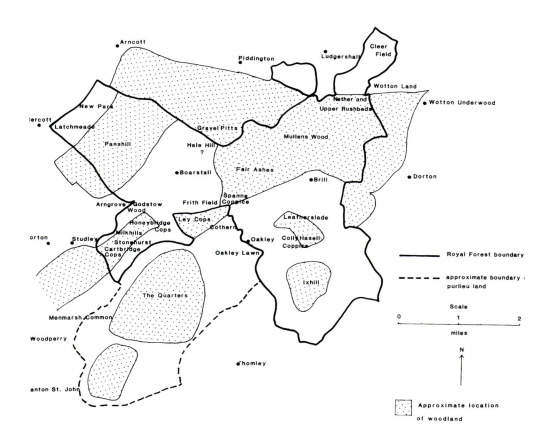

Fig. 19.5. Bernwood Forest *c.* 1600.

mercury, *Mercurialis perennis*, bramble, *Rubus fruticosus*, and honeysuckle, *Lonicera periclymenum*, creating woods with an open understorey and ground layer (Cook *et al.*, 1995; Kirby *et al.*, 1996). Grazing in the medieval coppices was likely to have similar effects on the understorey and ground flora such that the Lepidoptera, other invertebrates and birds, which make use of low cover, suffered during periods of heavy grazing. Conversely when grazing was reduced, expansion of bramble thickets with birch may have been one of the first responses, filling glades and developing on open land at the wood edge, benefiting the nectar feeding butterflies.

In contrast, some coppices in private hands were carefully protected from grazing stock by a fence line and ditch upon which much care and attention was lavished; in one case it cost about 25% of the revenue from sales to maintain. Here, underwood was cut according to a system which was to expand into other privately owned woods in later centuries and where the ground flora may have been more vigorous and varied.

Three species of deer, fallow, *Dama dama*, red, *Cervus elaphus*, and roe, *Capreolus capreolus*, have been recorded from the Forest although fallow appear to have been the most plentiful. Bernwood was unusual in the Midlands as it had a population of red deer in the thirteenth century. The number of deer, assessed from the number which were taken on royal command (on average, 23 fallow per year), appears low compared to the number culled today, possibly indicating that their impact in the Middle Ages was greater socio-economically than as browsers of vegetation. There are no references to deer after disafforestation, but their management is a major concern of Forest Enterprise today.

Over the bulk of the northern part of the medieval Forest, the land appears to have been managed as wood-pasture. Records of 522 *robora* (pollard bollings no longer able to produce a crop of poles (Latham, 1965; Rackham, 1980)) ordered to be taken from the Forest for firewood, between 1231 and 1314, indicate management of part of the Forest as wood-pasture. Grazing pressure is difficult to calculate, but by 1586 Commoners in the Forest 'towns' had at least 6050 cattle, pigs and sheep on the Forest while the hereditary forester had 7600 sheep. In addition, there were stock belonging to parishes outside the Forest as well as deer. The Forest covered about 7000 acres giving a minimum average grazing level of about two animals per acre. Although this is low by modern standards, when the enclosed coppice and additional stock and deer are taken into account, a picture of a very effectively grazed Forest emerges. Even so, habitats were probably expansive, grading into one another as enclosure and grazing levels fluctuated.

After disafforestation, land management developed along more compartmented lines. The remaining woods were managed as coppice-with-standards while stock were largely confined to pastures. In the eighteenth and nineteenth centuries timber and underwood were produced for a local market. Coppices were fenced to exclude stock and were cut on a sustainable 12–15 year cycle. Careful accounts were kept, sale and annual auction particulars were retained and their detail reveals the importance of coppice-wood in the local economy. Following the annual auction, coppicing occurred each winter and timber was felled each spring. Fixed extraction routes were agreed and mapped on coppice plans. It is possible to plot the route of the coppice rotation through an individual wood (Thomas, 1997). The underwood and timber probably fetched a good price; at least enough to ensure the continuity of the system for over 100 years from the 1770s to 1890s. In contrast to the heavy grazing of 150 years earlier, the coppice blocks were probably dense with underwood, creating conditions which favoured understorey species. Rotational cutting ensured that, at any one time, some compartments were recently cut while others were ready for cutting, creating structural variation within the wood.

Open space

Open space in the Forest occurs in several different forms. Arable and meadow have always been significant components. Arable was scattered across the Forest but was concentrated in the south. Today, much of the former arable is now agriculturally improved, species-poor, permanent pasture. Before agricultural drainage was well developed, heavy clay and flooding would have created conditions suitable for extensive areas of moist grassland, some sufficiently wet and acidic to support sphagnum moss (S.R.J. Woodell, pers. comm.). Today, moist grasslands are limited. The Berkshire, Buckinghamshire and Oxfordshire Naturalists' Trust manage some of the surviving examples as hay meadows. One of these has ridge and furrow, indicating a period under arable cultivation.

Within woods, temporary glades developed following timber extraction. These either remained open or grew over, according to usage. The medieval Forest may, therefore, have been a wooded landscape with a higher proportion of temporary open space, than in later years, and where canopy gaps came and went. Maps showing marked routes which were permanently open do not become available until the eighteenth century although there may have been permanent open space before then. Cutting coppice created temporary open space which closed in as the coppice grew. As a more structured system of coppice management extended across the Forest, the network of permanent and temporary open space in the woods became more formalized and predictable, and the extent of internal wood edge, especially around coppice coupes, increased. With the loss of systematic coppicing in the twentieth century this habitat largely disappeared, apart from a brief period after 1945, until glade and wood-edge management for insects developed as an important aspect of modern forestry.

Scrub/wood edge

The development of scrub, and its associated fauna, attracted entomologists to the woods in the early years of the twentieth century, earning Shabbington Wood its reputation as an important insect locality. Blackthorn, *Prunus spinosa*, and both species of hawthorn, *Crataegus* spp., are important elements in the underwood today and probably were historically. In 1232, 30 cartloads of thorn from the Forest were ordered for Studley Priory probably to be used either for fuel or dead hedging. Hedges, established to mark the allotments in lieu of common grazing in the early seventeenth century, were of thorn. As was indicated above, exploitation of the woods may have led to an increase in scrub but, where grazing pressure allowed, blackthorn thickets may have expanded out from the wood edge, as southwest of Finemere Wood today.

Eighteenth and nineteenth century documents give no account of the species composition of the underwood, but it is unlikely to have been substantially different from that recorded earlier or later, except through the influence of grazing. When, in 1947, prior to Forestry Commission acquisition, the Shabbington complex and Waterperry Wood were surveyed, the description of compartments included hawthorn, blackthorn and maple but the principal species were hazel, birch and oak. Once grazing was under control, hazel and ash, both susceptible to browsing but more commercial, may have become more abundant in the coppices. Internal wood-edge increased with systematic coppice management.

Today, arable or pasture usually occur right up to the wood edge, so external wood-edge scrub is restricted to the boundary of a few woods. The small meadow on the south side of Rushbeds Wood is one such boundary. This is managed by grazing, and occasionally cutting back, to retain the seral stages. This juxtaposition of habitats is very important for woodland edge invertebrates which require both mature trees and nectar sources to complete their development. The land gained by Forest Enterprise, southeast of Shabbington Wood, as a result of building the M40 motorway through the Forest, is also being managed as a scrub/wood-edge mosaic, rather than as commercial woodland or grassland.

Internal wood-edge is more extensive, and Bernwood has been one of the most influential areas in the development of internal wood-edge management for nature conservation. The importance of wood-edge and scrub in the Bernwood area has been acknowledged for most of the twentieth century and conservationists voiced considerable concern when it became apparent that it might be lost. In the 1950s, Charles Elton arranged for the management of ten 1 acre glades at the north end of Waterperry Wood to maintain suitable conditions, with sallow and oak, for the purple emperor butterfly, *Apatura iris*. Similarly, blackthorn management for the black hairstreak butterfly, *Strymonidia pruni*, was highlighted, initially by Marcus Goddard in the 1950s and 1960s, when special areas of blackthorn were designated. These were carefully protected in the 1960s when elsewhere in Shabbington Wood the broadleaf woodland was being converted to conifers. Some of these 'special areas' were protected by plastic sheeting when the chemical 2,4,5-T (trichlorophenoxyacetic acid) was sprayed to kill the understorey.

Box junction and ride management for butterflies, which have become standard management techniques in commercial woodland, were developed in Bernwood. However, deer browsing has prevented scrub regenerating on to the ride edges and intersections and the cutting regime, designed to create a herb-rich wood edge, led instead to the development of a coarse grass sward dominated by wood small-reed, *Calamagrostis epigejos*, and tufted hair-grass, *Deschampsia caespitosa*, to the exclusion of woodland herbs. Heavy browsing may have created the same conditions historically.

Recent work on the ecological importance of scrub (Hopkins, 1996) points out that we know little about the history of scrub management. This may be because there are no significant, direct economic measures to assess its abundance. Timber account books give a direct indication of woodland management, and stock levels indicate what was happening to grassland systems, but the condition and abundance of scrub is twice removed from the unit of measurement.

Dead wood

Bernwood has probably never been particularly important for dead wood fauna. Historically, small twigs and branches were collected for fuel. Timber trees may have been cut while still quite young, so never developed a significant dead wood element. The bollings of the pollards, a usual source of dead wood in wood-pasture systems, are referred to in relation to Bernwood only when they were being felled for fuel. Much of the wood-pasture was cleared in the seventeenth century. Only if a pollard was left as a boundary marker, or along the rivers, was it likely to acquire a dead wood fauna, although isolated trees probably had a poorer fauna than those in larger groups. Non-intervention areas in woodland nature reserves may now begin to develop a dead wood fauna although, because of isolation, these too will remain impoverished for many decades (Alexander, Chapter 7, this volume).

Habitat Fragmentation

The medieval Forest landscape was mainly wooded but changed to an agrarian one after disafforestation; a trend which continues today. Fragmentation of the landscape, at the whole Forest scale, occurred most markedly from about 1600, when woods and wood-pasture were cleared to make way for permanent pasture. The habitat mosaic had been extensive and the different elements – woodland, coppice, wood-pasture, clear fell, wooded meadows, scrub, pasture, arable, etc. – occurred over the whole Forest, grading into one another. After 1600, a tighter scale mosaic developed, where woodland habitats occurred within individual woods with greater structural variation, located within an agricultural landscape. This combination of habitats in close proximity is one reason why coppice-woods are so valuable for nature conservation.

Species Implications of Historical Changes in Forest Structure and Composition

Part of Bernwood Forest has been notified as a Site of Special Scientific Interest because of its Lepidoptera. The Lepidoptera of the Shabbington complex and Waterperry Wood have been studied extensively during the twentieth century (Waring, 1990) and ride and glade management for butterflies is now well established (Warren and Fuller, 1990). Different Lepidoptera, recorded from the Forest during the twentieth century and characteristic of particular conditions, can be used to illustrate the impact of the historical changes, although no assessment can be made for species confined to habitats lost before the records began, such as the wood-pasture specialists. These were probably present up to 1600, but not subsequently.

Light crimson underwing, *Catocala promissa*, and heart moth, *Dicyla oo*

The light crimson underwing is a moth of old oak forest; the caterpillars feed at night on oak leaves and hide by day on lichen-covered twigs and in crevices in oak bark, amongst which they are cryptic. It was recorded from Bernwood until 1947, was common in the 1920s and 1930s and was one of the species upon which Bernwood developed its entomological reputation (Waring, 1990). It may have been at its most abundant in the eighteenth and nineteenth centuries when, due to coppice-with-standards management, mature oaks were probably most frequent. The last large oaks in the Shabbington complex were felled in Hell Coppice during 1951–1952. Other Forest woods suffered similarly, and the moth is not known from the Forest today. The position regarding the heart moth is similar. Again, it is a species feeding on oak, preferring mature oaks in open airy conditions, so may have favoured oak standards growing over a coppiced understorey and, therefore, may have been particularly abundant in the eighteenth and nineteenth centuries. The last record of the moth in Bernwood was in 1940 (Waring, 1990).

Lead-coloured pug, *Eupithecia plumbeolata*, and speckled yellow, *Pseudopanthera macularia*

These are both moths of open conditions in woodland. The lead-coloured pug feeds on common cow-wheat, *Melampyrum pratense*, and flies in May and June. The moth was discovered in Bernwood between 1940 and 1959 but, as it can be confused with other species, there is a possibility that it had been overlooked previously (Waring, 1990). It is likely to be present as long as its food plant is present and may have been abundant in the eighteenth and nineteenth centuries when coppice management favoured *Melampyrum*. The speckled yellow is similar. The caterpillars feed on wood sage, *Teucrium scoradonia*, hedge woundwort, *Stachys sylvatica,* or dead nettle, *Lamium album*. It has not been recorded in the Forest since the 1920s and has probably always been restricted in Bernwood as wood sage is found only in the more acid areas (Waring, 1990). The moth is usually abundant where it occurs and, following coppicing, may have been localized in, for example, Furze Coppice in the centre of Oakley Wood where the soil is thinner and more acidic.

Barred rivulet, *Perizoma bifaciata*

This moth feeds on the seed capsules of red bartsia, *Odontites verna*, a distinctive plant of trampled rides. The moth may have occurred in disturbed areas and expanded its range along the permanent ride system as it became established and the food plant spread.

Narrow-boarded bee hawk-moth, *Hemaris tityus*

The larvae feed on devil's-bit scabious, *Succisa pratense*, a plant of damp grassland and the adults seek nectar often from bugle, *Ajuga reptans*, on woodland rides. Historically it was recorded from the damp rough meadows to the west of the Shabbington complex (Waring, 1990) and may have been widespread in other parts of the Forest when poorly drained grassland was abundant. As the caterpillar matures in July and August when hay was cut, it is more often found in permanent pasture adjacent to woodland than in hay meadows. It was last recorded in 1952.

Pine hawk-moth, *Hyloicus pinastri*

This species has been added to the Bernwood list in the 1990s (Waring, 1993). Scots pine has been present in Bernwood since the 1930s but the pine hawk-moth prefers mature pine trees and has expanded generally in the Oxford area, arriving in Bernwood recently.

Oecophorids

These are micro-moths which feed on fungi under bark and in dead wood. They are unlikely to have been a significant component of the Bernwood fauna but some species were probably always present.

Black hairstreak, *Strymonidia pruni*

This is one of the characteristic butterflies of the area and one of the species upon which the entomological reputation of the Forest was founded. It has a restricted distribution in the UK, largely confined to blackthorn thickets on heavy clay between Oxford and Peterborough. It prefers mature blackthorn at wood edges, in glades and in mature hedges. In recent years, much of Bernwood's scrub has been managed specifically for this butterfly. Blackthorn scrub appears to have been a significant historical component of the Forest and it is therefore likely that the butterfly has always been present. It may have suffered during periods of exploitation if blackthorn was cut or grazed hard, leaving few mature stands present. The sixteenth century may have been a good time for the black hairstreak when it may have been more abundant than today. Blackthorn remained as a wood edge and hedge species in the eighteenth and nineteenth centuries and as a component of the coppice underwood. However, at 15 years, the underwood was probably cut too young to be suitable for the black hairstreak butterfly. It probably remained in the hedges, especially in sheltered locations around small fields.

The above are just a few examples of the wide range of ecological deductions which can be drawn now that the available information on the Forest at different times during the millennium has been collected together.

Acknowledgements

The research for this chapter was carried out at Oxford Brookes University, then Oxford Polytechnic, in partial fulfilment of the requirements of the Council for National Academic Awards for the award of Doctor of Philosophy.

My thanks go especially to the late Dr Denis Owen and to Dr Nigel Heard for their encouragement and useful discussion; the staff of the Forestry Commission, former Eastern Conservancy, for access to the modern Bernwood Forest; to the staff of the Bodleian and Westgate Libraries, Oxford, the County Record Offices for Oxfordshire and Buckinghamshire and the British Library Department of Manuscripts. Dr Paul Waring helped with the section on species implications.

I should like to thank English Heritage for their support in preparation of this chapter.

References

Broad, J. and Hoyle, R. (eds) (1997) *Bernwood – the life and afterlife of a forest.* University of Central Lancashire, Preston.

Collins, E.J.T. (1985) Agriculture and conservation in England: an historical overview, 1880–1939. *Journal of the Royal Agricultural Society* 146, 38–46.

Cooke, A.S., Farrell, L., Kirby, K.J. and Thomas, R.C. (1995) Changes in abundance and size of dog's mercury, apparently associated with grazing in muntjac. *Deer* 9, 429–433.

Fuller, R.J. and Warren, M.S. (1990) *Coppiced woodlands: their management for wildlife.* Nature Conservancy Council, Peterborough.

Hopkins, J.J. (1996) Scrub ecology and conservation. *British Wildlife* 8, 28–36.

Kirby, K.J., Thomas, R.C. and Dawkins, H.C. (1996) Monitoring woodland changes in tree and shrub layers in Wytham Woods (Oxfordshire) 1974–1991. *Forestry* 69, 319–334.

Latham, R.E. (1965) *Revised medieval Latin word-list.* Oxford University Press, London.

Peterken, G.F. (1981) *Woodland conservation and management.* Chapman & Hall, London.

Rackham, O. (1975) *Hayley Wood – its history and ecology.* Cambridgeshire and Isle of Ely Naturalists' Trust, Cambridge.

Rackham, O. (1980) *Ancient woodland.* Arnold, London.

Spencer, J.W. and Kirby, K.J. (1992) An inventory of ancient woodland for England and Wales. *Biological Conservation* 62, 77–93.

Thomas, R.C. (1987) The historical ecology of Bernwood Forest. PhD thesis. Oxford Polytechnic, Oxford.

Thomas, R.C. (1997) Traditional woodland management in relation to nature conservation. In: Broad, J. and Hoyle, R. (eds) *Bernwood – the life and afterlife of a forest.* University of Central Lancashire, Preston, pp. 108–125.

Waring, P.M. (1990) Abundance and diversity of moths in woodland habitats. PhD thesis. Oxford Polytechnic, Oxford.

Waring, P.M. (1993) BENHS Field Meeting – Bernwood Forest, 31 July 1993. *British Journal of Entomology and Natural History* 6, 183–188.

Warren, M.S. and Fuller, R.J. (1990) *Woodland rides and glades: their management for wildlife.* Nature Conservancy Council, Peterborough.

20

Interpreting Present Vegetation Features by Landscape Historical Data: an Example from a Woodland–Grassland Mosaic Landscape (Nagykőrös Wood, Kiskunság, Hungary)

Zsolt Molnár

Institute of Ecology and Botany of the Hungarian Academy of Sciences, 2163 Vácrátót, Hungary

'Natural' vegetation fragments often play important roles in the reconstruction of the vegetation history and of the lost natural vegetation of a region. This approach involves the risk that even if the vegetation of an area seems highly natural, past human land use could have transformed it fundamentally.

The area (the Nagykőrös wood) is an historically important, 'seemingly natural', wooded sand steppe, where the following questions were studied: how natural is the pattern of the vegetation mosaic and how natural are the component habitats? A thorough vegetation survey and a detailed landscape, land use and historical vegetation reconstruction were made. Regionally determined indicator values of the plant species proved to be useful in estimating naturalness.

The habitats of the seemingly natural mosaic of the wooded sand steppes at Nagykőrös have been transformed by past human use. They differ in their degree of human disturbance (the larger openings are the more disturbed) and in their typicalness (woodland edges and grasslands near the edges have the most specialist species). Based on these results, the nature conservation priorities need to be re-evaluated: instead of closed woodlands the wooded steppe mosaic should be preserved.

Introduction

In Hungary, especially in the dry lowland regions, where pollen records are scarce or absent, the 'natural' vegetation fragments play an important role in the reconstruction of Holocene vegetation history and of the last natural vegetation in the Sub-Boreal (Rapaics, 1918; Soó, 1929; Zólyomi, 1945–1946, 1969, 1989; Boros, 1952; Járai-Komlódi, 1987). During our field visits to the sandy part of the plain (Kiskunság), we have, however, often had the

© CAB INTERNATIONAL 1998. *The Ecological History of European Forests* (eds K.J. Kirby and C. Watkins)

Fig. 20.1. View of a typical part of the wooded steppe in the Nagykőrös wood (Kiskunság, Hungary). Dry grasslands and more or less open oak woodlands form the dynamic vegetation mosaic. Though it seems natural, historical analysis pointed to some important human transformations to the wood.

impression that even the 'most natural' areas had some features which could not be of natural origin.

Jackson *et al.* (1988), Berglund (1991), Braun *et al.* (1993), Mitchell (Chapter 18, this volume) and Biró and Molnár (1998) showed that even if the vegetation of an area seems highly natural, past human land use could have transformed it fundamentally and thus it is now far from its natural state. Land use historical studies which interpret present vegetation features are therefore important.

Available historical data sources in Hungary limit the time-scale of detailed landscape historical studies to the last 250–300 years, but for this period there is a wide variety of information: maps, surveys, ethnographic data, local historical sources and so forth. The task for the botanist is to select those data that are relevant for vegetation history (Molnár, 1997). A large selection of botanical data is also available from the late eighteenth century onwards. In spite of the richness in data, landscape or land use historical studies with the aim of reconstructing past vegetation are quite scarce in Hungary (e.g. Rapaics, 1918; Zólyomi, 1945–1946, 1969; Somogyi, 1965, 1994; Majer, 1988; Frisnyák, 1990; Vidéki, 1993; Molnár, 1996; Biró and Molnár, 1998).

For the purpose of this study, a vegetation type was chosen which plays an important role in the reconstruction of the region's vegetation history: the oak-wooded sand-steppe (G. Fekete, pers. comm.). From several localities for this vegetation type a site was chosen where the vegetation seemed the most 'natural'. Another factor in selecting the site was that data on landscape, land use and vegetation history were available in detail for the last 300 years. The selected site is the Nagykőrös wood (Fig. 20.1). The area is very species rich, and the pattern of the vegetation mosaic 'seemed natural'. The main questions of our study were, first, how natural is the pattern of the vegetation mosaic and, second, how natural are the component habitats?

Others have noted the need to take account of human influence in this region. 'The vegetation of this sand region is already mixed and transformed by humans' (Borbás, 1886). 'It is difficult to study the ecological characteristics of the lowland woodlands since we might take man-made features to be the result of natural forces' (Bernátsky, 1901). 'Though we see the mixture of natural and man-made vegetation, a botanist can distinguish those species which are natural components of the vegetation from those which are weeds, and are only widespread because favoured by human disturbance' (Boros, 1952).

Study Site

The investigated area lies on the sand alluvial fan of the river Danube, west of the city of Nagykőrös (19°40', 47°05'). The central part of the fan is called Kiskunság. This region has, under a subcontinental wooded steppe climate, a dry sand vegetation with meadows, fens and alkali lakes in the depressions. The sand is highly calcareous, although in the investigated area the surface layers have been leached (pH 6.5–7). For general data for the region see Pécsi (1989) and Tóth (1996).

The Hungarian wooded steppe

The wooded steppe is one of the most typical vegetation types of Hungary (Zólyomi, 1989; Zólyomi and Fekete, 1994). This vegetation is a biome transitional between the temperate broad-leaved forests and the continental steppes (Soó, 1929). A characteristic feature of this vegetation type is that woodland and grasslands form a natural mosaic. Minor abiotic differences are responsible for the development of the wooded or non-wooded patches. The two parts of the vegetation are dynamically linked (Zólyomi and Fekete, 1994). In Hungary wooded steppes occur on loess, sand and also on alkali soils, as well as on the warm and dry slopes of the surrounding mountains (Zólyomi, 1989).

The sand wooded steppe of the Kiskunság region is made up of the following habitats (after Fekete, 1992):

- closed and open pedunculate oak woodland;
- white/grey poplar–juniper groves and other thickets;
- steppes and semi-desert-like grasslands; and
- in the depressions, wet meadows, fens and alkali lakes.

Vegetation of the Nagykőrös wood

The plant communities of the area (Boros, 1935; Hargitai, 1937, 1940) are as follows.

Steppe woodland (*Festuco rupicolae–Quercetum robori*) is the transition habitat between the steppes and the closed woodlands. The canopy is open, the dominant species is *Quercus robur*. In the 'grassy' herb layer, steppe and woodland species are codominant. The wooded steppe character is visible not only by the physiognomy of the stands but by the abundance of the so-called wooded steppe species: *Brachypodium pinnatum, Dictamnus albus, Geranium sanguineum, Iris variegata, Melampyrum cristatum, Peucedanum cervaria, Pulmonaria mollis, Quercus pubescens, Thalictrum minus* and *Vincetoxicum hirundinaria*. In recent years, as a consequence of drought, some of the closed oak woodland is turning into this habitat.

Closed oak woodland (*Convallario majalis–Quercetum robori*) is a closed, damp, high forest. The shrub layer is dense. Steppe species are usually absent. The present-day closed oak woodland includes partly dried out ash–elm–oak (riverine) woodland. Characteristic and dominant species are *Brachypodium sylvaticum, Convallaria majalis, Epipactis helleborine, Geum urbanum, Lapsana communis, Neottia nidus-avis, Poa nemoralis* and *Polygonatum latifolium*.

Closed steppe grassland (*Astragalo austriacae–Festucetum rupicolae*) is the species-rich closed grassland type of the humus-rich sand soils. In the study area it can be found near the edges of or in the place of oak woodland. Characteristic and dominant species are *Achillea pannonica, Aster linosyris, Carex humilis, Centaurea sadlerana, Dianthus pontederae, Festuca rupicola, Linum flavum, Medicago falcata, Origanum vulgare, Poa angustifolia, Scabiosa canescens, Seseli varium, Stachys recta, Stipa capillata, Teucrium chamaedrys* and *Veronica spicata*, and species that also occur in the steppe woodland: *Brachypodium pinnatum, Clinopodium vulgare, Dictamnus albus, Hieracium umbellatum, Melampyrum cristatum, Polygonatum odoratum, Sedum maximum, Silene nutans* and *Vincetoxicum hirundinaria*. Its more open subtype is the *Festucetum wagneri* community which is a transitional habitat between the open and closed sand grasslands with only few characteristic species of its own (e.g. *Festuca wagneri, Peucedanum arenarium, Iris arenaria*).

Open calcifuge sand grassland (*Festuco vaginatae–Corynephoretum*) forms the open grassland habitat of the Nagykőrös wood, derived from the next habitat by leaching of the topsoil (Hargitai, 1940). Characteristic species are *Corynephorus canescens* and *Jasione montana*. Open calciphilous sand grassland (*Festucetum vaginatae*) is the most characteristic habitat of the Kiskunság sand region: open, semi-desert like grassland with many endemic species. Characteristic species are *Achillea ochroleuca, Alkanna tinctoria, Bromus squarrosus, Dianthus serotinus, Euphorbia seguieriana, Festuca vaginata, Koeleria glauca, Onosma arenarium* and *Polygonum arenarium*. This habitat is rare in the wood.

Methods

Vegetation survey

Based on the recent woodland management plan and infra-red aerial photos, 41 areas (the 'best parts') of the wood were selected for a detailed field survey.

The character of the woodland–grassland mosaic and the edges, shape of the vegetation boundaries, species richness, health conditions and characteristic species of the component habitats were recorded and put in a database. The range of component habitats was studied by surveying the coenocline of the mosaic by plant sociological relevés (4 × 4 m). Cover values of each vascular species were estimated as percentages of the total area. Only the herb layer was surveyed to allow comparisons. Nomenclature follows Horváth *et al.* (1995).

The following habitats were studied:

1. Moist woodland interiors (two sites).
2. Dry woodland interiors (three sites).
3–4. The inner and outer parts of the woodland edge (three sites, respectively).
5. Grasslands in small openings where the centre of the opening is 5–10 m from the woodland edge (three sites).
6. Grasslands in medium sized openings where the centre of the opening is 15–20 m from the woodland edge (three sites).
7. Grasslands in large openings where the centre of the opening is 30–50 m from the woodland edge (two sites).

Vegetation analysis

Along the coenocline, the naturalness of the vegetation was measured by assuming that vegetation is more natural the smaller the proportion of the anthropogenic disturbance-tolerant species (degree of disturbance) and the greater the proportion of the characteristic species (for this habitat in this region) (typicalness).

This approach uses the indicator properties of plant species. With these proportions the naturalness of a habitat patch cannot be measured absolutely, but can be estimated relatively. When there is a complete lack of palynological data this method can be used successfully to detect coarse differences in naturalness. Though indicator values for Hungarian plant species are available (Simon, 1988; Borhidi, 1995; Horváth *et al.*, 1995), we think that for a detailed and specific botanical analysis the regional behaviour of the species has to be taken into consideration (cf. the regionally different behaviour of ancient woodland indicators, Peterken and Game (1984)). Each species that occurred in the vegetation samples was classified into the following categories based on its behaviour in the Kiskunság region using our 5 years' field experience and literature data. Local data (from Nagykőrös) were not considered to avoid circularity in the subsequent analysis. The categories used were:

- disturbance-tolerant species that mainly occur and flourish in habitats disturbed by humans;
- non-tolerant species that are forced back by anthropogenic disturbance and are more or less confined to non-disturbed habitats;
- specialists that occur in the region nearly always in steppe grasslands, in dry sand grasslands or oak woodland, respectively; and
- generalists that can be found in several different habitats or are accidental in the studied habitats.

Table 20.1. The plant sociological relevés used for the coenocline analysis; anthropogenic disturbance tolerants are marked with DT, strict specialists with 5, other specialists with 4, generalists with 3, indifferent species with 2, and other indifferent species with 1 (data show percentage cover values in 4 × 4 m quadrats).

DT	Indicator values	Species name	1	2	3	4	5	6	7	8	9	10	11	12	13	14	15	16	17	18	19
	5	Achillea ochroleuca																		10	
DT	5	Bromus squarrosus																		0.5	0.1
	5	Equisetum ramosissimum												0.2	0.3						0.4
	5	Euphorbia seguieriana																			1
	5	Festuca × wagneri							12	4	8	5	30	25	30	25	20	15	30	5	2
	5	Gypsophila paniculata																0.3			
	5	Holoschoenus romanus														0.5					
DT	5	Kochia laniflora																		0.1	0.2
	5	Koeleria glauca																20		8	5
	5	Linum flavum																			
	5	Origanum vulgare								3											
	5	Peucedanum arenarium								0.5				1.5		3					
DT	5	Polygonum arenarium												0.1					1.5	1.5	0.3
	5	Pulmonaria mollis						1													
	5	Artemisia pontica									1.5					0.2					
	5	Aster linosyris								0.3											
	5	Centaurea tauscheri																1	4		1
	5	Corynephorus canescens																		3	35
	5	Geranium sanguineum							0.5	0.1	1	0.5		0.5		4					
	5	Helianthemum nummularium											0.5		1.5						
	5	Iris variegata							0.5	0.5	0.5						0.5	0.5			
	5	Melampyrum cristatum						0.1	4	0.5	3		8	0.3	1	1	2				
	5	Neottia nidus-avis	1		0.1																

Code	Species	Values (across plots)
5	*Poa nemoralis*	4, 5, 4, 5
5	*Potentilla patula*	2
5	*Rosa elliptica*	0.1, 2
5	*Salix rosmarinifolia*	
5	*Seseli varium*	2, 3
5	*Silene nutans*	1.5, 0.3
4	*Achillea pannonica*	0.5, 0.5, 0.3, 0.5, 0.1, 0.1
4	*Alyssum tortuosum*	1, 0.5
DT 4	*Anthemis ruthenica*	1, 0.5
4	*Carex liparicarpos*	1
4	*Centaurea sadlerana*	0.5, 0.5
4	*Chamaecytisus ratisbonensis*	3
4	*Clinopodium vulgare*	1.5, 2, 0.5, 0.3, 1, 3, 5, 1.5
4	*Hieracium echioides* agg.	1, 0.3, 0.3, 1, 12, 5, 3
4	*Polygonatum odoratum*	7, 1.5
4	*Sedum maximum*	0.2, 1.5, 8, 1.5, 0.2, 0.1
4	*Stipa capillata*	1.5, 0.5, 0.2
4	*Thalictrum simplex*	2, 1
4	*Thesium linophyllon*	0.5, 0.5, 1
4	*Thymus glabrescens*	8, 8, 0.5, 0.5, 0.5
4	*Trifolium montanum*	0.3, 2, 2, 0.5, 20
4	*Vincetoxicum hirundinaria*	0.5, 0.3, 0.5
DT 4	*Artemisia campestris*	1, 0.2, 0.2, 0.2, 2, 2
4	*Asparagus officinalis*	1.5, 2, 0.2, 1, 3, 1
4	*Asperula cynanchica*	1.5, 2, 1, 3, 1.5, 0.5

continued over

Table 20.1. *Continued.*

Indicator values		Species name	No. of relevé																			
			1	2	3	4	5	6	7	8	9	10	11	12	13	14	15	16	17	18	19
4		*Betonica officinalis*							3												
4		*Bromus inermis*								1											
4		*Carex stenophylla*																	2		
4		*Dianthus pontederae*											1			2	1.5		0.5	0.3	
4		*Fragaria viridis*						0.5							1.5	0.5					
4		*Frangula alnus*					1					1									
4		*Inula salicina*							0.5	2	0.2				4	8					
4	DT	*Lithospermum officinale*								0.1											
4		*Phleum phleoides*						2				2						2		3	
4		*Ranunculus polyanthemos*								0.5	0.5		0.1								
4		*Silene otites*											0.2		0.5	1		1		1	
4		*Solidago virgaurea*																			
4		*Veronica chamaedrys*											0.5								
4		*Viburnum opulus*	1	4																	
3		*Filipendula vulgaris*							0.3	0.3											
3		*Linaria genistifolia*											0.3								
3	DT	*Ononis spinosa*									0.3				2	0.3	0.5				
3	DT	*Poa bulbosa*																2	0.2		
3		*Prunus spinosa*					2														
3		*Quercus robur*						0.1		0.2	1			0.1							
3		*Rosa canina*						0.5	0.5	0.3	0.3	1	2		1.5						
3		*Scabiosa ochroleuca*											0.5	1		2	3	10	2		
3		*Scrophularia nodosa*		0.5																	
3		*Scutellaria hastifolia*											0.1								

	3	*Seseli annuum*						1	1.5	0.5	2		1	0.5		0.2	1	0.3			
	3	*Teucrium chamaedrys*							0.2	10	20	2	1	12	5	3	30	0.4		15	
	3	*Verbascum austriacum*											1								
DT	3	*Verbascum lychnitis*							0.3	2	1	0.2	2	0.1	1	1					
	3	*Veronica spicata*							0.2		1.5		1.5	4	0.1	1.5	3		0.5	0.5	
	3	*Astragalus glycyphyllos*						1													
	3	*Brachypodium sylvaticum*	2	2	14			20		1		4									
	3	*Convallaria majalis*	5	3	8	15	8	10	8	5	2	7	0.2								
	3	*Crataegus monogyna*		8	5	2	5	2	2	6		3	1	0.5	3	0.5					
	3	*Epipactis helleborine s. str.*		0.1	0.2	0.5															
	3	*Euonymus europaea*			0.2	3	2					0.5									
	3	*Galium mollugo*			0.5	2		5	3	3	1	3	1								
	3	*Geum urbanum*	0.5	0.3		0.3		0.2				1									
	3	*Hieracium umbellatum* agg.							2			3									
	3	*Koeleria cristata s. str.*											0.3	2		0.5	0.5		0.2		
	3	*Ligustrum vulgare*	8	11	10	15	15	4	1	10		20									
DT	3	*Medicago minima*											0.1					2			
	3	*Polygala comosa*																0.2			
	3	*Polygonatum latifolium*	1		0.5	3	2	5	1	0.2		3									
	3	*Populus × canescens*																0.5			
	3	*Potentilla arenaria*											0.2	1		1		2	3	4	1
	3	*Pyrus pyraster*								2			3		1						
	3	*Rhamnus catharticus*	1.5					0.1		0.3		0.5									
DT	3	*Scleranthus annuus*																	0.5		
	3	*Ulmus minor*		1.5																	
	3	*Viola odorata*	1																		

continued over

Table 20.1. Continued.

Indicator values	Species name	1	2	3	4	5	6	7	8	9	10	11	12	13	14	15	16	17	18	19
DT 2	Ambrosia artemisifolia									1								0.1		
2	Carex flacca						4		5	6	0.5	1.5	0.5	2	0.3					
2	Coronilla varia																0.2		1.5	0.2
DT 2	Eryngium campestre							2									3		3	
DT 2	Euphorbia cyparissias									3	0.2	1	0.3	2			8			0.5
DT 2	Fallopia dumetorum		0.1		0.2						0.2		0.2		1					
2	Gallium verum						1					3	5	4	1	8			1	3
DT 2	Hypericum perforatum									0.5	1	0.1					0.3			
2	Populus tremula					4					1	1.5								
DT 2	Rubus caesius	3		2	3	3	0.1		3	1.5		1.5								
DT 2	Rumex acetosella										0.2					0.2	0.3		2	0.2
2	Agrostis stolonifera							1							0.5					
DT 2	Alliaria petiolata	2	1.5																	
DT 2	Bromus mollis												0.1					0.2		
DT 2	Bromus tectorum																	0.2		
2	Carex acutiformis	1																		
2	Carex spicata		0.3	3		2	3				4		8							
DT 2	Chondrilla juncea	3	1					0.2				0.5	0.2							
2	Cornus sanguinea				2						0.1									
DT 2	Cynodon dactylon											0.2		1.5	0.3		0.5	1.5	15	25
2	Knautia arvensis									0.5										
2	Lapsana communis		0.3						0.3											
DT 2	Leontodon hispidus									1.5										
DT 2	Lotus corniculatus															2				
DT 2	Pimpinella saxifraga										1		0.2							
2	Poa pratensis agg.						10				0.5		1							

	DT	Fresh interior	Dry interior	Inner edge	Outer edge	Small opening	Medium sized opening	Large opening
Prunus serotina	2	2			2			
Saponaria officinalis	2		2	2				
Silene vulgaris	2			0.1		0.5	0.5	
Tragopogon orientalis	2			0.3	0.2	0.1		
Trifolium arvense	2					0.5	0.5	0.1
Vicia cracca	2			1		0.1		
Agropyron repens	1				0.5	0.1	0.2	
Arenaria serpyllifolia	1					0.2		
Calamagrostis epigeios	1			0.5	2	0.5	10	1.5
Chenopodium sp.	1					1.5		
Crepis rhoeadifolia	1		1	2	2			0.1
Dactylis glomerata	1		0.3	3	3	0.5	4	0.2
Digitaria sp.	1			1	1	1	1.5	1.5
Erigeron canadensis	1						2	0.5
Galium aparine	1				1		1	0.5
Medicago lupulina	1						0.1	
Melandrium album	1			0.5	0.1	0.3		
Plantago lanceolata	1					0.3		
Potentilla argentea	1					0.1	0.3	
Robinia pseudo-acacia	1		1		1		5	
Setaria viridis	1							
Stellaria media	1	0.5					0.2	0.1
Tragus racemosus	1						0.2	
Habitats								

The degree of 'disturbance' was estimated by the proportion of anthropogenic disturbance-tolerant species in the vegetation samples, and typicalness by the proportion of specialist species. Both presence–absence and dominance data were used. This sort of classification always remains dependent on the expert's field experience. To test how robust the pattern was along the coenocline, we reclassified the species into five categories (from strict specialists to indifferent species) and repeated the analysis (Table 20.1).

Vegetation, landscape and land use history

Historical botanical data (floristic and coenological surveys, etc.; Hollós, 1896; Boros, 1935 and his travel diary 1918, 1919, 1920, 1922, 1934; Hargitai, 1937, 1940; Szentpéteri, 1990) provided clear data about the past flora and vegetation. This data and the detailed floristic and coenological survey of the present vegetation helped us decide what features and processes are the most useful to reconstruct. Other important historical sources were the maps, especially the military survey maps. The first was made in 1783 and is supplemented with a detailed Country Description. The second was made between 1861 and 1866, the third in 1884. Aerial photographs contributed to more recent reconstructions (1950, 1952, 1990 and 1992). Also helpful were a wide range of non-botanical historical documents and data which had been classified and interpreted by other specialists (historians, geographers, linguists, etc.), of which the most useful were Balla (1758), Nemcsik (1861), Galgóczy (1896), Hargitai (1940), Rédei (1978), and the woodland management plans from 1887 and the 1990s.

Results and Discussion

Vegetation

The most species-rich habitats are the woodland edges and the grasslands near the edge (Fig. 20.2e). There was a very big difference between the number of species in the inner part of the edge and the dry woodland interior. The grasslands of the larger openings are more disturbed than all the other habitats (Fig. 20.2a,b). If we use dominance data the same pattern remains, but the percentage of disturbance-tolerant species increases more sharply with increasing opening size.

Typicalness based on presence–absence data decreases from grasslands of small openings towards the woodland interiors (Fig. 20.2c,d). If we use dominance data we get the same pattern, but the differences are larger. It is difficult to estimate differences in typicalness between openings of different sizes since they show different patterns if we use binary or dominance data.

Comparison with the reference wooded steppe

Without a base of comparison we can only compare between habitats (openings, edge and woodland interior). Unfortunately there is no other wooded steppe fragment in the Great Plain which can be assumed to be more natural than the one we have studied, so we used a (probably) much more natural mosaic from the Central Hungarian Mountain Range for the comparisons (Morschhauser, 1990). This wooded steppe grows on limestone and has the three following habitats: closed dry oak woodland, open karst oak woodland and closed steppe grassland.

The low species number of the woodland interiors at Nagykőrös and the larger openings of the lowland mosaic compared to the mountain mosaic are striking (see the open squares on Fig. 20.2). The larger openings are much more disturbed in the lowland mosaic. Typicalness of habitats in the mountain mosaic decreases towards the grasslands, hence the characterless composition of the lowland woodland interiors is even more striking. The pattern remains the same if we use dominance values.

Testing the robustness of the results

When we divided the group of the specialists into two categories (strict specialists and other specialists) to test the robustness of the pattern we got two similar curves (Fig. 20.3, filled squares and plus signs). When the generalists' group was divided into three, we found that the category 3 species 'generalists' proper (stars) are responsible for most of the variation, the indifferent species playing a minor role. We can conclude that the patterns changed by the subdivision of the two groups but the botanical meaning of the patterns remained the same: specialist species became sparse towards the woodland interiors.

History of the wood

We have found big differences in the degree of disturbance and in typicalness between the component habitats of the wooded steppe mosaic and also between the lowland and mountain mosaic. This implies that human use has affected these habitats differently. The history of the wood has already been studied in detail by Galgóczy (1896), Hargitai (1940) and Rédei (1978). We give only a brief summary of those events that played an important role in the development of the present vegetation pattern.

In the sixteenth and seventeenth centuries this area was part of the Turkish Empire (Hanák, 1991) and the people of Nagykőrös had to pay a tribute of large quantities of wood to the Turks. Much wood was used for firewood, saltpetre production, fortification and tanning. Until the 1780s, people were allowed to use the woodland unrestrictedly (for cutting, grazing, mowing, etc.) (Galgóczy, 1896). Previously the wood must have been bigger if we consider the many toponyms in the surrounding landscape containing the words for woodland, oak and poplar (Hargitai, 1940).

The first data about the size of the wood come from 1783, from the Military Survey Map. At that time the wood occupied about the same area as it occupies today (Fig. 20.4), but it was more open as a consequence of over-cutting and grazing. The wood was surrounded by steppe grasslands and wet

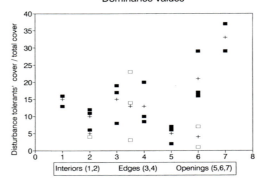

(a) **Degree of human disturbance**
Dominance values

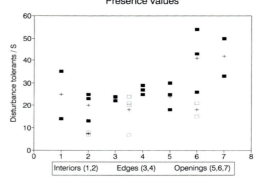

(b) **Degree of human disturbance**
Presence values

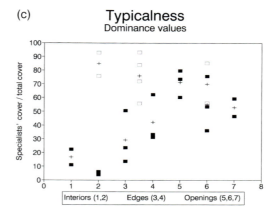

(c) **Typicalness**
Dominance values

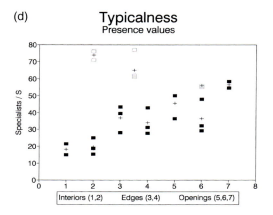

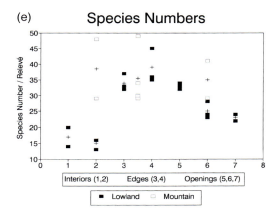

Fig. 20.2. (and opposite) The coenological features of the woodland–grassland coenocline in the Nagykőrös wood. The following habitats were surveyed: fresh and dry woodland interiors, the inner and outer parts of the woodland edge, grasslands of small, medium sized and large openings. Degree of disturbance and typicalness were estimated by the proportion of anthropogenic disturbance tolerant and specialist species of the vegetation samples, respectively. + signs indicate the averages. The habitats of the coenocline differ profoundly in their species number, typicalness and degree of disturbance. The differences could mainly be explained by local history. For comparison the data from a more natural mountain wooded steppe mosaic is also shown (open squares). The features of the two mosaics also differ.

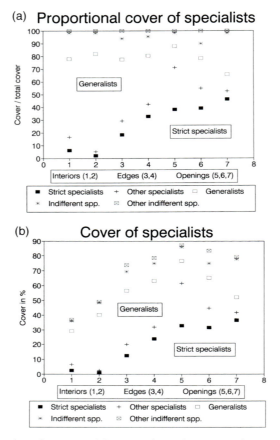

Fig. 20.3. To test the robustness of the coenological patterns of Fig. 20.2, the specialist/generalist species groups were subdivided into 2 and 3 categories, respectively. This figure shows the relative (a) and absolute (b) values of the 5 species groups. The patterns changed to some degree, but the botanical conclusions from the graphs changed little (see text for details).

meadows. Toponyms indicate that the dominant tree species of the wood were oak (*Quercus robur*), ash (*Fraxinus angustifolia*), birch (*Betula pendula*), elm (mainly *Ulmus minor*), alder (*Alnus glutinosa*), hazel (*Corylus avellana*), wild cherry (*Cerasus* sp.) and maple (*Acer tataricum*) (Hargitai, 1940).

As a consequence of the measures enforced by the city council in 1769, 1793 and 1796, the overuse gradually decreased from the beginning of the nineteenth century (Rédei, 1978). The canopy closure of the wood was still low, 40%, and some parts looked like a wooded pasture. The coppice cycle was 20 years. Oak trees were selected for and wild fruit trees were protected. In the longer run this must have caused a decrease in local tree diversity. The openings were mown, but some of them were also used as arable fields (Hargitai, 1940; Rédei, 1978). Grazing still continued in the wood in stands

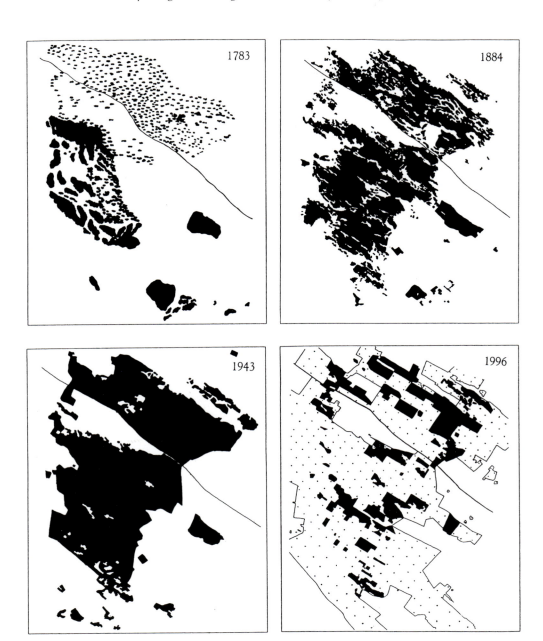

Fig. 20.4. The Nagykőrös wood (Kiskunság, Hungary) in the last 200 years. Black areas show the woodlands, white areas are grasslands or cultivated fields. On the 1996 map *Pinus* and *Robinia* stands with a totally degraded herb layer are indicated by dots. The opened up character of the wood in the eighteenth and nineteenth century is caused by overuse, later the wood closed. Notice the more or less unchanged shape of the wood and at the same time the profound changes in the pattern of the woodland–grassland mosaic.

older than 4 years, later, older than 15 years. The wood was used for fire-wood, tool making and for building. Shrubs were cut for brooms, harrows and for wickerwork (Galgóczy, 1896; Hargitai, 1940).

The first woodland management plan was completed in 1887 (Rédei, 1978). At that time 88% of the wood was oak (mainly *Quercus robur*, with some *Q. pubescens*), 9% was poplar (*Populus alba* and *tremula*, probably also *P. × canescens*) and 2.5% was occupied by the non-native black locust (*Robinia pseudo-acacia*). *Betula pendula, Fraxinus angustifolia, Ulmus minor, Pyrus pyraster* and *Malus sylvestris* also occurred, but only sporadically (Rédei, 1978). Openings made up one-third of the area (Fig. 20.4). The large extension of openings (sometimes 2–7 times bigger than on the cadastre map) meant that in the short run they were more profitable than the coppices (Galgóczy, 1896; Rédei, 1978). In these openings, steppe grass-lands and wet meadows were typical (Hargitai, 1940). Later (mainly after 1928) these openings were gradually planted with black locust. The coppice cycle was 25 years, later 30 years. The management plan made precise rules for coppicing, the use of the openings, and banned grazing. In the 1920–1930s, the only type of forestry was still coppicing (Rédei, 1978). From the late nineteenth century onwards we can assume that agricultural use of the wood ceased and the wood became more and more closed.

By the 1940s, black locust already occupied 20% of the area and many ancient oak woods were transformed into plantations for economic reasons (cf. Szabó, 1879). Later, pine (*Pinus nigra* and *sylvestris*) was also planted. By 1976 non-native species already occupied 58% of the total area (most of it was *Robinia*) (Rédei, 1978). From the 1950s, a new forestry practice started. Instead of simple coppicing, after cutting the trees, the stumps were bulldozed to the edge of the patch, then the area was ploughed, and the space between the newly planted tree lines was cultivated for a couple of years. This resulted in a total disappearance of the natural woodland flora (Molnár, unpubl. data). By the 1990s, more than 80% of the area experienced this total destruction.

As a consequence of the total abandonment of grazing after World War II, the openings began to close (this can be seen if we compare the aerial photos from 1950/52 and 1990/92) although they did not disappear totally. Birch become very rare. Since the 1930s some species became extinct (*Gentiana cruciata, Majanthemum bifolium, Acer tataricum*) but only a few species became more widespread (e.g. *Robinia pseudo-acacia* and woodland generalists (such as *Brachypodium sylvaticum* and *Polygonatum latifolium*). Weed species are not spreading in the remnant patches.

From the mid-1980s the lowering of the soil water table of the sand region became catastrophic (Pálfai, 1994; Ráth, 1994). The soil water table in the Nagykőrös wood was previously at a depth of *c.* 0.5–1 m, now it can be found 3–4 m below the surface (Szentpéteri, 1990). The lowering was caused by climatic drought, drinking water extraction and artificial drainage (Pálfai, 1994). Water shortage has caused a serious dieback of the oak, prevents its natural regeneration and has made wet habitats disappear.

The present-day distribution of the remaining wooded steppe patches can be seen in Fig. 20.4. These originated from coppices, and are about 40–70 years old. The 'high forest' physiognomy of the remnant stands is historically atypical, since in the last centuries the coppice physiognomy was typical. This may be one reason for the highly dynamic state of the wood. All the remaining openings are today dry steppe grasslands.

Relationship between history and present vegetation features

Mosaic structure

In the last 250 years the following factors have determined the size and pattern of the openings:

- grazing and mowing increased their area;
- planting with black locust decreased their area;
- ploughing of clear cuts decreased their area;
- abandonment of grazing decreased their area;
- water shortage increased their area.

With the knowledge of these facts, we cannot assume that the present pattern of the woodland and grasslands is a natural feature.

Woodland interiors

These habitats have the lowest typicalness and are most lacking in specialist species. This can be well explained by the fact that this habitat was not continuous in the past, many areas were once open and coppicing and grazing depleted the specialist flora further. Since the woodland flora of the region is not poor (see data in Szujkó-Lacza and Kováts, 1993), we think that the local flora has been depleted by human activities and is not the result of the subcontinental climate.

Woodland edges

The edge habitats show intermediate values: there are quite a few specialist and not too many disturbance-tolerant species. This habitat changed a lot in the last 250 years but has not been disturbed directly and heavily.

Grasslands of the small openings

These grasslands are the least disturbed and have the highest proportion of specialist species. This implies that they may be the most natural. This marginal habitat was continuous in the past and, being near to the forest edge, it was probably sheltered from disturbance.

Grasslands of the medium and large openings

The grasslands of the larger openings have the highest proportion of disturbance-tolerant species. In the past these openings were mown, grazed and used for traffic or as arable fields and so they became more degraded than the other habitats.

The above-mentioned 'explanations' are not necessarily cause–effect relationships, since in historical studies it is really very difficult to prove that a certain human effect caused a certain vegetational change. Questions which remain to be answered include:

- How poor was the woodland flora in the sixteenth and seventeenth centuries?
- How natural is the most natural habitat or the most natural part of the Nagykőrös wooded steppe?
- What would be the natural proportion of steppes and woodlands in this landscape?

Time-scale of some vegetational processes in the Nagykőrös wood

The following observations are derived from the general vegetation survey of the wood. They serve as hypotheses for future ecological research.

- Disturbance tolerant species disappear from the herb layer for several decades if the canopy closes. This is shown by the fact that present-day non-disturbed woodland interiors, though characterless, have very few disturbance indicator species.
- Secondary woodland patches are very poor in woodland interior specialist species. This implies that these specialist species spread very slowly, if at all, into the wood from refugia, even if circumstances improve (100–200 years was not enough). This observation corresponds with data from the Lincolnshire woods in England (Peterken and Game, 1984).
- If the canopy opens up (for example as a consequence of the recent drought), generalist steppe species are able to colonize these new habitats in 5–10 years. In some cases, however, the shrub layer becomes very dense, which prevents the invasion of the steppe species, but makes the survival of woodland interior species possible.

Recommendations for nature conservation

Nature conservation agencies responsible for this area emphasize the preservation of the closed oak woodlands even if the oak is dying back rapidly. The history and present-day vegetation of the wood suggests, however, that nature conservation should concentrate on the woodland edge habitat and the steppe grasslands near them. These parts of the wood should have the highest priority for conservation because it is here that the wooded steppe mosaic is in the best condition.

Conclusions

We can draw the conclusion that the habitats of the seemingly natural mosaic of the wooded sand steppes at Nagykőrös have been transformed by past human use. Future botanical and woodland ecological studies have to take this into consideration.

Without knowledge of the detailed history of the wood, the following features might have been misinterpreted:

- scarcity of woodland interior specialist species in the wood;
- the naturalness of the pattern of the wooded steppe mosaic; and
- the overwhelming dominance of *Quercus robur.*

By the use of the plant indicator species' properties the coarse estimation of naturalness became possible. When there is a complete lack of palynological data this method may prove to be useful in areas for which direct historical data is also lacking.

Acknowledgements

Gábor Fekete and Marianna Biró offered constructive comments on an earlier version of the manuscript. This study was supported by the Nature Conservation Directorate of Budapest, by the OTKA Grants Nos. T-16390 and T-14651.

Maps and Aerial Photographs

1st (1783), 2nd (1861–1866) and 3rd (1884) Military Survey Map, Museum of War History, Budapest.

Black and white aerial photographs from 1950, 1952 and 1990, Carthographic Institute of the Ministry of Defence, Budapest.

Infra-red aerial photograph from 1992, Institute of Geodesy and Remote Sensing, Budapest.

References

Balla, G. (1758) *Nagykőrös Krónika* (Chronicle of Nagykőrös). Nagykőrös.

Berglund, B.E. (ed.) (1991) The cultural landscape during 6000 years in southern Sweden – the Ystad Project. *Ecological Bulletin* 41, Copenhagen.

Bernátsky, J. (1901) Növényföldrajzi megfigyelések a Nyírségen. *Természettudományi Közlöny* 53, 203–216.

Biró, M. and Molnár, Zs. (1998) *Landscape types, their distribution, vegetation and land use history in the sand dunes of the Duna-Tisza köze (Kiskunság sensu lato) from the 18th century.* Történeti Földrajzi Füzetek, (in press).

Borbás, V. (1886) *A magyar homokpuszták növényvilága, meg a homokkötés.* Pesti könyvnyomda-részvény-társaság, Budapest.

Borhidi, A. (1995) Social behaviour types, naturalness and relative ecological indicator values of the Hungarian Flora. *Acta Botanica Hungarica* 39, 97–181.

Boros, Á. (1918–1934) *Utinapló* (Travel diary) – 1918, 1919, 1920, 1922, 1934. History of Science Collection of the Botanical Department of the Hungarian Natural Museum, Budapest.

Boros, Á. (1935) A nagykőrösi homoki erdők növényvilága. *Erdészeti Kísérletek* 37, 1–24.

Boros, Á. (1952) A Duna-Tisza köze növényföldrajza. *Földrajzi Értesítő* 1, 39–53.

Braun, M., Sümegi, P., Szűcs, L. and Szöőr, Gy. (1993) The history and development of the Nagy-Mohos fen at Kállósemjén: man induced fen formation and the 'archaic' fen concept. *Jósa Múzeum Évkönyve* 33–34, 335–366.

Fekete, G. (1992) The holistic view of succession reconsidered. *Coenoses* 7, 21–30.

Frisnyák, S. (1990) Magyarország történeti földrajza. Tankönyvkiadó, Budapest.

Galgóczy K. (1896) *Nagykőrös város monográfiája*. Nagykőrös.

Hanák, P. (ed.) (1991) *The Corvina history of Hungary – from the earliest times until the present day*. Corvina Books, Budapest.

Hargitai, Z. (1937) Nagykőrös növényvilága. I. A flóra. *Debreceni Református Kollégium Tanárképző Intézet Dolgozatai* 17, 1–55.

Hargitai, Z. (1940) Nagykőrös növényvilága. II. A homoki növényszövetkezetek (Vegetation of Nagykőrös II. The plant communities on sand). *Botanikai Közlemények* 37, 205–240.

Hollós, L. (1896) Kecskemét növényzete. In: Bagi, L. (ed.) *Kecskemét múltja és jelene*. Tóth L. Nyomdája, Kecskemét, pp. 77–147.

Horváth, F., Dobolyi, K., Morschhauser, T., Lőkös, L., Karas, L. and Szerdahelyi, T. (1995) *Flóra adatbázis 1.2, Taxonlista és attributumállomány*. Flora Workgroup, Institute of Ecology and Botany, Botanical Department of the Hungarian Natural Museum, Vácrátót.

Jackson, S.T., Futyma, R.P. and Wilcox, D.A. (1988) A palaeoecological test of a classical hydrosere in the Lake Michigan dunes. *Ecology* 69, 928–936.

Járai-Komlódi, M. (1987) Postglacial climate and vegetation in Hungary. In: Pécsi, M. and Kordos, L. (eds) *Holocene Environment in Hungary*. Geographic Research Institute, Budapest, pp. 37–47.

Majer, A. (1988) *Fenyves a Bakonyalján*. Akadémia; Kiadó, Budapest.

Molnár, Zs. (1996) A Pitvarosi-puszták vegetáció- és a tájtörténete az Árpád-kortól napjainkig. *Natura Bekesiensis* 21, 65–97.

Molnár, Zs. (1997) The land use historical approach to study vegetation history at the century scale. In: Tóth, E. and Horváth, R. (eds) *Research, Conservation, Management Conference*. Agglelek, pp. 345–354.

Morschhauser, T. (1990) A Remete-szurdok flórája, vegetációja és degradáltsági állapotának felmérése. MSc Thesis, Eötvös Lóránd University, Budapest.

Nemcsik, J. (1861) *A nagykőrösi erdő, annak kezelése és haszna*. Ballagi Nagykőrösi naptár, Nagykőrös.

Pálfai, I. (1994) Összefoglaló tanulmány a Duna-Tisza közi talajvízszint-süllyedés okairól és a vízhiányos helyzet javításának lehetőségeiről. In: Pálfai, I. (ed.) *A Duna-Tisza közi Hátság vízgazdálkodási problémái*. Nagyalföld Alapítvány, Budapest, pp. 111–123.

Pécsi, M. (ed.) (1989) *National atlas of Hungary*. Kartográfiai Vállalat, Budapest.

Peterken, G. and Game, M. (1984) Historical factors affecting the number and distribution of vascular plant species in the woodlands of central Lincolnshire. *Journal of Ecology* 72, 155–182.

Rapaics, R. (1918) Az Alföld növényföldrajzi jelleme. *Erdészeti Kísérletek* 21, 1–146.

Ráth, I. (1994) Kritikus vízháztartási helyzet a Duna-Tisza közi hátságban. *Ö.K.O.* 5, 29–36.

Rédei, K. (1978) *Adatok Nagykőrös város erdőgazdálkodásának történetéhez*. Kecskemét.

Simon, T. (1988) A hazai edényes flóra természetvédelmi érték-besorolása. *Abstracta Botanica* 12, 1–23.

Somogyi, S. (1965) A szikesek elterjedésének időbeli változásai Magyarországon. *Földrajzi Közlemények* 11, 41–55.

Somogyi, S. (1994) Az Alföld földrajzi képe a honfoglalás és a magyar középkor időszakában. *Észak- és Kelet-Magyarországi Földrajzi Évkönyv* 1, 61–75.

Soó, R. (1929) Die Vegetation und die Entstehung der ungarischen Puszta. *Ecology* 17, 329–350.

Szabó, A. (1879) Tölgyesek irtása és ákáczosok telepítése a Kecskemét városi erdőkben. *Erdészeti Lapok* 18, 14–26.

Szentpéteri, S. (1990) Ökoszisztéma rekonstrukciós terv a Nagykőrös határában lévő, Strázsadomb" környéki területre. MSc Thesis, Sopron.

Szujkó-Lacza, J. and Kováts, D. (eds) (1993) *The flora of the Kiskunság National Park*. Hungarian Natural Museum, Budapest.

Tóth, K. (ed.) (1996) 20 éves a Kiskunsági Nemzeti Park 1975–1995. In: *A tudományos konferencia előadásai és hozzászólásai. Tudományos kutatási eredmények*. Kiskunság National Park, Házinyomda Kft, Kecskemét.

Vidéki, R. (1993) *A társadalmi beavatkozások hatása a Duna-Tisza köze geomorfológiai, vízrajzi, növénytani viszonyaira*. Kiskunfélegyháza.

Zólyomi, B. (1945–1946) Természetes növénytakaró a tiszafüredi öntözőrendszer területén. *Öntözésügyi Közlemények* 7–8, 62–75.

Zólyomi, B. (1969) Földvárak, sáncok, határmezsgyék és a természetvédelem. *Természet Világa* 100, 550–553.

Zólyomi, B. (1989) Natural vegetation of Hungary. In: Pécsi, M. (ed.) *National atlas of Hungary*. Kartográfiai Vállalat, Budapest, pp. 89.

Zólyomi, B. and Fekete, G. (1994) The Pannonian loess steppe: differentiation in space and time. *Abstracta Botanica* 18, 29–41.

Researching Forest History to Underpin the Classification of Dutch Forest Ecosystems

21

R.J.A.M. Wolf

*DLO Institute for Forestry and Nature Research, PO Box 23,
NL-6700 AA Wageningen, the Netherlands. Present Address: Johan
Fabriciuskade 8, NL-6708 SG Wageningen, the Netherlands*

The role of forest history research within the classification of Dutch forest ecosystems is illustrated by a case study of the Amerongen Forest. Although the Dutch forest ecosystem typology is based on combining site types with vegetation types, data on forest history are systematically collected to underpin this ecological classification. These data are used to obtain an overview of the general historical background of woodland and forests in different Dutch landscapes and to provide historical foundations for the ecological basis of the classification system. Certain standard data on forest history are linked to the ecological data set for the latter purpose.

The integration of the two approaches results in knowledge of the history and former management of woodland being used to improve our understanding of the present-day variety of forest ecosystems and to predict their development. Historical knowledge contributes substantially to the provision of an ecological frame of reference to underpin forest management decisions.

Introduction

All the present-day forest ecosystems in the Netherlands are currently being classified in a nationwide project. Although the classification system is based on ecological data, considerable attention is being paid to forest history in this project. This chapter discusses the role of forest history research within the project and how this integrated historical–ecological approach contributes to the usefulness of the classification system in forest management, using a case study of the Amerongen Forest.

Classifying Forest Ecosystems

The classification of Dutch forest ecosystems is being carried out by a team of scientists from the DLO Institute for Forestry and Nature Research (IBN-DLO) and the DLO Winand Staring Centre (SC-DLO). The aim is a survey of the Dutch forest which:

- is based on integration of biotic and abiotic characteristics of the ecosystem;
- describes the present-day situation;
- includes recently planted and intensively managed forests as well as woodland several hundreds of years old;
- makes it possible to predict the development of the different forest ecosystem types; and
- is suitable for forest management up to the level of forest stand (Clerkx *et al.*, 1994).

A multidisciplinary approach which combines knowledge of soil, geomorphology, hydrology, vegetation, forestry and forest history is being used to achieve the five points outlined above. Some 1500 field descriptions (relevés) of Dutch forests and woodlands are being made to establish the forest ecosystem classification. Each relevé covers 50–400 m² and consists of three components:

1. A description of abiotic features (land forms, soil profile, humus profile, groundwater levels, pH, etc.).
2. A vegetation description (relevé according to the French–Swiss school, see Schaminée *et al.*, 1995).
3. Notes on the visible signs of forest history and management practices.

   Field descriptions are made by a team; field discussions between scientists from different disciplines result in the direct exchange of information and insights. The data from the relevés are supplemented with information from the literature. Additional data on historical aspects are derived from old topographic maps, archives and interviews with local forest managers.

   These data are used to prepare a site typology and vegetation typology. Each site type is characterized by specific stable abiotic conditions, determined by factors which are important for the vegetation and for the forest stand, but are not influenced by the development of these (Clerkx *et al.*, 1994; Wolf *et al.*, 1996). The ecosystem model of Jenny (1941, 1980), which has proved its usefulness in several ecosystem studies and manuals (e.g. Spurr and Barnes, 1980; Fanta, 1982; Vos and Stortelder, 1992; Otto 1994), serves as a guideline for working out the site classes.

   Vegetation types are derived from differences in species composition. The methods used for drawing up the vegetation classification are based on those of Schaminée *et al.* (1995). The typology is founded on vegetation tables, based on the relevés collected supplemented by a large number of relevés derived from literature.

   The basic unit of the classification project, the forest ecosystem, is a characteristic combination of a site type and a vegetation type (Fig. 21.1). Various vegetation types may occur on the same site type. One of these usually corresponds to the potential natural vegetation (PNV): the vegetation of the final succession stage at the site type in question. All other vegetation types differ from the PNV as a result of their young age (pioneer communities) or silvicultural treatments (replacement communities).

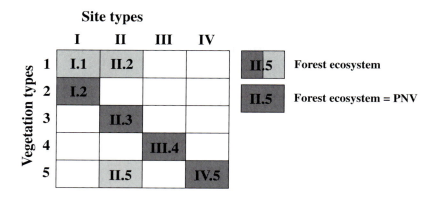

Fig. 21.1. Matrix forest ecosystems. Each existing combination of a site and vegetation type forms a forest ecosystem.

Historical Data on Forests

Historical data are used in two ways within the classification project (Wolf, 1992, 1995; Clerkx *et al.*, 1994). The data found in the literature enable a general outline to be made of developments which have led to the present-day variety of forest ecosystems in each Dutch landscape. This historical overview covers forest and woodland areas and landscapes as a whole. The approach is similar to that used by Rackham (1980, 1986) and Peterken (1981) in the UK, Tack *et al.* (1993) in Belgium, and Pott and Hüppe (1991) in Germany. The overview is used as an introduction to the classification, to help understand the general historical background of woodland and forest in the various landscapes.

Data on forest history are also used to improve the foundations of the ecosystem classification. For this, historical forest data are linked to the ecosystem relevés. A standard set of historical data is prepared from each relevé, mainly on aspects which have substantially influenced the development of the present-day forest ecosystems. These standard data are to do with the origin and age of the woodland and the current stand, and with human interference in the forest ecosystem. The following data are collected from each relevé:

- Land use before the development of the woodland/forest.
- Approximate date of development of the first generation of woodland/forest (if after *c.* 1840).
- Approximate date of development of the current woodland generation/forest stand.
- Soil treatments.
- Dominant tree species in current and former woodland generations.
- Origin of the current woodland generation/forest stand.
- Former and present-day silvicultural management systems.

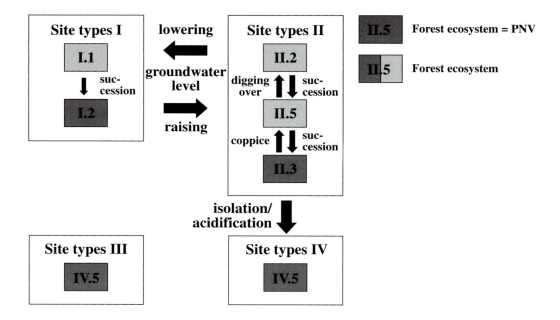

Fig. 21.2. The principle of a forest development diagram.

Any additional historical data specific to certain forests or woodland are also collected.

Collecting information on forest history and management for each relevé separately, helps elucidate the reasons for differences between various forest ecosystem types (Clerkx *et al.*, 1994). The presence of various vegetation types on one particular site is usually the result of differences in their historical backgrounds. Thus, insights into the interrelationships between several forest ecosystem types can be obtained by integrating historical and ecological data collected in the relevés. These interrelationships are summarized in a forest development diagram (Fig. 21.2), which shows for each site type how one forest ecosystem can evolve into another, for example as a result of succession or certain silvicultural treatments. Sites may also change category as a result of major changes in site characteristics, such as groundwater level or soil acidity.

Forest Ecosystems in the Amerongen Forest

The Amerongen Forest (Amerongse bos) is a woodland of *c.* 1000 ha, on the transition from the ice-pushed ridge of the Utrechtse Heuvelrug to the Rhine floodplain (Fig. 21.3). Fourteen relevés have been described in the Amerongen Forest as part of the forest ecosystem classification project. In addition, the history of this woodland has been studied. The relevés do not depict the complete scope of ecosystem variation within the Amerongen Forest. However, combined with the historical study of this wood, they are

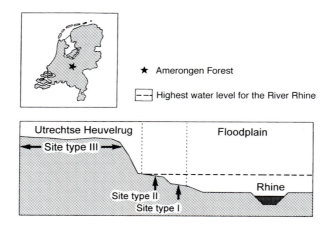

Fig. 21.3. Location of the Amerongen Forest and the site types distinguished.

suitable for illustrating the role of forest history research within the classification project.

Using the criteria applied in the forest ecosystem classification system, the relevés in the Amerongen Forest can be assigned to three site types (Table 21.1). Site type I is located on the transition from the ice-pushed ridge to the floodplain (Fig. 21.3). When the Rhine floods, this footslope of the ice-pushed ridge is inundated with calcareous, nutrient-rich river water. Floods happen quite regularly; the mean duration of flooding per annum is *c.* 3–4 days. Hence the topsoil consists of a mixture of coarse-grained sand from the ice-pushed ridge, and river clay deposited by the Rhine. Site type II is situated slightly higher than site type I, hence floods are less frequent and do not last as long. Since inundation by Rhine water happens only incidentally (mean duration of flooding per annum < 1 day), the clay content and pH value are lower than those of site type I. Site type III can be found on top of the ice-pushed ridge and consists of nutrient-poor loamy sand with low pH.

Table 21.1. Amerongen Forest site types.

	Situation	Inundation	Texture topsoil	Acidity topsoil
Site type I	On footslope of ice-pushed ridge	Fairly regularly (3–4 days p.a.)	Light clay (20–30% < 2μm)	pH-KCl *c.* 5
Site type II	On gentle slope on edge of ice-pushed ridge	Incidentally (< 1 day p.a.)	Light sandy clay (c. 10% < 2μm)	pH-KCl 4–4.5
Site type III	On top of ice-pushed ridge	Never	Loamy sand (15–25% < 50μm)	pH-KCl *c.* 3

Table 21.2. Vegetation table, Amerongen Forest.

Relevé number	1	2	3	4	5	6	7	8	9	10	11	12	13	14
Vegetation type	1	2	3		4				5		6			
Tree layer														
Quercus robur	8	8	2	1	–	–	6	–	6	5	–	–	3	1
Fraxinus excelsior	7	6	–	–	–	–	–	–	–	–	–	–	–	–
Alnus glutinosa	5	–	–	–	–	–	–	–	–	—	–	–	–	–
Betula pendula	–	–	7	5	6	–	6	3	7	5	–	–	–	1
Fagus sylvatica	–	–	–	5	2	–	–	–	–	–	–	–	–	–
Quercus petraea	–	–	–	–	8	–	–	–	–	–	–	–	–	–
Pinus sylvestris	–	–	–	5	–	–	8	–	–	–	–	–	–	–
Larix kaempferi	–	–	6	–	–	–	–	7	6	8	9	–	–	–
Pinus nigra v. maritima	–	–	–	–	–	9	–	–	–	–	–	–	–	–
Picea abies	–	–	–	–	–	–	–	–	–	–	–	8	–	–
Pseudotsuga menziesii	–	–	–	–	–	–	–	–	–	–	–	–	7	9
Natural regeneration														
Prunus padus	2	–	–	–	–	–	–	–	–	–	–	–	–	–
Quercus robur	–	2	–	1	–	–	2	–	–	–	–	–	–	–
Quercus petraea	–	–	–	–	7	–	–	–	–	–	–	–	–	–
Prunus serotina	–	–	–	–	–	2	2	–	–	–	–	–	–	–
Fagus sylvatica	–	–	–	–	–	–	1	2	–	–	–	–	2	–
Sorbus aucuparia	–	–	–	–	–	2	2	1	–	2	–	2	4	1
Betula pendula	–	–	–	–	–	–	–	–	–	–	–	2	2	–
Pseudotsuga menziesii	–	–	–	–	–	–	–	–	–	–	–	–	3	2
Herb/shrub layer														
Alliaria petiolata	5	–	–	–	–	–	–	–	–	–	–	–	–	–
Geum urbanum	5	–	–	–	–	–	–	–	–	–	–	–	–	–
Urtica dioica	5	–	–	–	–	–	–	–	–	–	–	–	–	–
Allium scorodoprasum	3	–	–	–	–	–	–	–	–	–	–	–	–	–
Festuca gigantea	3	–	–	–	–	–	–	–	–	–	–	–	–	–
Poa nemoralis	3	–	–	–	–	–	–	–	–	–	–	–	–	–
Rubus caesius	3	–	–	–	–	–	–	–	–	–	–	–	–	–
Senecio fluviatilis	2	–	–	–	–	–	–	–	–	–	–	–	–	–
Chaerophyllum temulum	2	–	–	–	–	–	–	–	–	–	–	–	–	–
Heracleum sphondylium	2	–	–	–	–	–	–	–	–	–	–	–	–	–
Veronica hederifolia	2	–	–	–	–	–	–	–	–	–	–	–	–	–
Viola riviniana	2	–	–	–	–	–	–	–	–	–	–	–	–	–
Ribes rubrum	2	–	–	–	–	–	–	–	–	–	–	–	–	–
Galium aparine	7	2	–	–	–	–	–	–	–	–	–	–	–	–
Glechoma hederacea	7	2	–	–	–	–	–	–	–	–	–	–	–	–
Sambucus nigra	5	5	–	–	–	–	–	–	–	–	–	–	–	–
Crataegus monogyna	7	8	–	–	–	–	–	–	–	–	–	–	–	–
Anemone nemorosa	2	7	–	–	–	–	–	–	–	–	–	–	–	–
Ranunculus ficaria	2	7	–	–	–	–	–	–	–	–	–	–	–	–
Ornithogalum umbellatum	2	3	–	–	–	–	–	–	–	–	–	–	–	–
Euonymus europaeus	1	3	–	–	–	–	–	–	–	–	–	–	–	–
Corylus avellana	–	5	–	–	–	–	–	–	–	–	–	–	–	–
Lonicera periclymenum	–	2	–	–	–	–	–	–	–	–	–	–	–	–
Cornus sanguinea	–	2	–	–	–	–	–	–	–	–	–	–	–	–
Pteridium aquilinum	–	–	9	8	–	–	–	–	–	–	–	–	–	–
Vaccinium myrtillus	–	–	–	2	7	6	9	7	–	–	3	1	–	–
Ceratocapnos claviculata	–	–	–	1	5	5	3	–	3	–	–	–	–	–

Table 21.2. *Continued.*

Relevé number	1	2	3	4	5	6	7	8	9	10	11	12	13	14
Vegetation type	1	2	3				4			5			6	
Galium saxatile	–	–	–	–	–	2	4	–	–	–	–	–	3	–
Dryopteris dilatata	–	–	–	–	2	2	–	–	–	1	–	5	5	3
Dryopteris carthusiana	–	–	–	–	2	5	–	–	2	1	–	2	–	2
Deschampsia flexuosa	–	–	3	7	6	8	7	8	**9**	**9**	2	2	3	2
Carex pilulifera	–	–	–	–	–	–	–	–	–	–	–	1	3	1
Moss layer														
Mnium hornum	2	2	–	–	1	–	–	–	–	–	2	2	4	–
Brachythecium rutabulum	2	–	3	3	3	–	2	–	2	4	4	–	3	–
Hypnum jutlandicum	1	–	3	–	5	7	7	–	5	6	7	–	–	–
Pleurozium schreberi	–	–	4	–	4	4	3	6	5	6	3	–	–	–
Pseudoscleropodium purum	–	–	5	–	4	–	–	–	4	5	3	–	–	3
Lophocolea heterophylla	2	–	2	2	–	2	–	–	2	–	2	–	2	2
Eurhynchium praelongum	3	2	4	–	3	3	–	–	4	5	7	8	3	2
Atrichum undulatum	–	–	2	–	–	–	–	–	–	–	3	–	–	2
Dicranella heteromalla	–	–	2	–	–	–	–	–	–	–	2	–	3	2
Dicranum scoparium	–	–	–	–	–	2	2	3	3	2	2	–	3	2
Plagiothecium curvifolium	–	–	–	–	2	–	–	–	–	4	5	5	–	5
Polytrichum formosum	–	–	–	–	–	–	–	–	–	–	4	–	–	–
Brachythecium oedipodium	–	–	–	–	–	–	–	–	–	–	3	–	–	–
Eurhynchium striatum	–	–	–	–	–	–	–	–	–	–	3	–	–	–
Campylopus flexuosus	–	–	–	–	–	–	–	–	–	–	–	–	3	–

Vegetation types:
1 *Allium* type (*Ulmion carpinifoliae*)
2 *Anemone* type (*Ulmion carpinifoliae*)
3 *Pteridium* type (*Quercion robori-petraeae*)
4 *Vaccinium* type (*Quercion robori-petraeae*)
5 *Deschampsia* type (*Quercion robori-petraeae*)
6 Moss type (*Quercion robori-petraeae*)

Number of individuals/cover:
1 1–3 individuals; negligible cover
2 4–20 individuals; cover < 5%
3 20–100 individuals; cover < 5%
4 > 100 individuals; cover < 5%
5 cover 5–12%
6 cover 13–25%
7 cover 25–50%
8 cover 50–75%
9 cover 75–100%
Species which gave their name to the vegetation type are shown in bold figures.

The relevés in the Amerongen Forest have been classified into six vegetation types. Their species composition is shown in Table 21.2. The *Allium* type (1) occurs on site type I and is designated forest ecosystem I.1. The *Anemone* type (2) has been found on site type II (forest ecosystem II.2), the other vegetation types (3–6) on site type III (forest ecosystems III.3-III.6).

Fig. 21.4. The irregular, crooked stems of the oaks in this woodland have developed from old degraded coppice.

History of the Amerongen Forest

Burial mounds in the Amerongen Forest indicate that even in prehistoric times people were using the woodland present at that time. In the early Middle Ages the site of the present-day Amerongen Forest was part of an extensive woodland area with large trees. Gradually, the use of the woodland intensified, resulting in a growing interest in recording forest use rights based on customary laws. From the thirteenth to the fourteenth centuries these rights were divided among commoners: the members of the 'marke' of Amerongen, called 'markegenoten'. The use of the forest by these commoners resulted in extensive oak coppices replacing the original high forest. As a result of excessive woodcutting, over-grazing, removal of litter and lack of protective measures during the fifteenth, sixteenth and seventeenth centuries, the oak coppices gradually degraded to vast heathlands, inhabited by flocks of sheep. Only a small amount of degraded oak coppice was left as a remnant of the original woodlands (Fig. 21.4) (Buis, 1985; Oosterhuis and Rademaker, 1988).

By the beginning of the eighteenth century, all heathlands near Amerongen became the property of the Lord of Amerongen. The commoners thereby had lost their rights to the forest. During the second half of the eighteenth century the Lord of Amerongen started a systematic re-afforestation programme. He afforested some lower parts of his estate, mainly with ash (*Fraxinus excelsior*), alder (*Alnus glutinosa*), elm (*Ulmus* spp.) and common oak (*Quercus robur*). However, the vast majority of the new forest was established on the heathlands, on the dry sandy soils of the ice-pushed ridge. Scots pine (*Pinus sylvestris*) and common oak were mainly used to afforest the heathlands, the latter for recreating coppice compartments. In addition, a limited area of broad-leaved high forest (mainly beech, *Fagus sylvatica*) and a considerable number of avenues of beech or oak were

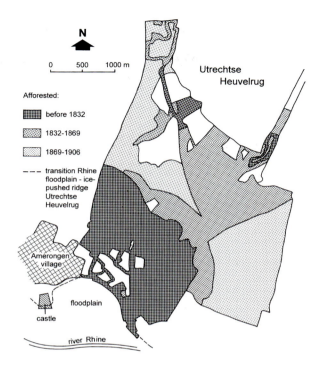

Fig. 21.5. Increase of the wooded area in the present-day Amerongen Forest between 1832 and 1906.

planted. The progress in afforestation between 1832 and 1906 is depicted in Fig. 21.5, which shows that the oldest parts (woodland before 1832) are concentrated in the southwestern part of the estate, close to the castle of the Lords of Amerongen. Moreover, most of the estate had been re-afforested by the beginning of this century.

The first Scots pine stands in the Amerongen estate, dating from the second half of the eighteenth century and the beginning of the nineteenth century, were generally created by broadcasting seed by hand after burning or mowing the heather. After sowing, the pine seeds were covered with sand. Wooded banks were created to protect the new pine stands from livestock and game. Later in the nineteenth century, the soil was turned over completely to depths varying from 20 to 60 cm prior to sowing. Around 1900, this method changed to local and superficial soil treatment, followed by sowing or planting. During the nineteenth and early twentieth centuries, the pine timber was mainly used for pit props. Rotations of *c.* 55–60 years were used (van der Jagt, 1976; Oosterhuis and Rademaker, 1988).

Parcels of oak coppice were planted after the soil had been dug over to 60–80 cm depth. Initially, the coppice was used for firewood and other domestic uses by the local people. Later, during the nineteenth century, the oak bark was used for tanning, and gradually this became the most important market product for the coppices. This led to oak coppice areas being re-established in the Amerongen Forest in the late nineteenth century. Rotations of the oak coppice were *c.* 8–10 years (Buis, 1985; Oosterhuis and Rademaker, 1988).

Most of the Amerongen Forest was sold to the Unitas agency in 1935, whose main purpose was to raise the productivity and value of the property. To achieve this, the remaining heathland was afforested, virtually all oak coppice was converted into coniferous stands, stands were enlarged and poorly growing Scots pine stands were replaced by stands of other coniferous species. The planting of Douglas fir (*Pseudotsuga menziesii*), Japanese larch (*Larix kaempferi*), Norway spruce (*Picea abies*), sitka spruce (*Picea sitchensis*) and Corsican pine (*Pinus nigra* var. *maritima*), resulted in the proportion of Scots pine within the coniferous woods falling from 97% in 1935 to 59% in 1950. During this afforestation the soil of the whole stand was dug over and manured. The stands were not thinned until the thinnings could be sold. Wood from the first thinnings was mainly sold to local people for domestic use; subsequent thinnings and the final felling yielded pit props (van der Jagt, 1976; Staatsbosbeheer, 1985).

About 100 ha of the Amerongen Forest remained the property of the Lords of Amerongen in 1935. In this part, many parcels of oak coppice and Scots pine stands were also converted to new coniferous stands. However, much more oak coppice survived than in the Unitas part of the Amerongen Forest. Most of these remaining oak coppices have not been cut since the Second World War, and because of this have developed into today's over-grown coppice (Klingen and Litjens, 1985).

In 1974, Unitas sold its properties in the Amerongen Forest to the state. Since then, these have been managed by the State Forest Service (Staatsbosbeheer). The land still owned by the Lords of Amerongen was sold to the regional nature management organization, Utrechts Landschap, in 1977. These transactions brought to an end a long period in which woodland management had been chiefly concentrated on timber production. Gradually, more opportunities were created for natural processes. Clear felling and artificial forest regeneration have been replaced by small-scale forest management and natural regeneration. In addition, the area of coniferous forest is being reduced to benefit indigenous deciduous tree species such as beech, common oak and silver birch (*Betula pendula*).

Fertilization has ceased, and only locally is some superficial soil treatment applied to encourage natural regeneration. Today, timber production plays a minor role in the management of the Amerongen Forest, which is focusing on maintaining and developing the forest's natural, cultural–historical and recreational value (Klingen and Litjens, 1985; Staatsbosbeheer, 1985). A forest reserve was created in 1983 (Galgenberg, 42 ha), in which no human interference is allowed except for the maintenance of the paths. Its natural forest development is being monitored as part of a comprehensive nationwide research programme (Koop, 1989; Koop and Clerkx, 1995; Clerkx *et al.*, 1996).

Table 21.3. Historical backgrounds of Amerongen Forest relevés.

Relevé no.	Forest ecosystem	Former land use	Depth soil treatments	First generation		Current generation		
				Period of establishment	Main tree species	Period of establishment	Main tree species	Origin
1	I.1	Woodland*	–	before 1832	co/as	1940–1950	co/as (al)	Former coppice
2	II.2	Woodland*	–	before 1832	co/as	1940–1950	co/as	Former coppice
3	III.3	Woodland*	25 cm	before 1832	sp	1939	jl/sb	Planting (jl), nat. regeneration (sb)
4	III.3	Heathland	30 cm	1832–1869	sp	1880	sp/be/sb	Sowing (sp), nat. regen. (be/sb)
5	III.4	Heathland	superficial	1832–1869	sp	1890–1900	do/sb	Nat. regeneration, degraded coppice
6	III.4	Heathland	20 cm	1869–1906	sp	1959	cp	Planting
7	III.4	Heathland	30 cm	1869–1906	sp	1894 (1st gen.)	sp/sb/co	Sowing (sp), nat. regen. (sb/co)
8	III.4	Heathland	45 cm	1832–1869	sp	1940	jl (sb)	Planting (jl), nat. regen. (sb)
9	III.5	Woodland*	25 cm	before 1832	sp	1939	jl/sb/co	Planting (jl), nat. regen. (sb/co)
10	III.5	Woodland*	75 cm	before 1832	co†	1940	jl (sb/co)	Planting (jl), nat. regen. (sb/co)
11	III.6	Heathland	25 cm	1869–1906	sp	1967	jl	Planting
12	III.6	Woodland*	superficial	before 1832	be/co	1940	ns	Planting
13	III.6	Heathland	80 cm	1869–1906	co†	1938	dg (co)	Planting (dg), nat. regen. (co)
14	III.6	Woodland*	90 cm	before 1832	sp	1940	dg	Planting

* Before 1832; † coppice.
be, Beech; cp, Corsican pine; dg, Douglas fir; as, Ash; ns, Norway spruce; sp, Scots pine; jl, Japanese larch; sb, Silver birch; co, Common oak; al, Alder; do, Durmast oak.
Main tree species: brackets = cover < 12.5%.

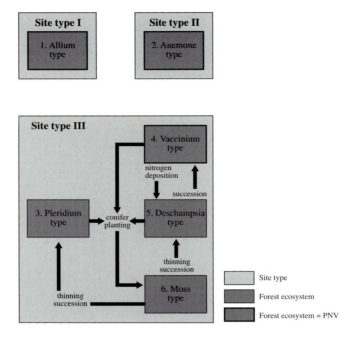

Fig. 21.6. Forest development diagram for Amerongen Forest.

Forest Development in the Amerongen Forest

Table 21.3 shows a number of standard data on the historical background of the Amerongen Forest relevés, collected by the methods used in the classification project. This has enabled various items from the preceding general historical outline of the Amerongen Forest to be transformed into characteristics of the relevés. The relevés have been grouped according to forest ecosystem type in the table. Combining these historical data with the information derived from the site and vegetation descriptions reveals the interrelationships and possible development of the different forest ecosystems (Fig. 21.6).

Relevés 1 and 2, corresponding to forest ecosystems I.1 and II.2 respectively, contain ancient stored oak–ash coppice. No soil treatment has been applied here. The coppice has not been cut since the Second World War, and there has been hardly any forest management. Because of this, the forest floor has become darker, which has probably slightly decreased the number of plant species in the understorey. The vegetation corresponds to the potential natural vegetation (PNV) of the sites in question, which both belong to the *Ulmion carpinifoliae* alliance (Westhoff and den Held, 1969; van der Werf, 1991). The species composition of the vegetation is mainly the result of processes related to flooding by the river Rhine. At site type I, a considerable supply of nutrients from the river is responsible for the presence of a lush, species-rich understorey belonging to the *Allium* type (Table 21.2). This

Fig. 21.7. Wood anemone (*Anemone nemorosa*) and lesser celandine (*Ranunculus ficaria*) dominate the herb layer under the canopy of abandoned oak coppice on a site incidentally flooded by Rhine water (forest ecosystem II.2).

nutrient-rich situation has encouraged an active soil fauna to develop. Hence, organic substances decompose rapidly, preventing litter and slightly decomposed humus from accumulating on top of the mineral soil. Site type II is less nutrient-rich, which causes a slower decomposition of organic matter than on site type I (Wolf *et al.*, 1997). However, the organic substances that have accumulated on top of the mineral soil are eroded during river floods approximately once every 15 years (de Waal, 1996). This process results in very favourable conditions for geophytes such as wood anemone (*Anemone nemorosa*) and lesser celandine (*Ranunculus ficaria*) just after each flood. These poor competitors survive the winter as rhizomes or bulbs and sprout in early spring. They are able to expand rapidly on the bare mineral soil and thus create a dense, species-poor herb layer (Fig. 21.7) (Wolf *et al.*, 1996).

The other relevés (3–14) belong to four forest ecosystems situated on top of the ice-pushed ridge of the Utrechtse Heuvelrug (site type III; Table 21.3). These are woods on former heathland, with ages varying from *c.* 100 years (1869–1906) to more than 160 years (woodland before 1832). In all relevés, the soil has been dug over to depths varying from 15 to 90 cm. The first woodland generation generally consisted of Scots pine stands, with oak coppice or deciduous high forest in only a few relevés. Except for relevé 7, the first generation has since been replaced, mostly by plantations of Japanese larch, Douglas fir, Norway spruce and Corsican pine, planted *c.* 1940 or later. In some relevés, a substantial part of the present-day canopy and understorey originates from natural regeneration of deciduous tree species such as silver birch, common oak and beech. The durmast oaks (*Quercus petraea*) in relevé 5 are probably relicts of medieval degraded oak coppice (Knoppersen, 1995). Table 21.3 shows that differences in the historical backgrounds of the four forest ecosystems of site type III are not much larger than those within each forest ecosystem. However, some trends are visible. For example, all relevés in which the current woodland generation is more than 60 years old belong to forest ecosystems III.3 and III.4; forest ecosystems III.5 and III.6 only contain relatively recent coniferous stands. Nevertheless, to understand how different forest ecosystem types have developed on the same site, these historical data need to be combined with the ecological characteristics of the woodland in the relevés, such as the plant species composition, woodland structure and humus profile. The insight provided by this integration of historical and ecological knowledge reveals the interrelationships and possible development of forest ecosystems as a result of succession and various silvicultural measures (Fig. 21.6).

The Moss type (forest ecosystem III.6) occurs in fairly recent, dark coniferous stands. Initially, in the thicket stage and early pole stage, ground vegetation is virtually absent in the dark, dense stands. After the first thinning, at the age of *c.* 15–25 years, mosses start dominating the ground vegetation, resulting in development of the Moss type (Table 21.2). As these coniferous stands grow older and are thinned further, herb species take over the dominant position in the ground vegetation. Depending on which herb species becomes dominant, the *Pteridium* (forest ecosystem III.3) or the *Deschampsia* type (forest ecosystem III.5) arises. Replanting or dense natural regeneration may result in reversion to dark coniferous stands belonging to the Moss type.

In the Dutch woodlands, bracken (*Pteridium aquilinum*) depends on vegetative reproduction under a light canopy. Because of this, the occurrence of the *Pteridium* type (forest ecosystem III.3) is related to existing bracken populations as well as to favourable light conditions. In this ecosystem, bracken rhizomes form a dense ground cover and thus suppress other plant species, including regenerating trees and shrubs. So, the *Pteridium* type forms a sub-climax whose vegetation usually remains virtually unchanged for many years. In this way, further succession towards the PNV is delayed or obstructed.

Like the *Pteridium* type, the *Deschampsia* type (forest ecosystem III.5) can be found in the light coniferous, deciduous or mixed woods. This type is characterized by a dense mat of wavy hair-grass (*Deschampsia flexuosa*) covering the ground (Fig. 21.8). Both relevés belonging to this ecosystem type in the Amerongen Forest are 50–60 year-old stands of Japanese larch, on soils dug over completely before planting. As a result of natural regeneration, the larches are currently intermixed with indigenous deciduous trees. In both relevés the *Deschampsia* type has developed from the Moss type. However, when the young stand is light enough, the *Deschampsia* type may also develop without a preceding Moss stage. As a result of succession, the *Deschampsia* type will gradually turn into the *Vaccinium* type (forest ecosystem III.4).

The vegetation of the *Vaccinium* type approximately corresponds to the PNV on this site, belonging to the alliance of oak woodlands on nutrient-poor soils (*Quercion robori-petraeae*; Westhoff and den Held, 1969; van der Werf, 1991). The ground vegetation is dominated by wavy hair-grass and bilberry

Fig. 21.8. The understorey in many light coniferous and deciduous stands on ice-pushed ridges is dominated by a thick mat of wavy hair-grass (*Deschampsia flexuosa*; forest ecosystem type III.5).

(*Vaccinium myrtillus*). As the stands age, this ecosystem develops from the former, because bilberry gradually ousts the wavy hair-grass. Nowadays, however, air pollution – especially deposition of nitrogen – triggers development in the opposite direction: wavy hair-grass suppresses bilberry and the *Deschampsia* type may return (Wolf *et al.*, 1996).

Conclusion

The example of the Amerongen Forest illustrates that collecting historical data on the Dutch forest as part of the nationwide Dutch ecosystem classification project results in an overview of the historical background of Dutch forests in general, as well as of the individual forest ecosystem types.

The general historical overview describes the interactions between human activities and natural developments through the ages. It gives an impression of the age and relative naturalness of present-day woodlands and forests in different Dutch landscapes. Furthermore, the context in which cultural–historical and natural–historical relicts should be put, is made clear.

Historical data on the Amerongen Forest show that the woodland, in the transition zone between the Rhine floodplain and the ice-pushed ridge of the Utrechtse Heuvelrug, is old, overgrown coppices, mainly of common oak and ash, on untreated soils. The woods situated on top of the ice-pushed ridge originate from the eighteenth to twentieth century afforestation of the heathlands which had replaced the original woodland as a result of centuries of overexploitation, a historical background typical of much Dutch woodland on sandy soils. Burial mounds and avenues of beech or oak are cultural relicts; groups of durmast oaks with irregular, crooked stems are considered natural-historical relicts, originating from highly degraded original woodland.

Knowledge of the historical backgrounds of the forest ecosystem types distinguished provides a foundation for the ecological basis of the classification system. The historical and ecological knowledge are integrated by collecting data on vegetation and on the abiotic environment as well as data on forest history in each relevé made to establish the classification. This integration of knowledge provides added value to the classification system. It elucidates the combinations of natural development and human interference in the past which have led to the composition of each forest ecosystem occurring today.

The relevés in the Amerongen Forest show that the composition of the two forest ecosystems on site types I and II is mainly determined by natural processes triggered by occasional inundation by river water. In both ecosystems, the vegetation corresponds to the potential natural vegetation (PNV). The composition of the four ecosystems described on the ice-pushed ridge (site type III), on the other hand, largely results from previous silvicultural measures; only one of the ecosystems approximately corresponds to the PNV. Combining the ecological and historical data from the relevés made on this site type has clarified the interrelationships and possible development of these forest ecosystems.

The results of the historical research conducted within the forest ecosystem classification project may contribute considerably to improving the foundations of forest management decisions. On the one hand, a historical frame of reference is drawn up by means of a description of the general woodland history of each Dutch landscape. On the other hand, historical data are integrated in an ecological frame of reference: the forest ecosystem classification. The historical research concentrates on aspects of forest and woodland history which have substantially influenced the development of the present-day Dutch woodlands and are most important for forest managers. It underpins the ecosystem classification by improving our understanding of the variety of the present-day forest ecosystems and indicating which woodland types can be expected on each site as a result of succession or certain management measures. Forest managers can use this knowledge to determine which targets can be achieved on a certain site.

References

Buis, J. (1985) *Historia forestis, Nederlandse bosgeschiedenis. Deel 1. Bosgebruik, bosbeheer en boswetgeving tot het midden van de negentiende eeuw.* HES, Utrecht.

Clerkx, A.P.P.M., Dort, K.W. van, Hommel, P.W.F.M., Stortelder, A.H.F., Vrielink, J.G., de Waal, R.W. and Wolf, R.J.A.M. (1994) *Broekbossen van Nederland.* DLO Instituut voor Bos – en Natuuronderzoek/DLO Staring Centrum, Wageningen.

Clerkx, A.P.P.M., Broekmeyer, M.E.A., Scabo, P.J., Hees, A.F.M. van, Os, L.J. van and Koop, H.G.J.M. (1996) *Bosdynamiek in bosreservaat Galgenberg.* DLO Instituut voor Bos – en Natuuronderzoek, Wageningen.

Fanta, J. (1982) *Natuurlijke verjonging van het bos op de droge zandgronden.* Rijksinstituut voor onderzoek in bos – en landschapsbouw De Dorschkamp, Wageningen.

Jagt, J.L. van der (1976) Boswachterij Amerongse berg. *Nederlands Bosbouwtijdschrift* 48, 218–224.

Jenny, H. (1941) *Factors of soil formation*. McGraw-Hill, New York/London.

Jenny, H. (1980) *The soil resource. Origin and behaviour.* Springer, New York.

Klingen, L.A.S. and Litjens, G.J.J.M. (1985) *Beheersplan Amerongse Bos 1985–1995.* De Bilt, Stichting Het Utrechts Landschap.

Knoppersen, G. (1995) *Algemene informatie van het bosreservaat 3, Amerongen Galgenberg.* IKC-natuurbeheer, Wageningen.

Koop, H. (1989) *Forest dynamics (SILVI-STAR: a comprehensive monitoring system).* Springer-Verlag, Berlin.

Koop, H.G.J.M. and Clerkx, A.P.P.M. (1995) *De vegetatie van bosreservaten in Nederland. Deel 3. Bosreservaat Galgenberg, Amerongen.* DLO Instituut voor Bos- en Natuuronderzoek, Wageningen.

Oosterhuis, J.W. and Rademaker, B.F.M. (1988) *Historisch onderzoek naar het beheer van bossen en natuurterreinen. Deelproject 11.1: 19e-eeuwse droge heidebebossing Boswachterij de Amerongse Berg.* Staatsbosbeheer, Utrecht.

Otto, H.J. (1994) *Waldökologie.* Ulmer, Stuttgart.

Peterken, G.F. (1981). *Woodland conservation and management.* Chapman & Hall, London.

Pott, R. and Hüppe, J. (1991) *Die Hudelandschaften Nordwestdeutschlands.* Westfälisches Museum für Naturkunde, Münster.

Rackham, O. (1980) *Ancient woodland. Its history, vegetation and uses in England.* Arnold, London.

Rackham, O. (1986) *The history of the countryside.* Dent, London.

Schaminée, J.H.J., Stortelder, A.H.F. and Westhoff, V. (1995) *De Vegetatie van Nederland. Deel 1. Inleiding tot de plantensociologie, grondslagen, Methoden, toepassingen.* Opulus, Uppsala/Leiden.

Spurr, S.H. and Barnes, B.V. (1980) *Forest ecology.* Wiley, New York.

Staatsbosbeheer (1985) *Beheersplan voor de boswachterij De Amerongse Berg over de periode 1984–1993. Deel 1. Inventarisatie.* Staatsbosbeheer, Utrecht.

Tack, G., Bremt, P. van den and Hermy, M. (1993) *Bossen van Vlaanderen. Een historische ecologie.* Davidsfonds, Leuven.

Vos, W. and Stortelder, A.H.F. (1992) *Vanishing Tuscan Landscapes. Landscape ecology of a submediterranean-montane area.* Pudoc, Wageningen.

Waal, R.W. de (1996) De dynamiek van strooisellagen in bosecosystemen op de overgang van kalkrijk naar kalkarm. In: Kemmers, R.H. (ed.) *De dynamiek van strooisellagen.* DLO Staring Centrum, Wageningen, pp. 67–79.

Werf, S. van der (1991) *Natuurbeheer in Nederland. Deel 5. Bosgemeenschappen.* Pudoc, Wageningen.

Westhoff, V. and Held, A.J. den (1969) *Plantengemeenschappen in Nederland.* Thieme, Zutphen.

Wolf, R.J.A.M. (1992) *Ontstaansgeschiedenis en beheer van de Nederlandse elzen – en berkenbroekbossen.* DLO Instituut voor Bos – en Natuuronderzoek, Wageningen.

Wolf, R.J.A.M. (1995) *Geschiedenis en beheer van de Nederlandse ooibossen.* DLO Instituut voor Bos – en Natuuronderzoek, Wageningen.

Wolf, R.J.A.M., Dort K.W. van and Vrielink, J.G. (1996) Groeiplaatsen als basis voor bostypologie. Vaste grond onder de voeten van bosbeheerders. *Nederlands Bosbouwtijdschrift* 68, 177–189.

Wolf, R.J.A.M., Vrielink, J.G. and Waal, R.W. de (1997) Riverine woodlands in the Netherlands. *Global Ecology and Biogeography Letters*, 6, 287–295.

Historical Ecology of Woodlands in Flanders

22

G. Tack[1] and M. Hermy[2]

[1]Ministry of the Flemish Community, Administration of
Monuments and Sites, Gebr. Van Eyckstraat 2–6, B–9000 Gent,
Belgium; [2]Catholic University of Leuven, Department of Land
Management, Laboratory for Forest, Nature and Landscape Research,
Vital Decosterstraat 102, B–3000 Leuven, Belgium

Conclusions from 13 years' research into the historical ecology of woodland in Flanders are presented. The region is one of the most densely populated in Europe. Detailed multidisciplinary studies of the Wood of Ename have been combined with broader-scale surveys of more than 500 woods in the region. In the Middle Ages, demographic pressure led to woodland clearance to create more farmland. Around 1250 there was only about 10% woodland cover. A period of afforestation brought the woodland cover back to about 16% by 1775. In the late eighteenth and early nineteenth centuries coal tended to replace firewood, and woodland was again cleared, but this was followed by a further period of afforestation. Most woods in Flanders are therefore a complex mosaic of differing histories, with only small areas that can be considered ancient. Woodland uses such as pannage, sod cutting and pasturage gradually declined from the Middle Ages onwards. Large areas were coppiced and in the south of Flanders standards for timber production were abundant. In the north, underwood was more important. Flemish woods remain rich in woodland plants, including distinctive ancient woodland species. Some of these are more abundant in large than in small woods. Abandonment of coppicing has also led to losses of species, particularly species of the wood edge.

Introduction

'Historical ecology seeks to interpret the natural and artificial factors that have influenced the development of an area of vegetation to its present state' (Rackham, 1980) and has a particularly strong tradition in the UK (Tubbs, 1968; Rackham, 1975, 1976, 1986, 1989; Peterken, 1981, 1996). However, the idea that woodland history is important for nature conservation and forestry is much older, usually from the nineteenth century (for the UK, see Watkins, 1988; for Belgium, see Goblet D'Alviella, 1927–1930; for Germany, see Ellenberg, 1963). Partly because of the stimulating British work, the increased

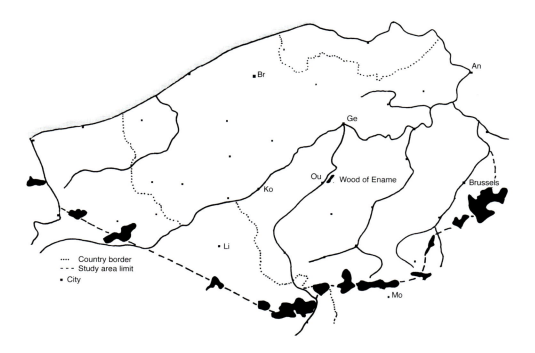

Fig. 22.1. The study area includes the historical county of Flanders, with adjacent parts of Brabant and Hainaut. It now encloses parts of Belgium, France and the Netherlands.

global environmental problems and the increased public awareness of the value of woodland, historical ecology – as a multidisciplinary and integrating science – has recently gained renewed scientific interest (e.g. Pott and Hüppe, 1991; Foster, 1992; Tack *et al.*, 1993; During and Schreurs, 1995; Wulf, Chapter 24, this volume).

Here we present the main conclusions of 13 years of research concerning the historical ecology of woodlands in Flanders (Tack *et al.*, 1993). By Flanders, is meant not the actual political region of Flanders in Belgium, but the historical county of Flanders with adjacent parts of Brabant and Hainaut, now enclosing parts of Belgium, France and the Netherlands, between Dunkirk and Antwerp, with cities such as Brussels, Lille, Ghent and Bruges (Fig. 22.1). This region, of about 11,000 km², has always been one of the most populated areas of Europe, together with the valley of the Po in northern Italy. This has resulted in a variety of anthropogenic pressures on woodland.

The research has focused on two spatial levels. On the one hand, a detailed multidisciplinary investigation of a particular, well-documented wood with a high example value, the Wood of Ename, and on the other hand, research on a broader scale concerning more than 500 woods in the whole region.

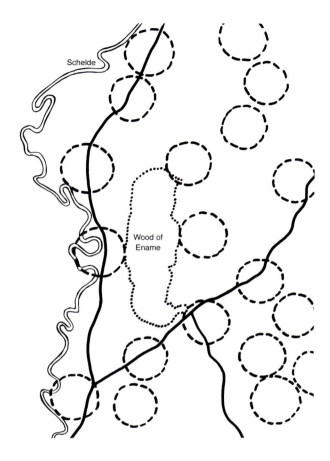

Fig. 22.2. Impression of the Gallo-Roman occupation (circles) around the Wood of Ename (…). Each circle corresponds with a settlement of 50 ha.

Here we will summarize the main results for the three parts of the research, namely the spatial evolution of woodland in the past, the traditional management of woodlands and trees outside woods, and finally today's woodland flora and vegetation as a result of their history. For further results we refer to Tack *et al.* (1993).

Woodland Evolution

The history of woodland in Flanders is a sequence of periods of woodland regression, but there were also important periods when woodland expansion occurred. Because of the ecological importance of ancient woods, we focus on medieval and post-medieval times. The events in some periods are remarkable in a European context.

For Roman times no detailed analysis is possible. Yet from scattered data (Fig. 22.2) we can infer that considerable alteration to the woodland occurred through grazing and management (coppicing). Figure 22.2 shows the average size (c. 50 ha) of the Roman villas and their localization around the Wood of Ename, suggesting a high density of human disturbance and fragmentation of the forest area.

In the Middle Ages, demographic pressure meant that there was a high priority for more arable fields and pasture. Medieval woodland deforestation reached its maximum in Flanders around 1250, leaving only about 10% of land as woodland. In the same period, peat supplies were exhausted, and firewood became the only remaining fuel. This led to a massive reafforestation of a large area of wood-pasture, heathland, former peat-grounds and marginal arable fields that had only recently been cleared. This movement started at the end of the thirteenth century and continued until the second half of the eighteenth century. From 1250 till 1775, the woodland area increased from 10 to 16%. The Flemish landscape was scattered with thousands of small woods. Especially in the north of Flanders, fragmentation and isolation of woods was extreme. In 1775, in the whole region, only 50 woods were larger than 500 ha. Except in a few restricted areas, heathland and wood-pasture almost totally disappeared. Concurrently, the northern half of Flanders became a very dense bocage-region with hedges, not only around meadows but also around most of the arable fields. Firewood-supply was the main reason for the enclosing of the landscape in this very densely populated area, with the large medieval cities of Ghent, Bruges and Ypres nearby.

In the second half of the eighteenth and the first half of the nineteenth century, coal became a real alternative for firewood, and population growth reached such a level that expansion of arable fields was again necessary. The woodland cover collapsed from 16 to 6% in 1880. At that moment, only 50 woods larger than 200 ha remained. In the period between 1851 and 1868, the Wood of Ename was also deforested for agricultural purposes. It was divided into about 430 parcels of 3100 m². A lot of small, linear hedges survived the deforestation. These hedges probably functioned as sources for colonization after abandonment of the arable land by the end of the nineteenth century.

From about 1880 until World War I, reafforestation took place in parts of woodland cleared only decades earlier because of an agricultural crisis due to grain imports and the need for mining wood. So, coal-mining that had initially caused clearance now caused reafforestation. This picture for the whole region is also reflected in the changes in the area of the Wood of Ename (Fig. 22.3). After World War II, urbanization again reduced the remaining area of woodland. Through an explosive expansion in road infrastructure, house building and industries, the fragmentation of the forest area increased tremendously in the last century. This fragmentation process stopped at the end of the sixties, but the edge effects on the woodland greatly increased, due to the environmental problems of air, soil and water pollution caused by urbanization and industry, and intensification in agriculture. In the last decade

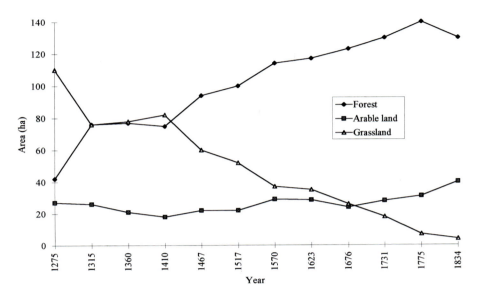

Fig. 22.3. Evolution of the area of various land use types in the Wood of Ename from 1275 to 1834. This picture corresponds quite well with that of the whole region.

there has been a renewed trend towards increasing the woodland area, due to an active governmental policy of forest protection and expansion.

This very turbulent history means that most of the present woods in Flanders consist of a mosaic of parts with a different past, and only small parts of many woods can be called ancient. Concurrent with the forest area fluctuations there were of course management changes as well.

Traditional Management of Woods

Flemish woods are extremely influenced by human interference. From the Middle Ages until World War I, management focused mostly on firewood and timber production. Other (originally common) woodland uses, such as pasturage, pannage, sod cutting (for fuel), etc., which were not compatible with wood production, were progressively banned during the Middle Ages by abolishing common rights. They only remained in the larger woods after 1250. In a lot of smaller woods, wood-pasture was limited to a period starting from the third or fourth year after coppicing, to certain parts of a wood and to specific farmers. Alternatives developed such as mowing woodland vegetation (for fodder and manure) and gathering acorns. Mowing of brambles was permitted and very common because it was favourable to woodland regeneration, and brambles were used as a fuel in bakeries.

With the exception of the greater part of the Forest of Zoniën near Brussels, in the second half of the thirteenth century probably all woods in Flanders were coppice or coppice-with-standards woods. Apart from gathering dead wood, common rights on the use of coppice were abolished in the Late Middle Ages. Woodland management was already rationalized by having fixed boundaries to the woods themselves (narrow woodbanks), and their panels, divided in small plots, fixed by surveyors; short coppice cycles (8–10 years); replanting of open places with short planting distances (80–120 cm), sometimes after deep digging; planting in the coppices after the cut; digging out of less productive stools; digging out of stumps of standards; removal of brambles, climbers etc.; emptying drainage ditches and spreading out the mud into the coppice. This intensive management provided a continued and very high production.

The yield ratio of underwood to timber (from the standards) varied from 6 : 4 to 8 : 2. In the north, management concentrated on the underwood production because of the import of timber through the ports of Bruges and Antwerp. In that area, wood and timber production of coppiced hedges, pollard and tree rows was also very important. In the south of Flanders, coppices were filled with a lot of standards for timber production. Management intensity was more important in smaller woods than in the 50 larger ones. However during the longer war periods, regular management was heavily disturbed, especially the desirable age structure of the standards.

Wood demand not only influenced the woodland cover and the extent of the coppice-with-standards-system, but also the species composition. Until the sixteenth century, oak (*Quercus* spp.) was favoured because of the demand for its timber and its multiple uses. From the seventeenth century onwards, black alder (*Alnus glutinosa*), sweet chestnut (*Castanea sativa*) and sycamore (*Acer pseudoplatanus*) as underwood and grey poplar (*Populus* sp.), beech (*Fagus sylvatica*), pine (*Pinus* spp.) and hybrids of poplar were planted as standards on a large scale for economic reasons.

Woodland management did not allow hunting and shooting to become a problem for woodland production. Apart from in the longer war periods, when large mammals, wolves included, invaded from the south, wild swine, red deer and roe deer were restricted to the larger woods, of which some were legal hunting forests. Here management was less intensive anyway. The introduction of rabbits from 1200 onwards caused some trouble, but the production of meat and fur had some economic benefits too.

By the Late Middle Ages, woodland management was committed to specialized managers, called 'bosmeesters' in Dutch (woodmasters). For that reason, we called the very intensive wood management practice in Flanders 'bosmeesterschap', comparable with woodmanship in English. The financial value of an average wood was little lower than the value of an average arable field or meadow.

Woodland Flora as a Reflection of the Past

Woodland evolution and woodland management are both reflected in the flora and vegetation of woods today. In spite of their turbulent history, Flemish woods are still fairly rich in woodland plants. The higher diversity in woodland plants in the south compared with the north is mainly caused by

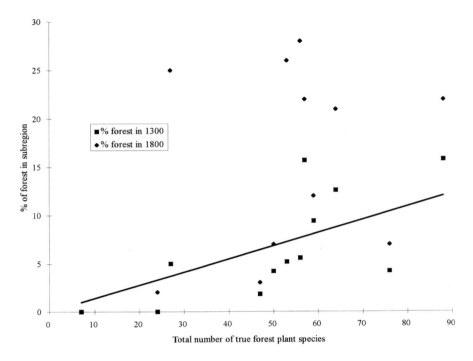

Fig. 22.4. Relationships between the total number of true woodland plant species and the woodland index (% woodland of the total subregion area) in 1300 and 1800 based on data from 13 subregions within the study area. A strong correlation is observed for the early fourteenth century (fitted linear regression line; $R^2 = 0.46$).

differences in soil texture and fertility (silty soils in the south, sandy soils in the north). When we divide Flanders into districts, based on: (i) the historical density of woodland in 1300, after the great medieval clearings and before the reafforestation; and (ii) the density of woodland in 1775, after the reafforestation and before the modern clearings; we observe a strong correlation between the richness in woodland plants today and the woodland index in 1300 (Fig. 22.4). In 1300, due to degradation and clearing, the few remaining woods in the north of Flanders were very small and isolated. The possibilities to sustain, colonize and exchange species were greater in the south.

As in other parts of Europe (Hermy *et al.*, 1993), ancient woodland in Flanders has a fairly high number of so-called ancient woodland plants (Peterken, 1974) (Table 22.1). (The threshold date used here for ancient woodland is the land shown as wooded on the maps of the Ferraris (1770–1779).) These ancient woodland plant species have been described from the UK, Belgium, the Netherlands, Denmark, Poland, Sweden, Germany and Czech Republic (Hermy *et al.*, 1998). They have a low colonization capacity in common. This may be due to short distance dispersal, a low

Table 22.1. True forest plant species having an affinity for ancient woodland in Flanders.

Species	Strength of association	Dispersal type
Acer campestre	*	ANEw
Anemone nemorosa†	*	MYR
Carex pallescens	*	UNSP
Chrysosplenium oppositifolium	*	UNSP
Cornus sanguinea	*	END
Corydalis solida	*	MYR
Corylus avellana	*	END
Crataegus laevigata^	*	END
Dryopteris carthusiana	*	ANEd
Euonymus europaeus	*	END
Euphorbia dulcis	*	MYR
Lamium galeobdolon	*	MYR
Malus sylvestris	*	END
Mercurialis perennis	*	MYR
Mespilus germanica	*	END
Milium effusum	*	ANE
Neottia nidus-avis	*	ANEd
Phyteuma spicatum‡	*	ANE
Platanthera chorantha	*	ANEd
Polygonatum multiflorum§	*	END
Polystichum aculeatum§	*	ANEd
Primula vulgaris§	*	MYR
Pteridium aquilinum	*	ANEd
Pulmonaria officinalis	*	MYR
Pyrus commune	*	END
Ranunculus auricomus	*	EPI
Sorbus torminalis	*	END
Tamus communis	*	END
Vaccinium myrtillus†	*	END
Viola reichenbachiana	*	MYR
Allium ursinum	**	MYR
Asperula odorata	**	EPI
Carex pendula	**	HYD
Carex sylvatica¶	**	MYR
Convallaria majalis	**	END
Equisetum sylvaticum	**	ANEd
Hyacinthoides non-scripta	**	BAR
Luzula sylvatica	**	MYR
Lysimachia nemorum¶	**	ANE
Maianthemum bifolium	**	END
Narcissus pseudonarcissus‡	**	BAR
Orchis mascula	**	ANEd
Oxalis acetosella	**	AUT
Paris quadrifolia	**	END
Sanicula europaea	**	EPI
Veronica montana	**	MYR
Vinca minor	**	MYR
Carex strigosa	***	HYD
Euphorbia amygdaloides	***	MYR
Luzula pilosa	***	MYR
Melica uniflora	***	MYR
Stellaria nemorum	***	HYD

Association with ancient woodland:
* weak association; ** moderate association; *** strongly associated.
Dispersal type:
MYR, dispersal by ants (myrmecochores); END, dispersal by animals and birds via digestion (endozoochores and ornitho-chores); EPI, dispersal by adhesion on animals (epizoochores); BAR & AUT, passive and active dispersal by plant itself (baro- and autochores); UNSP, unspecified, uncertain or unknown; HYD, dispersal by water (hydrochores); ANE; dispersed by wind; ANEw, diaspores winged or flattened; ANEd, diaspores minute (orchids, ferns & horsetails); ANE, heavier than ANEd and not winged.
† sometimes in old grasslands or heathlands, on former woodland sites.
‡ sometimes in grasslands, particularly in Central Europe.
§ sometimes in wood edges.
¶ often along woodland rides.

reproductive capacity and/or restricted recruitment (germination and/or establishment). Most of this group of species do not have a persistent seed bank. The consequence is that even temporary deforestation for agricultural land use has dramatic effects on the potential for recolonization after abandonment of the agricultural land use.

Yet woods that were ancient at the moment of their modern clearance, mostly in the nineteenth century, and were (partly) replanted after a short period of agricultural use, can retain a considerable part of their original woodland flora, when enough points of recolonization, such as hedges, remain. When coppicing stopped, woods in Flanders became poorer in flowering plants, a phenomenon also described by Peterken (1981).

Conclusion

Although Flemish woods still have most of their woodland plants, a lot of species connected with wood edges, small lawns, verges of rides, small ponds and so forth disappeared locally or became very rare as a result of changes in woodland management. Coppicing stopped and the canopy closed, mowing stopped the edge of woodland at its surroundings which thus became sharp as a razor with the widespread use of barbed wire.

References

During, R. and Schreurs, W. (1995) *Historische ecologie*. Stichting Uitg. Kon. Ned. Natuurhist. Ver., Utrecht.

Ellenberg, H. (1963) *Vegetation Mitteleuropas mit den Alpen*. Ulmer, Stuttgart.

Foster, D.R. (1992) Land use history (1730–1990) and vegetation dynamics in central New England, USA. *Journal of Ecology* 80, 753–771.

Goblet d'Alviella (1927–1930) *Histoire des bois et forêts de Belgique*. Lechevalier P., Paris and Lamertin M., 4 parts, Brussels.

Hermy, M., Bremt, P., van den and Tack, G. (1993) Effects of site history on woodland vegetation. In: Broekmeyer, M.E.A., Vos, W. and Koop, H. (eds) *European forest reserves*. Pudoc, Wageningen, pp. 219–231.

Hermy, M., Honnay, O., Firbank, L., Grashof-Bokdam, C.J. and Lawesson, J. (1998) An ecological comparison between ancient and other forest plant species of Europe, and the implications for forest conservation. *Conservation Biology* (in press).

Peterken, G.F. (1974) A method for assessing woodland flora conservation using indicator species. *Biological Conservation* 6, 239–245.

Peterken, G.F. (1981) *Woodland conservation and management*. Chapman & Hall, London.

Peterken, G.F. (1996) *Natural woodland. Ecology and conservation in northern temperate regions*. Cambridge University Press, Cambridge.

Pott, R. and Hüppe, J. (1991) *De Hudelandschaften Nordwest-deutschlands*. Westfalischer Museum für Naturkunde, Munster.

Rackham, O. (1975) *Hayley Wood. Its history and ecology*. Cambridge and Isle of Ely Naturalists Trust.

Rackham, O. (1976) *Trees and woodland in the British Landscape*. Dent, London.

Rackham, O. (1980) *Ancient woodland. Its history, vegetation and uses in England*. Arnold, London.

Rackham, O. (1986) *The history of the countryside.* Dent, London.

Rackham, O. (1989) *The last forest: the story of Hatfield Forest.* Dent, London.

Tack, G., van den Bremt, P. and Hermy, M. (1993) *Bossen van Vlaanderen. Een historische ecologie.* Davidsfonds, Leuven.

Tubbs, C.R. (1968) *The New Forest: an ecological history.* David & Charles, Newton Abbot.

Watkins, C. (1988) The idea of ancient woodland in Britian from 1800. In: Salbitano, F. (ed.) *Human influence on forest ecosystems development in Europe.* Pitagora Editrice, Bologna, pp. 237–246.

Occurrence of Woodland Herbs in an Area Poor in Woodland: NW Zealand, Denmark

23

Peter Milan Petersen

Department of Plant Ecology, University of Copenhagen,
Ø. Farimagsgade 2 D, DK-1353 Copenhagen K, Denmark

The herb flora of ancient woods, fragments of ancient and supposed ancient woodland, and open, north-facing coastal slopes in NW Zealand are compared. Several true woodland herbs, such as *Circaea lutetiana, Convallaria majalis, Luzula pilosa, Majanthemum bifolium* and *Oxalis acetosella* are identified. Other woodland herbs, elsewhere considered indicators of woodland continuity, occur on open, north-facing slopes. These findings and probable causes are discussed. The term ancient woodland species should only be used about woodland herbs in relation to a specified geographical area.

Introduction

The discussion of the factors determining the flora of recently established woods often focuses on immigration and dispersal ability of the species. The dispersal ability of most woodland herbs is very limited. In areas poor in woodland the probability of recent dispersal to woodland fragments of species belonging to this group is low. Hence, studies of the occurrence of woodland plants in areas poor in woodland may contribute to our knowledge of the persistence of populations of these species, for even woodland plants may survive a period of deforestation and become part of the flora of more recently established woodland at that point.

The Investigation Area

NW Zealand (Fig. 23.1) is situated in the moist, temperate woodland zone. The climax vegetation is deciduous broad-leaved forest. The climate is temperate–subcontinental and precipitation is about 500 mm year^{-1}. The landscape is characterized by large terminal moraines which form the core of a number of peninsulas with steep recent and raised coastal slopes. These

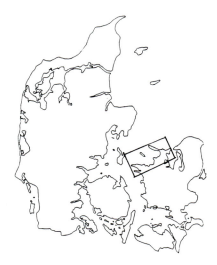

Fig. 23.1. Map of Denmark, showing the investigation area, NW Zealand.

slopes have never been cultivated. Geologically the area is dominated by glacial till, ranging from gravelly moraine to moraine clay, but marine deposits and raised beach formations are also represented.

NW Zealand has been poor in woodland for at least 1000 years and probably longer, as indicated by numerous graves from the Stone and Bronze Ages as well as by a pollen diagram from Lake Tissø (Aaby, 1992).

Old maps, archives and cadasters may give some information about the occurrence, extent and exploitation of woodland. However, it is impossible to obtain an extensive and reliable picture of the occurrence of woodland until 1769 when the maps of Videnskabernes Selskab appeared. Therefore ancient woodland is here defined as woodland older than 225 years. More recently established woodland – planted or self-sown – is called secondary woodland. The area of woodland in NW Zealand reached a minimum about 1850. Since then the wooded area has increased, especially in the form of coniferous plantations on sandy marine deposits.

Material and Methods

The material consists of species records of woodland herbs from five large ancient woods (22–200 ha), ten fragments of ancient woodland (0.04–6.5 ha), six fragments of supposed ancient woodland or scrub characterized by the occurrence of true forest trees (0.1–6.5 ha), and open, north-facing coastal slopes (Fig. 23.2). For larger woods and some woodland fragments (>*c.* 5 ha) the maps (Videnskabernes Selskabs Konceptkort) have been used to decide whether they should be considered ancient or not. For smaller fragments of woodland and scrub not shown on the maps the occurrence of true native forest tree species (*Acer platanoides, Fagus sylvatica, Tilia cordata* and *T.*

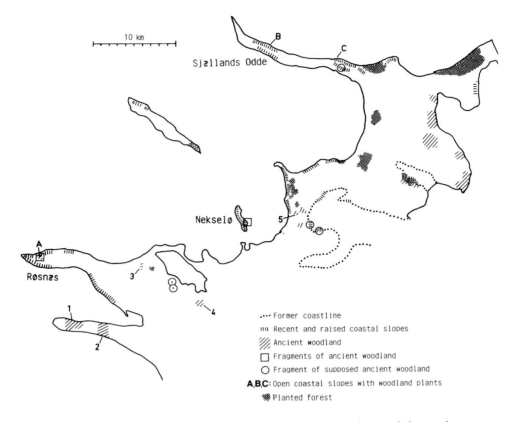

Fig. 23.2. The investigation area. The map shows recent and raised coastal slopes; the present coastline and the coastline prior to the reclamation of land at the end of the nineteenth century; and ancient and more recently planted woodland. 1–5, ancient woods investigated.

platyphyllos) has been used as an indicator of age (supposed ancient woodland) (Fig. 23.3).

The surveys were made in 1982–1996 (Petersen, 1988, 1994). A few records from Grøntved (1936) were included in the material. The majority of sites were visited at least twice – in spring and late summer – but some of the larger woods were visited in summer only. Woodland herbs are defined as shade-tolerant herbs which in Denmark are normally restricted to woodland. Species growing on wet soil have been omitted from the investigation because the majority are also able to grow on wet soil in the open.

A total of 42 species were found which fulfilled these criteria (Table 23.1). They can all be characterized as shade plants or shade-tolerant plants (with a light value according to Ellenberg *et al.* (1991) ranging from 1 to 7, mode 4) and are mesic (with a moisture value according to Ellenberg *et al.* (1991) from 4 to 7, mode 5). Species definition and nomenclature is in accordance with Hansen (1981).

Fig. 23.3.
Raised
coastal slope
with fragment
of supposed
ancient
woodland,
with *Acer
platanoides,
Corylus
avellana,
Fraxinus
excelsior,
Quercus
robur, Tilia
platyphyllos*
and *Ulmus
glabra.*

Results

Forty-two species were found in the five ancient woods. Thirty-three species were found in fragments of ancient woodland on Røsnæs and the island of Nexelø, 26 in supposed ancient woodland fragments and scrub and 11 on north-facing coastal slopes (Table 23.1).

The number of species in the individual ancient woodland fragments ranged from 7 to 17 (Table 23.2), and in the supposed ancient woodland fragments from 8 to 17 (out of the 42 species mentioned in Table 23.1). There is a significant positive correlation between area and number of woodland species for ancient and supposed ancient woodland fragments together (Spearman Rank Correlation Coefficient r_s = 0.557, 0.05>P>0.01).

Discussion

The records from woodland fragments are more complete than those from the woods, in particular as regards the vernal species. This is mainly because the woodland fragments are more easily surveyed than the woods, due to their limited size, but it is also due to the fact that some of the woods were visited only once.

The flora of woodland herbs of ancient woodland fragments is only slightly less rich in species than that of ancient woods, despite the fact that the total area of the 16 fragments is only 12 ha, compared to 600 ha for the five woods investigated. This agrees with Dzwonko and Loster (1989) who found that groups of small ancient woods support more woodland species than do single woods of equal area. Certain woodland herbs have not been found in the woodland fragments or are less frequent here than in the woods. There may be several reasons for this:

Table 23.1. Occurrence of woodland herbs in various site types in NW Zealand.

Species	a (N =5)	b (N = 10)	c (N = 6)	d (N =3)
Adoxa moschatellina	4	3	1	1
Allium ursinum	4		3	1
Anemone nemorosa	5	10	6	3
Anemone ranunculoides	2	2	3	2
Hedera helix	4	2	3	1
Hepatica nobilis	3	2	3	1
Mercurialis perennis	5	1	2	1
Orchis mascula		2		2
Ranunculus auricomus	3	1	4	1
Ranunculus ficaria	5	10	6	3
Scrophularia nodosa	5	4		1
Brachypodium sylvaticum	5	1	2	
Campanula trachelium	1	4	2	
Convallaria majalis	4	1	1	
Corydalis intermedia	2	5	3	
Festuca gigantea	5	1	2	
Gagea lutea	3	7	5	
Gagea spathacea	2		1	
Melica uniflora	5	7	2	
Milium effusum	5	8	2	
Moehringia trinervia	5	4	4	
Poa nemoralis	5	8	6	
Polygonatum multiflorum	4	9	2	
Roegneria canina	4	5	3	
Sanicula europaea	4	1	1	
Stachys sylvatica	5	7	2	
Stellaria holostea	5	9	4	
Corydalis bulbosa	2	4	1	
Ajuga reptans	3	1		
Carex sylvatica	5	2		
Luzula pilosa	2	1		
Oxalis acetosella	5	6		
Paris quadrifolia	2	5		
Pulmonaria officinalis	4	1		
Veronica montana	5	1		
Bromus ramosus	4			
Campanula latifolia	2			
Circaea lutetiana	5			
Galium odoratum	4			
Lamiastrum galeobdolon	3			
Majanthemum bifolium	4			
Primula vulgaris	2			

a, larger ancient woods (area 10–200 ha); b, fragments of ancient woodland on the peninsula Røsnæs and the island Nekselø (0.04–6.5 ha); c, fragments of supposed ancient woodland or scrub, not shown on the maps from 1769, but characterized by the occurrence of true native forest tree species (0.1–6 ha); d, herbaceous vegetation on north-exposed coastal slopes.

Table 23.2. Number of woodland species (out of the 42 species mentioned in Table 23.1) found in the individual woodland fragments (the sites have been arranged according to increasing area).

Fragments of ancient woodland		Fragments of supposed ancient woodland	
area (ha)	number of species	area (ha)	number of species
0.04	10	0.1	11
0.19	9	0.2	13
0.25	7	0.8	8
0.25	9	1.0	17
0.32	7	2.5	11
0.39	12	6.0	14
0.49	13		
0.81	10		
2.7	17		
6.5	14		

- Most ancient and supposed ancient woodland fragments are situated on slopes, with an inclination of up to 50°. On steep slopes soil pH is generally higher and organic matter lower than in woodland on more stable ground. Hence species restricted to acid soils, such as *Convallaria majalis, Majanthemum bifolium, Luzula pilosa* and *Oxalis acetosella* are absent or rare.

- For the same reason, hydrological conditions are more varied in woods than in fragments of woodland. In particular in woodland fragments on south-facing slopes, habitats suitable for species preferring moist soil (e.g. *Circaea lutetiana*) may be lacking.

- Exploitation has been different, at least for the last 150–200 years. The woods are today managed as high forest, often with clear felling. In woodland fragments selected felling is most often practised. Cattle grazing was forbidden in the woods in 1805, but continued for a longer period of time in the woodland fragments. These differences may be of importance to some species characteristic of high forest, such as *Bromus ramosus* (Brunet, 1993).

The occurrence of woodland herbs in open herbaceous vegetation on north-facing coastal slopes (Fig. 23.4) may be explained by the microclimate being fairly similar to that of woodland. The local influence of seepage water on some of these slopes (as indicated by the occurrence of *Filipendula ulmaria* and even *Epilobium hirsutum*) may further contribute to a cool and moist microclimate. According to Hermy *et al.* (1993), eight of the species found on open, north-facing coastal slopes have, in European literature, been mentioned as ancient woodland species and are thus indicators of woodland continuity (*Adoxa moschatellina, Allium ursinum, Anemone nemorosa, A. ranunculoides, Hepatica nobilis, Mercurialis perennis, Orchis mascula, Ranunculus auricomus*) (Fig. 23.5). In other parts of Denmark the woodland

Fig. 23.4. N-exposed coastal slope on Røsnæs with open herbaceous vegetation (site 3 on Fig. 23.2).

herbs *Anemone nemorosa* and *Ranunculus auricomus* grow in meadows – *A. nemorosa* even along the shores of lakes – while *Majanthemum bifolium* can be found in heathland (Hansen, 1981). Peterken and Game (1984) also mention hedges as an alternative habitat for woodland species.

Conclusions

There is evidence that populations of many woodland herbs have persisted in isolated fragments of ancient woodland in NW Zealand for centuries. Even

Fig. 23.5. *Anemone nemorosa* and *A. ranunculoides* in open herbaceous vegetation on the slope shown on Fig. 23.4.

very small fragments may serve as habitat for quite a number of species. However, the specific topography, soil conditions, and management regime through the ages of the woodland fragments in a given area may exclude other species from growing there. Similarly, certain woodland herbs may survive a period of deforestation in non-wooded habitats, in NW Zealand typically on north-exposed coastal slopes with neutral–slightly acid soil.

If these fragments of ancient woodland and non-wooded habitats are included in secondary woodland – planted or natural – the species may become part of the flora of that secondary woodland. Thus the preservation of small woodland remnants in the agricultural landscape may be of great importance for the maintenance of woodland species diversity, as pointed out by Dzwonko and Loster (1989).

The observation that in NW Zealand some woodland herbs do not occur in small fragments of woodland and that others grow in non-wooded habitats, may contribute to the discussion of woodland herbs as indicators of woodland continuity in time (Peterken, 1981; Hermy *et al.*, 1993). Area and ecological conditions in remnants of ancient woodland, as well as alternative habitats, may vary between regions. Consequently the term ancient woodland species should only be used about woodland herbs in relation to a specified geographical area.

References

Aaby, B. (1992) Sjællands kulturlandskaber i jernalderen. *Arkæologiske skrifter* 6, 209–236.

Brunet, J. (1993) Environmental and historical factors limiting the distribution of rare forest grasses in southern Sweden. *Forest Ecology and Management* 61, 263–275.

Dzwonko, Z. and Loster, S. (1989) Distribution of vascular plant species in small woodlands on the Western Carpathian foothills. *Oikos* 56, 77–86.

Ellenberg, H., Weber, H., Düll, R., Wirth, W., Werner, W. and Paulissen, D. (1991) Zeigerwerte von Pflanzen in Mitteleuropa. *Scripta Geobotanica* 18, 1–248.

Grøntved, P. (1936) Om floraen i nogle nordvestsjællandske skove. *Bot. Tidsskr* 43, 325–356.

Hansen, K. (ed.) (1981) *Dansk feltflora*. Gyldendal, Copenhagen.

Hermy, M., van den Bremt, P. and Tack, G. (1993) Effects of site history on woodland vegetation. In: Broekmeyer, M.E.A., Vos, W. and Koop, H. (eds) *European forest reserves*. Pudoc, Wageningen, pp. 219–232.

Peterken, G.F. (1981) *Woodland conservation and management*. Chapman & Hall, London.

Peterken, G.F. and Game, M. (1984) Historical factors affecting the number and distribution of vascular plant species in the woodlands of central Lincolnshire. *Journal of Ecology* 72, 155–182.

Petersen, P.M. (1988) En botanisk beskrivelse af ni småskove på Røsnæs med præg af tidligere tiders drift. *Flora og Fauna* 94 (1), 15–22.

Petersen, P.M. (1994) Flora, vegetation, and soil in ancient and planted woodland, and scrub on Røsnæs, Denmark. *Nordic Journal of Botany* 14, 693–709.

Distribution of Ancient Woodland, Afforestation and Clearances in Relation to Quaternary Deposits and Soil Types in Northwestern Brandenburg (Germany)

24

Monika Wulf

*Center for Agricultural Landscape and Land Use Research e. V.
Müncheberg, Institute of Land Use Systems and Landscape Ecology,
Eberswalder Str. 84, 15374 Eberswalde, Germany*

About 65% of the present woodland area of Brandenburg has been wooded for at least the last 200 years. The rates of deforestation and afforestation have been almost equal since 1780 and reach 6.5% and 8.3%, respectively. But the proportion of deforestation and afforestation differs between distinct landscape units. Despite this, there is a tendency to deforest the sites with higher nutrient level whether influenced by ground water or not, whereas dry soils with low nutrient levels have been afforested. This is demonstrated in detail for a part of Brandenburg which was intensively cleared before 1780 and also during the last 200 years. In the northwestern part of the Prignitz, the distribution of afforestation and clearances relates closely to the soil and geological characteristics of the area. Seventy per cent of the afforestation covered well-drained and sandy soils whereas about 52% of the cleared woodland had been growing on loamy sites.

The shifting of woodland from soil types rich in nutrients and often poorly drained, to soil types poor in nutrients and without any influence of ground water during the last 200 years has led to increased uniformity of the countryside. The predominance of Scots pine in afforestation has further strengthened the uniform impression of the countryside. The consequences of these changes for the herb layer of forest stands are discussed, particularly with respect to species diversity.

Introduction

In most countries of the northwestern European lowlands, 60–70% of the area is used for intensive agriculture and about 10% is urban land. The remaining

20–30% consists of lakes, forests, and other areas of a more or less semi-natural vegetation. Most of these semi-natural areas are small, often less than 100 ha, or even less than 10 ha (Kalkhoven, 1993).

Changes in the distribution of wooded and non-wooded areas are determined both by the natural conditions and by social influences, while changes in woodland management are mainly affected by changes in society (Berglund, 1991). Thus, the present distribution of agricultural and forest areas and recent changes within the landscape have evolved historically and were subject to numerous influences, including, for example, past woodland management and atmospheric deposition.

In Brandenburg, the greatest human-induced change to the extent of wooded and non-wooded areas occurred between the 700s and 1300s, when large areas of the woodland were cleared (Ministerium für Ernährung, Landwirtschaft und Forsten, 1994). Then in the 1600s, large woodland areas were cleared due to country-wide colonization (Mantel, 1990). Thus, a more or less natural and semi-natural vegetation was replaced by other vegetation. Considerable parts of the deciduous woodland area were converted to coniferous forest plantations from about 1750 (Hesmer, 1938). This has changed the appearance of many parts of the countryside and has had profound effects on plant species composition over large areas of woodland. For example, in Brandenburg, large areas of Scots pine stands are dominated by the grass *Calamagrostis epigeios,* showing the influence of nitrogen fertilization (Hofmann, 1994).

In general, woodland on soils with good nutrient supply has been cleared to form arable fields (Thomasius, 1973). This is often described, but there is considerable potential to use geographical information systems to quantify this tendency. Despite its importance for landscape ecology there are few studies from Germany about the size of clearances in relation to site conditions (Leuschner and Immenroth, 1994). Therefore, this study has the following three major objectives:

1. To describe long-term changes in the distribution of wooded and non-wooded areas in Brandenburg using data from digitized historical maps.
2. To document spatial and temporal patterns of ancient woodlands, afforestation and clearances in the northwestern part of Brandenburg in detail and to relate them to soil types and geological substrates using a geographical information system.
3. To draw conclusions on the species diversity of the herb layer in woodlands at the landscape level with regard to the preferential afforestation of sandy soils during the last 200 years.

Study Area

Brandenburg is located in the northeastern part of Germany (Fig. 24.1) and comprises an area of about 3 million hectares of which woodland covers nearly 970,000 ha, about 33% (Table 24.1). About 85% of the wooded sites are Scots pine stands (*Pinus sylvestris*).

Fig. 24.1. Location of the investigated area.

The Prignitz is situated in the northwestern part of Brandenburg (Fig. 24.1) and has an area of about 3400 km². The region is one of the most intensively cultivated regions of Brandenburg; therefore woodland is relatively rare and most woods are extremely fragmented. About 15% (61,500 ha) of the total woodland area can be classed as ancient woodland (Wulf and Schmidt, 1996), and about a twentieth of this could be classed as semi-natural. All of this had been treated as coppice, coppice-with-standards or less commonly as wood-pasture. Although much woodland in the Prignitz has been cleared, the number of surviving semi-natural ancient woods is relatively high, with at least 350 stands. Most belong to the phytosociological order *Fagetalia* (Passarge, 1966).

The landscape of the Prignitz is characterized by a hummocky relief ranging in height from about 25 to 100 m. The climate is sub-Atlantic to sub-continental with a mean annual precipitation ranging from 665 mm in the north to 525 mm in the south. The mean annual temperature does not vary much and ranges from 7.8 to 8.2°C. Compared with Brandenburg, the mean temperature of the warmest month is lower and in July reaches no more than 17.5°C (Krumbiegel and Schwinge, 1991).

After the Saalian glaciation, the present land surface of the Prignitz was covered by boulder clay, with some small areas of sandy deposits and end-moraines mainly in the southern, eastern and northern part (Fischer, 1963;

Table 24.1. Part of ancient woodlands, afforestations and cleared woodlands in Brandenburg, in the Prignitz and in the northwestern part of the Prignitz.

	Brandenburg (1780 to 1990)	Prignitz (1780 to 1990)	Northwestern part of the Prignitz (only 1780 to 1937)
Size total [ha]	2,971,100	410,000	147,700
Woodland areas total [ha]	968,100	89,900	16,600
% of size	32.6	22.0	11.2
Ancient woodlands total [ha]	721,200	61,500	10,100
% of size	24.3	15.1	6.8
Afforestations total [ha]	246,900	28,400	6500
% of size	8.3	7.0	4.4
Cleared woodlands total [ha]	193,100	44,000	24,500
% of size	6.5	10.8	16.6

Akademie der Wissenschaften der DDR, 1981). The boulder clay soils are deeper and consist mainly of sand, loam and clay (Gellert, 1954). The potential natural vegetation types on the boulder clay are mixed oak–beech woods and on the sandy deposits more or less mesotrophic oak woods (Scamoni, 1964). In the Prignitz, slightly loamy brown earths are about equally common as sandy brown podsolic soils, but true podsols only occur mostly in the southern, eastern and northern edges of the Prignitz (Akademie der Wissenschaften der DDR, 1981).

Methods

The origin of each wood was determined using historical maps of the eighteenth century, recent maps from the late twentieth century, and using the maps from detailed studies such as Schauer (1957). The most important historical maps of Brandenburg are the Schmettausche Karte (scale 1:50,000; Wulf and Schmidt, 1996), drawn between 1767 and 1787 by the Preußische Regierung (Prussian government). So far, ancient woodland, afforestation and clearances in the northwestern part of the Prignitz have been digitized only for the period from 1780 to 1937, but the tendencies within this period are the same as in the period from 1937 to 1990.

Information about soil types and geology was obtained from maps at a scale of 1:100,000 ('mittelmaßstäbige Kartierung') and 1:500,000, respectively. The 'mittelmaßstäbige Kartierung' is the mapping of topsoils used as arable fields (Schmidt and Diemann, 1991). The small scale of geological maps is a problem because it limits the accuracy of the results, but for the Prignitiz no other sources were available. Due to the coarse scale, errors could be great, but in practice are ignored because of the very small amount of afforestations in the study area.

All maps have been digitized at the scale 1:250,000. Using the geographical information system ARC-Info, the spatial overlap between soil types and clearances was examined. In the same way, the spatial overlap between geological substrates and afforestation was examined. The digitization enabled me to quantify the areas of afforestation and cleared woodland in relation to the Quaternary deposits and soil types.

Results

The area of woodland in Brandenburg has been nearly constant over the present century and now covers about 35% of the total land area. Compared with other states of the FRG, however, the proportion of woodland is still high (Statistisches Bundesamt, 1996). Within Brandenburg, the woodland areas are distributed very unevenly. In some landscape units, such as in parts of southeastern Brandenburg, there has been massive afforestation. In other areas, such as parts of northwestern Germany, there has been much clearance. Although there is considerable regional variation in forest gain and loss during the last 200 years, the area of woodland has increased over this period by about 54,000 ha.

In Brandenburg, about 65% of the woodland is ancient woods (Richter, 1957; Wulf and Schmidt, 1996). Overall, ancient woodland was reduced from about 910,000 ha to about 720,000 ha between 1780 and 1990 (Richter, 1957; Schauer, 1957). Between 1780 and 1937 about 7% of the woodlands in the whole state, mostly on rich nutrient sites, were cleared. At the same time 8% of the land surface has been afforested mostly on poor sandy soils. This trend has continued since 1937 (Wulf and Schmidt, 1996).

Within Brandenburg there are differences in the distribution of ancient woodland, afforestation and cleared woodland during the last 200 years between distinct landscape units (Wulf and Schmidt, 1996). In the northern region of Brandenburg, most of the woodland consists of ancient woods whereas in the Spree Wood (in the southeast) and in the Oder Valley (in the east) the woodland area is equally ancient and recent.

Studies of woodland loss and gain at the local scale show that the detailed patterns of woodland change are very complex. The Prignitz, in general, was more agriculturally orientated than other parts of Brandenburg and, thus, the rate and extent of forest clearing are greater than the state average (Table 24.1) and reached about 11%. Most of the woodland in the Prignitz has been fragmented and is now found scattered in small pieces. Only on sandy soils are there large areas of predominantly coniferous stands.

Between 1780 and 1937 clearances were concentrated particularly in the northwestern part of the Prignitz. From 1780 onwards the whole northwestern part of the Prignitz has changed to a more and more open, sparsely wooded, cultural landscape, dominated by semi-natural vegetation types and cultivated fields in varying proportions. The very small areas of afforestation of about 4% (6500 ha) of the land surface do not change this general picture (Table 24.1). Therefore, only about 7% of the northwestern part of the Prignitz has had woodland cover continuously since at least 1780 and thus can be defined as ancient (Fig. 24.2). About 17% has been cleared (Table 24.1), especially on soils with high nutrient supply and sufficient water conditions (Table 24.2).

Fig. 24.2.
Areas of ancient
woodlands,
afforestations
and clearances
in the
northwestern
part of the
Prignitz
between 1780
and 1937.

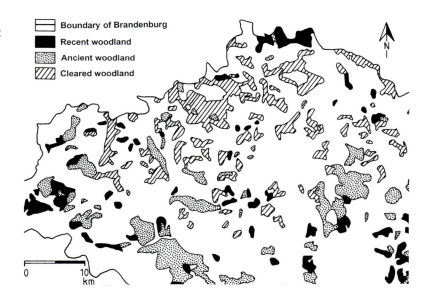

Boundary of Brandenburg
Recent woodland
Ancient woodland
Cleared woodland

0   10 km

Sandy and loamy soils, whether well watered in the top soil or not, contributed about 70% of all areas of cleared woodlands. Little more than 4% of
such soils have been afforested (Table 24.1) usually where they lie on sandy
deposits (Table 24.3). Afforestation on boulder clay and end moraines, both
with a relatively good nutrient supply, extends to less than 1000 ha.

Discussion

Historical changes in the size and duration of wooded and non-wooded areas
reflect the type and intensity of land use. This was shown for different
countries, for example by Berglund (1991), Bunce *et al.* (1993) and Leuschner
and Immenroth (1994), and seems also to be valid for Brandenburg. One
major conclusion derived from studies in different European countries is that,
at any scale, land use practices and the resulting woodland patterns have
changed continuously (see Bunce *et al.*, 1993). For distinct landscape units in
Brandenburg the land use pattern is exceedingly complex and variable in
scale and intensity over time. Over the past 200 years there are some stable
landscape units in which the extent of forest and open land was relatively
constant, for example the Lower Lusatia (in the southeast) and the Mittlere
Mark (in the southwest) (Wulf and Schmidt, 1996).

In the whole state, the long-term trend until the eighteenth century was
woodland clearance, but there were periods when land that had been cleared
became wooded again as in other European regions (Mantel, 1990; Bunce *et
al.*, 1993). The actual distribution of afforestation and clearance is closely
related to the Quaternary deposits and the soil types, which has also been
shown by Berglund (1991) for the Ystad-region in southern Sweden; by

Table 24.2. Soil types of the areas of cleared woodlands.

Soil types	Sandy soils with influence of backwater or ground water	Loamy and sandy sites (loam at least in deeper layers) with influence of backwater or ground water	Riverside meadow on loam	Mor with underlying sandy or loamy horizons	No data available
Total [ha]	4620	12,625	2.0	3310	3961
% of cleared woodlands	18.8	51.5	<1	13.5	16.2

Foster (1992) for Worcester County in central New England in the USA; and by Leuschner and Immenroth (1994) for the Lüneburger Heathland in north-western Germany.

A geographical information system has been used for the Prignitz to determine the rate of loss and gain of woodland and to monitor the changes in woodland patterns in relation to soil types and geological substrates. At least one-third of the study area is covered with boulder clay, mostly rich in sand and loam. Consequently, great areas of the Prignitz have been cleared and were mainly used as arable fields (equivalent to 140,000 ha or 80%) (Landesamt für Datenverarbeitung und Statistik, 1996).

The continued deforestation of the Prignitz is one of the most remarkable aspects of the landscape during the last 200 years, continuing the tendency established before 1780. Moreover, despite modern techniques in land use and concurrent high yields, the principle of clearing woodland sites on nutrient-rich soils and afforestation on poor sandy soils has also continued until recent times. The major reason may be that the study area is a very old area of settlement for German people, and the greatest part of the Prignitz was traditionally used as arable land (Ministerium für Umwelt, Naturschutz und Raumordnung, 1995). In other words, arable farming is the ancient use of this landscape back to at least the 1300s (Ellenberg, 1990). Ancient woodland has survived mainly on soils which are not suitable as arable land.

Table 24.3. Geological substrates of the areas of afforestations.

Geological substrate	Peats	Boulder clay	End moraine	Fluvial sand deposits	Mixed sand and gravel fluvial deposits	Sand of glens	Aeolian sands/ dunes
Total [ha]	200	220	69.0	4450	120	240	530
% of the afforested woodlands	3.1	3.4	10.7	69.0	1.9	3.7	8.2

The displacement of the woodland from richer to poorer soils is one of the important factors affecting the actual species composition of woodland and the distribution patterns of typical woodland plant species. Meso- and eutrophic woodland in the Prignitz has become rare and threatened because these woodland types have to a large extent been cut down and converted into arable land. The importance of retaining the remaining areas of semi-natural ancient woodland on meso- and eutrophic sites is emphasized when the massive loss of this type of habitat since 1780 is recognized (Wulf and Schmidt, 1996).

The red list of plants in Brandenburg now contains 660 species (Ministerium für Umwelt, Naturschutz und Raumordnung, 1993, 1995), of which about 68 plant species grow especially in deciduous woodland. Of these, 39 endangered species are characteristic of mesotrophic woodlands and eutrophic woodland. Many of these woodland plants depend on ancient semi-natural woodland. Thus, relicts of semi-natural vegetation, especially deciduous ancient woodland, are a refuge for woodland plant species which have disappeared elsewhere. Those habitats have not only decreased with increased clearances, but also with the transformation of deciduous stands into coniferous stands. This emphasizes the importance of protecting the remaining semi-natural woodland. From a biological perspective, landscapes are always patchy (Finke, 1994). This lack of uniformity is one of the main reasons for the biological variation and diversity in the landscape. The biological diversity of a landscape also depends on its biogeographical history, that is, on the number of plant species that have spread into the area. Furthermore, the effect of dispersal movements is influenced by the distance between the habitat patches, by the resistance of the landscape in between, and by the behaviour of the organisms concerned. Peterken and Game (1984) have shown that many woodland species have a very low capacity to disperse over long distances. These plant species do not have much chance to colonize isolated forest stands (Wulf, 1995), often they have to survive in their small habitats. This could be seen as a further decrease of diversity, but at the species level. Logically, small habitats can only support small populations of plant species, and small populations have a higher risk that they will lose genetic variation or even become extinct locally (Kalkhoven, 1993). Another important point for species diversity at the landscape level is the degree to which all forest stands were subjected to varying agricultural use, including grazing.

Some woodland plant species appear to have depended on the effects of different methods of woodland management to survive (Brunet, 1994). In former times the different methods of woodland management formed a mosaic of land use both in space and time. Now woodland management tends to be much simpler being driven primarily by the need to produce timber.

Conclusions

The concentration of afforestation on sandy soils, the change of deciduous woodland into coniferous stands, and the predominance of woodland management for timber have led to relatively uniform woodland being created across the whole landscape. The lack of diversity in the woodland vegetation has led to a decrease of species diversity at the landscape level.

Acknowledgements

Investigations were supported by the BMIL (Federal Ministry for Food, Agriculture and Forestry, FRG) and the MELF (Ministry for Food, Agriculture and Forestry, Brandenburg). Many thanks are given to Ute Kunter for much cartographic help.

References

Akademie der Wissenschaften der DDR (ed.) (1981) *Atlas der DDR*. Gotha, Haack, 53 map sheets.

Berglund, B.E. (1991) The cultural landscape during 6000 years in southern Sweden. The Ystad project. *Ecological Bulletin* 41, 495 pp.

Brunet, J. (1994) *Der Einfluß von Waldnutzung und Waldgeschichte auf die Vegetation südschwedischer Laubwälder NNA-Ber. 7. Jg., H. 3*, 96–101.

Bunce, R.G.H., Ryszkowski, L. and Paoletti, M.G. (1993) *Landscape ecology and agro-ecosystems*. Lewis Publishers, London.

Ellenberg, H. (1990) *Bauernhaus und Landschaft in ökologischer und historischer Sicht*. Eugen Ulmer, Stuttgart.

Finke, L. (1994) *Landschaftsökologie*. Das Geographische Seminar – 2. Aufl., Braunschweig.

Fischer, W. (1963) *Flora der Prignitz*. Heimatmuseum des Kreises Pritzwalk, Pritzwalk, 135 pp.

Foster, D.R. (1992) Land-use history (1730–1990) and vegetation dynamics in central New England, USA. *Journal of Ecology* 80, 753–772.

Gellert, J.H. (1954) *Bemerkungen zur Karte der physisch-geographischen Gliederung der Deutschen Demokratischen Republik im Maßstab 1: 1,000,000*. Petermanns Geographische Mitt., Jg. 1954, H. 1.

Hesmer, H. (1938) *Die heutige Bewaldung Deutschlands*. Paul Parey, Berlin.

Hofmann, G. (1994) *Der Wald. Sonderheft Waldökosystem-Katalog*. Deutscher Landwirtschaftsverlag, Berlin.

Kalkhoven, J.T.R. (1993) Survival of populations and the scale of the fragmented agricultural landscape. In: Bunce, R.G.H., Ryszkowski, L. and Paoletti, M.G. (eds) *Landscape ecology and agroecosystems*. Lewis Publishers, London, pp. 83–90.

Krumbiegel, D. and Schwinge, W. (1991) *Witterung – Klima, Datenzusammenstellung für Mecklenburg-Vorpommern, Brandenburg und Berlin*. Potsdam.

Landesamt für Datenverarbeitung und Statistik Brandenburg (Hrsg.) (1996) *Statistische Berichte. Bodennutzung im Land Brandenburg 1995*. Potsdam.

Leuschner, Chr. and Immenroth, J. (1994) Landschaftsveränderungen in der Lüneburger Heide 1770–1985. Dokumentation und Bilanzierung auf der Grundlage historischer Katen. *Arch. für Natursch. u. Landschaftspfl*. 33, 85–139.

Mantel, K. (1990) *Wald und Forst in der Geschichte: ein Lehr- und Handbuch*. Alfeld-M. und H. Schaper, Hanover.

Ministerium für Ernährung, Landwirtschaft und Forsten (Hrsg.) (1994) *Wald und Forstwirtschaft im Land Brandenburg*. UNZE-Verlagsgesellschaft, Potsdam.

Ministerium für Umwelt, Naturschutz und Raumordnung (Hrsg.) (1993) *Rote Liste. Gefährdete Farn- und Blütenpflanzen, Algen und Pilze im Land Brandenburg.* UNZE-Verlagsgesellschaft, Potsdam.

Ministerium für Umwelt, Naturschutz und Raumordnung (Hrsg.) (1995) *Landschaftsprogramme Brandenburg. Erläuterungen.* Selbstverlag, Potsdam.

Passarge, H. (1966) Waldgesellschaften der Prignitz. *Archiv f. Forstwesen* 15, 475–504.

Peterken, G.F. and Game, M. (1984) Historical factors affecting the number and distribution of vascular plant species in the woodlands of Central Lincolnshire. *Journal of Ecology* 72, 155–182.

Richter, A. (1957) Zur Entwicklung der Waldverbreitung im Gebiet der DDR während der letzten 150 Jahre. *Arch. Forstwes. Bd. 6, Heft 11/12,* 802–810.

Scamoni, A. (1964) *Vegetationskarte der Deutschen Demokratischen Republik (1:500,000).* Akademie-Verlag, Berlin.

Schauer, W. (1957) Untersuchungen zur Entwicklung der Waldverbreitung in den Bezirken des ehemaligen Landes Brandenburg (1780–1937). Diss. Forstwiss, Fakultät Eberswalde, Humboldt-Univ., Berlin.

Schmidt, R. and Diemann, R. (eds) (1991) *Erläuterungen zur mittelmaßstäbigen landwirtschaftlichen Standortkartierung.* Nachdruck im Selbstverlag der Akademie der Landwirtschaftswissenschaften der DDR, Bereich Bodenkunde/Fernerkundung, Eberswalde.

Statistisches Bundesamt (ed.) (1996) *Statistisches Jahrbuch 1996 für die Bundesrepublik Deutschland.* Metzler & Poeschel, Wiesbaden.

Thomasius, H. (1973) Wald. In: *Landeskultur und Gesellschaft.* Theodor Steinkopff, Dresden.

Wulf, M. (1995) Sollten Erstaufforstungen an kontinuierlich bewaldete Flächen angrenzen? Ein Beitrag zur Migrationsproblematik typischer Waldpflanzen. *Zalf-Bericht* 52, 52–60.

Wulf, M. and Schmidt, R. (1996) Die Entwicklung der Waldverteilung in Brandenburg in Beziehung zu den naturräumlichen Bedingungen. *Beitr. Forstwirtsch. u. Landschaftsökol* 30, 125–131.

Opportunities to Protect the Biodiversity of Ancient Woodland in England

$\boxed{25}$

S.A. Bailey, R. Haines-Young and C. Watkins

*Department of Geography, University of Nottingham,
Nottingham NG7 2RD, UK*

**This chapter considers the potential offered by afforestation to help
conserve ancient woodland in England by exploring the criteria for
locating new woodland and the need to consider the ancient woodland
resource in a regional context. A number of frameworks, some designed
specifically to represent ecologically similar regions, are used to quantify
England's ancient woodland resource; their ability to describe ancient
woodland pattern is analysed.**

Introduction

Recent government proposals have been to increase substantially the woodland cover of England in the next 50 years (DOE and MAFF, 1995). Much of
this afforestation will take place on former agricultural land in lowland
England. At the same time, although the rate of ancient woodland loss is
declining, there is still a need to protect this valuable semi-natural habitat. We
consider whether landscape ecological principles can be used to help decide
where to establish new woodland with the aim of protecting ancient woodland
and yet maximizing the potential for species from semi-natural sites to colonize
the new woodland. Regions with similar ancient woodland character need to
be identified to improve decisions about afforestation. We examine the distribution of ancient woodland at the county level and then assess the different ways
Natural Areas and Joint Character Areas (JCAs) represent the resource. The
chapter concludes with a discussion of the effectiveness of the JCAs as a framework in which to represent the character of ancient woodland.

Fragmentation as a Threat to Ancient Woodland

Ancient woodlands are recognized by many ecologists as the most important
type of woodland for nature conservation being generally far more diverse in

the species they support than recent woods (Rackham, 1980, 1990; Peterken, 1981; Watkins, 1990; Spencer and Kirby, 1992). Although great advances have been made in their protection and conservation over the past 20 years, they are still threatened with clearance and damage. At the site level they can be protected by their designation as a Site of Special Scientific Interest (SSSI) or they can be purchased by a conservation body (Thomas *et al.*, 1997). However, ancient woodlands are highly fragmented and as such the species associated with ancient woodland are threatened by a number of processes associated with fragmentation (Kirby, 1995). For example, a reduction in habitat size can lead to loss of internal habitat diversity and relative increase in edge habitats. Species composition changes at woodland edges and an increase in invasive species may push out forest-interior species (Dawson, 1994).

Fragmentation inhibits the dispersal of species of ancient woodland either because the distance between woods is increased or because modification of the surrounding land cover matrix makes it more difficult for species to move through it. Some woodland plant species have been found to be highly vulnerable to isolation (Peterken and Game, 1984; Verkaar, 1990), as have some sedentary bird species such as nuthatch (*Psitta psitta*) and marsh tit (*Parus palustris*) (Fuller *et al.*, 1995). Dormice (*Muscardinus avallanarius*) tend to travel through continuous upper story woodland cover. Their dispersal is inhibited by breaks in woodland cover and they are very unlikely to cross agricultural land. Van Dorp and Opdam (1987) found that it was the forest-interior species with poor powers of dispersal that were significantly affected by habitat isolation. Thus fragmentation of ancient woodland can reduce species diversity within ancient woodland and be sufficient to hinder dispersal to other potential habitats such as recent woodland.

Opportunities to Protect Ancient Woodland Using New Planting

An appreciation of landscape-scale ecological processes is integral to the protection of ancient woodland (Peterken, 1992). The consideration of landscape structure when locating new woodland could be used to reverse the detrimental effects of fragmentation and enhance the biodiversity of ancient woodland. However, empirical work on both the effects of landscape structure on the movement of species associated with ancient woodland and the quantification of fragmentation of ancient woodland at scales appropriate to species, is deficient. Government policy for England is to encourage an increase in woodland cover (DOE and MAFF, 1995). If this new woodland were established in association with ancient woodland then not only could further species loss from the ancient woods be reduced but the new woods themselves could be more easily colonized by woodland species.

Various schemes are available to encourage farmers and landowners to afforest agricultural land. Currently the level of planting is very low because farming is profitable, but changes in the agricultural support mechanism are likely to change this situation (Watkins *et al.*, 1996). Forestry Commission grants are awarded for both the establishment and management of woodlands

(Forestry Authority, 1993). These schemes are open to landowners and tenants across the whole country although there is an increasing tendency for woodland establishment to be preferentially encouraged in special areas, such as Community Forests and the New National Forest. Before the location of new planting is agreed and the relevant grants awarded, any 'significant' planting proposals are investigated to ensure no conflicts arise with other interests such as agriculture, landscape, public recreation, nature conservation and archaeology. However, no special provision is currently made to encourage new woodland where it could enhance and protect the existing ancient woodland or where the new woodland itself could benefit from being colonized by species from ancient woodland.

The Ancient Woodland Inventory

English Nature's Ancient Woodland Inventory (AWI) identifies and records all ancient woodlands above 2 ha in area in England. The inventory records the extent and location of each ancient wood; the area of ancient semi-natural woodland; the area of ancient semi-natural woodland that has been converted to plantation since the 1930s; and the area of ancient semi-natural woodland that has been cleared since the 1930s. The designations that are associated with each site and ownership by public bodies are also recorded (Spencer and Kirby, 1992). The spatial reference provided for every ancient woodland enables detailed examination of the pattern of different types of ancient woodland across England. It also allows the fitting of various spatial frameworks to the ancient woodland pattern.

National Woodland Frameworks

Ancient woodland varies in distribution from very dense areas such as the Weald (Kent and Sussex) and parts of Herefordshire to areas almost completely cleared of ancient woodland, for example in the Fens and Lincolnshire. New woodland planting policies should take account of this variation. Which frameworks most appropriately represent ancient woodland pattern in England? To date the woodland resource of England has been represented at the county level, and more recently by Natural Areas and Joint Character Areas (English Nature, 1993). County boundaries are essentially historical administrative boundaries which pay little heed to topographical features. English Nature's Natural Areas, in contrast, are specifically designe and designated to represent some form of natural region and English N? is shifting the context of its conservation work from administrative re such as counties and districts, to Natural Areas (Reid *et al.*, 1996).

The Countryside Commission and English Nature published ? Map' of the English countryside in December 1996. The integrates with and will supersede the present Natur (Countryside Commission, 1994), has been created to 'ident cultural elements which contribute to the distinctive c' parts of the county' (Forestry Authority and Countrysi'

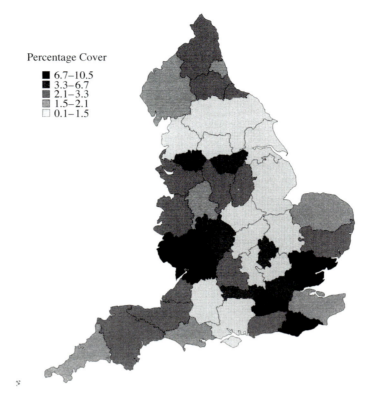

Fig. 25.1. Area of ancient woodland *c.* 1980 described by county. Source: Ancient Woodland Inventory.

It has been suggested that the Character units could help identify areas where new woodland might 'enhance, restore or create landscape character' (Forestry Authority and Countryside Commission, 1996).

Ancient Woodland Distribution

The map of the ancient woodland resource based on the county framework shows broad trends in the distribution of ancient woodland (Fig. 25.1). Counties in the South East and along the Welsh Border have a large share of England's ancient woodland resource. Northamptonshire, Oxfordshire and Wiltshire are also identified as having a high proportion of ancient woodland. Counties that have retained the least ancient woodland are in clearly defined regions in the East of England, the West Midlands and the North West, with the exception of Cumbria. This precise assessment of the distribution of ancient woodland by county, however, provides little advance over existing descriptions (Marren, 1992; Spencer and Kirby, 1992). The use of counties as a framework obscures considerable variation within counties. The disparity in

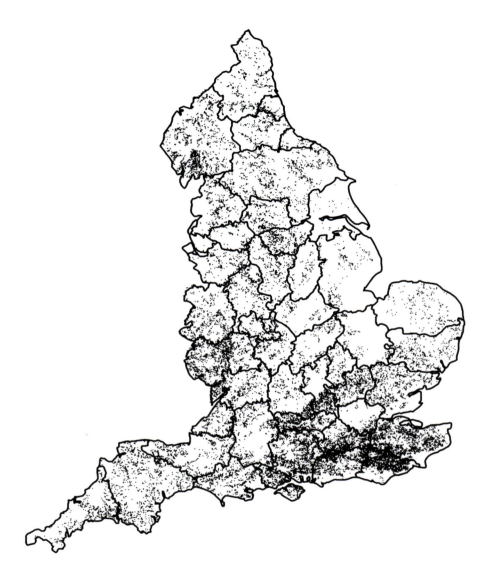

Fig. 25.2. The distribution of ancient woodland centroids across England overlain with county boundaries. Source: Ancient Woodland Inventory.

overall ancient woodland distribution and county boundaries is made clear if Fig. 25.1 is compared to Fig. 25.2 which shows the distribution of centroids recording the location of each ancient woodland held in the AWI for England. This provides a better impression of the fit of the ancient woodland pattern to particular landscape regions.

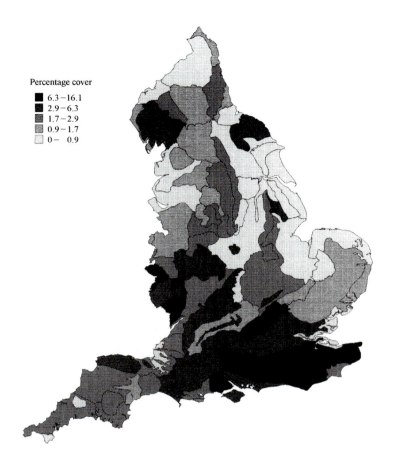

Fig. 25.3. Area of ancient woodland *c.* 1980 described by Natural Areas. Source: Ancient Woodland Inventory.

Gloucestershire, for example, contains dense areas of ancient woodlands within the Cotswolds and Forest of Dean and a sparse cover in the Severn Valley. In Buckinghamshire, the heavily wooded Chilterns lie adjacent to the Clay Vale which is almost devoid of ancient woodland. The South East, particularly the Weald, stands out as one of the areas most dense in ancient woodland. It is bounded by a clearly defined line running from the New Forest along the north edge of the Chiltern Hills into Suffolk. Within the South East, Greater London stands out as a noticeably sparsely wooded area. Other densely wooded areas are in the Wye Valley and uplands in Herefordshire and the Lake District in Cumbria. Often such regions cover a fraction of county area or are dissected by county boundaries which consequently mask the underlying distribution of ancient woodland.

Description of the ancient woodland resource within the framework of counties is of value to local authorities seeking to put their ancient woodland

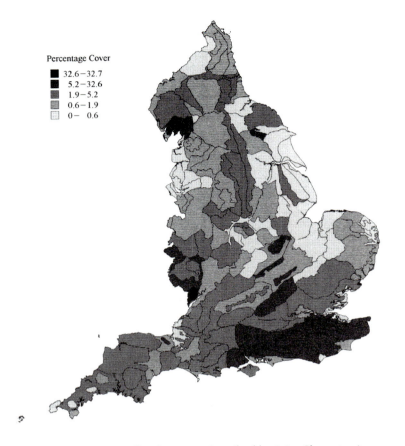

Percentage Cover
- 32.6–32.7
- 5.2–32.6
- 1.9–5.2
- 0.6–1.9
- 0– 0.6

Fig. 25.4. Area of ancient woodland *c.* 1980 described by Joint Character Areas.
Source: Ancient Woodland Inventory.

resource into a wider context and to those organizations involved in making policy decisions regarding woodland within administrative units (such as MAFF, the Countryside Commission and the Environment Agency). However, the extent of within-county variation justifies the use of other spatial units and the area of ancient woodland will now be considered within the context of Natural Areas and Joint Character Areas (JCAs) which give a finer resolution of the data (Figs 25.3 and 25.4). The large areas of ancient woodland in the Weald, the Lake District, Dean Valley and Wye Plateau, North York Moors and Chilterns, referred to above, are clearly depicted using these two frameworks. Regions with sparse ancient woodland cover are also apparent. The boundaries defining Natural Areas and JCAs appear to parallel the broad pattern of ancient woodland centroids shown in Fig. 25.2.

Both the Natural Areas and JCA frameworks pick out areas where there is a strong discontinuity in the pattern of ancient woodland. Contrasts are particularly noticeable between the Weald and surrounding area; the

Herefordshire Plain and the surrounding well-wooded areas; and in Oxfordshire where the Oxfordshire Heights, Bedfordshire Greensand and Chilterns are clearly defined. The North York Moors are clearly distinguished from the sparsely wooded North East and Charnwood Forest is separated from the surrounding Trent Valley and Levels. Areas with sparse ancient woodland, such as the Northumberland Coastal Plain, the coast of East Anglia and the Thames Estuary, and the inland upland regions of Bodmin and the Borders, are also more clearly defined.

Although Natural Areas are better than counties in reflecting the pattern of ancient woodland, some regions with clusters of ancient woodlands, or that are devoid of ancient woodlands, are not defined by Natural Areas (Reid *et al.*, 1996). The JCAs present the data at a finer resolution than provided by Natural Areas. Additional heavily wooded regions that are defined include, for example, the Yardley-Whittlewood Ridge, Rockingham Forest, the Yorkshire Southern Pennine Fringe, and the Durham Coalfield Pennine Fringe. At the other extreme, regions with few ancient woods such as Bowland Fells and moorlands, and the Cornish Killas (Bodmin, Hensbarrow and Carnmallis) are also picked out. The accurate definition of such areas is important to ensure that the ecological integrity of a region is protected when new woodland is being located. In addition, between these extremes of ancient woodland character, the finer resolution of JCAs allows for a more accurate description of the ancient woodland pattern than is true for Natural Areas.

Effectiveness of Natural Areas and JCAs in Representing Ancient Woodland

Methods

The comparison of maps suggests that the JCAs provide the best framework for developing an understanding of the pattern of ancient woodland in England. The best framework should be the one that partitions the information available on ancient woodlands in the most effective way. Analysis of variance (ANOVA) is used to describe how well variance is partitioned by the frameworks and indicates whether there is more variation between the regions described by the frameworks than within the regions themselves. The null hypothesis is that the samples (regions) are taken from a common, normally distributed population, or identical populations (Ebdon, 1995). The test determines whether there is more variation between the samples than within them. The framework that describes the variation in ancient woodland character most effectively is that which has proportionally higher between-group variation explained by the ANOVA. The F statistic resulting from one-way ANOVA can only be compared to another F ratio where the degrees of freedom used in the calculation are the same. To enable frameworks with different numbers of regions, and therefore different degrees of freedom to be compared, a general factorial ANOVA model that adjusts for the different degrees of freedom was used. The ANOVA with this adjustment provides an

Table 25.1. R_a^2 for ANOVA on ancient woodland attributes within three frameworks.

Framework	Size of ancient wood (c. 1980)	Distance to nearest wood	Increased distance to nearest wood	Unmodified woods	Reduction of wood area	Area of wood replanted	Area of wood cleared	Removal of woods
County	44	8.7	0.4	3.1	4.4	3.8	4.1	0.6
Natural Area	40	10.4	0.8	3.5	5.8	5.7	4.6	1.0
JCA	52	12.1	1.3	4.0	6.3	6.2	6.2	1.3

indication of the variation explained by each framework's representation of the data; the framework that has the largest variability between its samples (the largest R_a^2) will provide the most effective partition of the ancient woodland feature that is being considered.

A table was constructed which related the information about every ancient woodland in England to the various frameworks of which it was a part and the distance to its nearest neighbour. This table was imported into the Statistical Package for Social Sciences (SPSS) (Norušis, 1993). ANOVA assumes that the data has a normal distribution, so, before the ANOVA could be performed the values of each attribute were converted to log natural values to force the data into a normal distribution. General factorial analysis of variance (ANOVA) was then performed on each unit's representation of the AWI attributes.

Results

All frameworks explain very small proportions of the total variation in the distribution of each attribute (Table 25.1) and are therefore ineffective at representing the attributes that describe the reduction of the ancient woodland resource to its current highly fragmented state. The attribute that describes the variation in the distribution of ancient woodland size is more effectively explained within the frameworks. The JCA framework, which was expected to explain the most variation within the attributes considered, does give the best representation of the distribution of the majority of the attributes. But JCAs still only account for just over 6% of the variation in distribution of the attributes that describe the change in ancient woodland between *c.* 1930 and *c.* 1980.

In particular, all frameworks are very ineffective in explaining the distribution of the removal of ancient woodlands and the increased distance to nearest ancient wood following the removal of ancient woods between *c.* 1930 and *c.*1980. The failure of the frameworks to explain the distribution of cleared ancient woods is because they are being considered by the ANOVA within the context of the whole population of ancient woods. The models are only explaining the partition of 730 values from a population of over 21,000 records. The removal of ancient woodlands between *c.* 1930 and *c.* 1980 has increased the nearest neighbour distance for 645 ancient woodlands (3% of all ancient woodlands). The average increase in distance to the nearest neighbour of these ancient woodlands is almost half a kilometre. These figures are an underestimate as they only account for an increase of distance from an ancient woodland to its nearest neighbour, not all surrounding ancient woods.

Effectiveness of JCAs in Representing the Pattern of Ancient Woodland

The poor results of the ANOVA suggest that it is more difficult to link the pattern of ancient woodland to regional geographical and cultural processes at this scale than was anticipated. The description of the distribution of ancient woodland within the three frameworks indicates that JCAs provide the most effective framework within which to represent ancient woodland distribution because they pick out the extremes: the regions that are particularly noted for being well wooded and at the other end of the spectrum the regions that are devoid of ancient woodlands.

The JCAs were defined with the intention of describing regions according to their landscape, wildlife and natural features and as such are better tailored to the distribution of ancient woodland than counties, and are more effective in this respect than Natural Areas. Both the Countryside Commission and English Nature are exploring how to incorporate the use of JCAs within their woodland planning policy to: 'help in decisions about the siting and composition of new woodland, with a view to achieving the target of doubling England's tree cover ... in ways that respect local character' (Countryside Commission and English Nature, 1997).

One way forward would be to aggregate JCAs based on ancient woodland character, woodland change and complete woodland cover through profiles holding information on the amount and nature of the reduction of the ancient woodland resource for each JCA. An index could be calculated describing the isolation of all ancient woods from both recent and other ancient woodland. These profiles could then be developed into a typology of JCAs within which groups of JCAs with similar profiles would be aggregated.

Acknowledgement

We wish to thank English Nature for permitting us to use the AWI, Natural Area and JCA boundaries.

References

Countryside Commission (1994) *Countryside Character Programme.* John Dower House, Cheltenham.
Countryside Commission and English Nature (1996) *The character of England: landscape, wildlife and natural features.* English Nature, Peterborough.
Dawson, D. (1994) *Are habitats conduits for animals and plants in a fragmented landscape? A review of the scientific evidence.* English Nature (Research Report 94), Peterborough.

DOE and MAFF (1995) *Rural England: A nation committed to a living countryside.* Department of the Environment and Ministry of Agriculture, Fisheries and Food. HMSO, London.

Ebdon, D. (1995) *Statistics in geography*, 2nd edn. Blackwell Scientific Publishers, London.

English Nature (1993) *Strategy for the 1990's – natural areas.* English Nature, Peterborough.

Forestry Authority (1993) *Woodland Grant Scheme.* April 1993. Forestry Authority, Corstorphine Road, Edinburgh.

Forestry Authority and Countryside Commission (1996) *Woodland creation: needs and opportunities in the English countryside. A discussion paper.* Forestry Authority and Countryside Commission, Northampton.

Fuller, R.J., Gough, S.J. and Marchant, J.H. (1995) Bird populations in new lowland woods: landscape design and management perspectives. In: Ferris-Kaan, R. (ed.) *The ecology of woodland creation.* Wiley and Sons, London.

Kirby, K. (1995) *Rebuilding the English countryside: habitat fragmentation and wildlife corridors as issues in practical conservation.* English Nature (English Nature Science 10), Peterborough.

Marren, P. (1992) *The wild woods.* David & Charles, Newton Abbot.

Norušis, M.J. (1993) *SPSS for Windows*, Release 6.0. Chicago, USA.

Peterken, G.F. (1981) *Woodland conservation and management.* Chapman & Hall, London.

Peterken, G.F. (1992) Woodland connectivity and design. In: Haines-Young, R. (ed.) *Landscape ecology in Britain.* Department of Geography, University of Nottingham, Nottingham.

Peterken, G.F. and Game, M. (1984) Historical factors affecting the number and distribution of vascular plant species in the woodlands of central Lincolnshire. *Journal of Ecology* 72, 155–182.

Rackham, O. (1980) *Ancient woodland.* Edward Arnold, London.

Rackham, O. (1990) *Trees and woodland in the British landscape*, revised edition. Dent, London.

Reid, C.M., Kirby, K.J. and Cooke, R.J. (1996) *A preliminary assessment of woodland conservation in England by natural areas.* English Nature (Research Report 186), Peterborough.

Spencer, J.W. and Kirby, K.J. (1992) An inventory of ancient woodlands for England and Wales. *Biological Conservation* 62, 77–93.

Thomas, R.C., Kirby, K.J. and Reid, C.M. (1997) The conservation of a fragmented ecosystem within a cultural landscape – the case of ancient woodland in England. *Biological Conservation* 82, 243–252.

Van Dorp, D. and Opdam, P. (1987) Effects of patch size, isolation and regional abundance on forest bird communities. *Landscape Ecology* 1, 59–73.

Verkaar, H.J. (1990) Corridors as tools for plant species conservation. In: Bunce, R.G.H. and Howard, M.C. (eds) *Species dispersal in agricultural habitats.* Belhaven Press, London.

Watkins, C. (1990) *Woodland management and conservation.* David & Charles, Newton Abbot.

Watkins, C., Lloyd, T. and Williams, D. (1996) Constraints on farm woodland planting in England: a study of Nottinghamshire farmers. *Forestry* 69, 167–176.

The Ancient Woodland Inventory in England and its Uses

26

K.J. Kirby, C.M. Reid, D. Isaac and R.C. Thomas
English Nature, Northminster House, Peterborough PE1 1UA, UK

Ancient woodland (land continuously wooded since AD 1600) is the most important category for nature conservation in England. In 1981 the Nature Conservancy Council started to list the ancient woodland sites in each county. The resulting inventories have been used in developing national forestry policies, promoting local action to defend sites that were under threat, and as a basis for exploring how woods and their conservation are affected by changes at a landscape level. While in some places increased fragmentation of woodland has occurred during the last 100 years, elsewhere recent woodland development has reduced the isolation of ancient woods. Conservation of ancient woods may be promoted through site designation, sympathetic ownership or wider land use policies. These have different significance in different parts of the country and this must be allowed for in an overall conservation strategy.

Introduction

We describe the use of historical classification of woodland sites to promote woodland conservation in Great Britain that has enabled the conservation movement to have an influence well beyond the statutorily protected Sites of Special Scientific Interest and nature reserves.

The work stems from the identification of ancient semi-natural woodland as the most important category for nature conservation in Great Britain (but particularly in England) by Peterken (1977, 1981) and Rackham (1976, 1980) because of the species, communities and historical features that it contains. Ancient woodland sites in Britain are those that are believed to have remained wooded continuously since AD 1600. Semi-natural woods are those composed of trees and shrubs native to the site, which have grown from stump regrowth (coppice, pollards) or natural regeneration, as opposed to having been planted.

Peterken and Rackham also realized that only a minority of the ancient woods were within (or likely to be brought into) statutorily protected sites and that the value of many of these woods had been greatly reduced through clearance and conversion to coniferous plantations (Peterken and Harding, 1975; Rackham, 1976). Conservation of the ancient woodland resource would therefore depend on conservation being approached at a landscape scale, and an important tool in this would be an ancient woodland inventory. The four major elements in developing this inventory approach have been:

1. Creation of provisional lists of ancient woods for each county using relatively simple methods based on old maps, aerial photographs and existing field surveys (Goodfellow and Peterken, 1981; Kirby *et al.*, 1984; Walker and Kirby, 1987, 1989; Roberts *et al.*, 1992; Spencer and Kirby 1992).
2. Gaining acceptance of the importance of ancient woods by the Forestry Commission and the private forestry sector, and the incorporation of this idea in national forestry policy (Forestry Commission, 1985a; HMSO, 1994) and in local authority and voluntary conservation movement thinking.
3. Increasing the level of understanding of the management needs of ancient woods through training, advice and publications of different sorts (Peterken, 1981; Forestry Commission, 1985b, 1994; Watkins, 1990).
4. Refinement of our ideas of how ancient woods function, their history and role in the landscape and for the species that live in them.

The first three elements are only briefly reviewed below as they have been well described elsewhere (see above) and so we concentrate on this last subject and on results from England.

Creation of the Ancient Woodland Inventory

The county ancient woodland inventories for England were started in 1981 and completed in 1992. In practice results for all counties were available by 1988 but early drafts for some counties were revised and reissued over the next four years to take account of changes in the way the data were being presented and used as the project developed (Spencer and Kirby, 1992). Some counties have been subject to further revision and periodic amendment reports produced (Thomas and Phillips, 1994; Isaac and Reid, 1996). The main conclusions to be drawn from this first round of inventories were that:

- the areas of ancient and ancient semi-natural woodland in England are 340,158 ha and 198,445 ha, respectively (about 2.6 and 1.5%, respectively, of the land surface area);
- the sites are mainly small (47% less than 5 ha and 83% less than 20 ha) and widely distributed across the countryside, but with both notable gaps and local concentrations of sites (Fig. 26.1);
- replanting (mainly with conifers) and clearance (mainly for agriculture) have affected about 39% and 7% respectively of the ancient woodland present in the 1930s; and
- only about 15% of ancient and 21% of ancient semi-natural woods are within Sites of Special Scientific Interest and nature reserves.

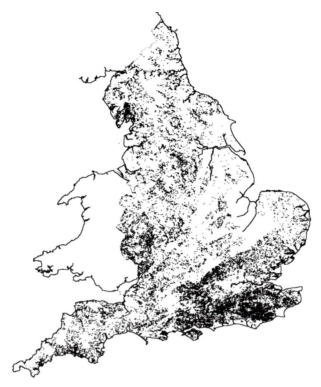

Fig. 26.1. Ancient woodland distribution across England.

Gaining Acceptance of the Use of the Inventories in Forestry Policy and Practice

The keys to this lie in:

- the sound theoretical and empirical evidence for the importance of ancient woods for nature conservation laid down by Rackham and Peterken;
- recognition that even in England where most ancient semi-natural woods occurred, they were a minority of the total woodland cover (Fig. 26.2);
- recognition by the forestry industry (both state and private) that it needed to demonstrate a commitment to maintaining woodland biodiversity and that, in this context, restrictions on timber production in ancient woods were not unreasonable;
- widespread (largely free) distribution of the inventories to forestry, conservation and Government bodies as well as to landowners on request;
- willingness by the Nature Conservancy Council (and subsequently English Nature) to use the inventories in a flexible way with respect to sites that might have been wrongly included or classified on the inventory. Openness and honesty about the limitations of the methods used to produce the inventories have, in the long run, proved very valuable in defending them against criticism.

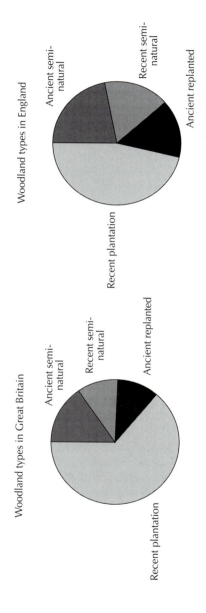

Fig. 26.2. Proportion of different woodland types.

Dealing with conflicts and debate about whether sites should or should not be on the inventory, and NCC's willingness to revise the inventory database in the light of new evidence, took up much staff time in the period 1985–1988 but helped to gain acceptance of the inventories as a tool in forest management planning.

Improving the Management Advice Available for Ancient Woods

This has come about through increasingly close working between the Forestry Commission (latterly the Forestry Authority) and the statutory nature conservation bodies in both published advice and in training. The Forestry Commission showed its commitment to improving its ability to provide advice on the conservation value and management of ancient woods by running nine courses in 1991, each a week long and attended in total by about 150 of their staff. The bulk of the technical input on these courses was provided by the staff of the conservation agencies. Since 1991 further courses have been arranged when necessary.

To go with the changing approach and advice on management there have been changes to the grants and incentives available to woodland owners and managers to try to encourage sympathetic treatment of ancient semi-natural woods. In 1996, for example, a grant was introduced to cover 50% of the costs of restoring long-neglected coppice in areas where it should benefit endangered butterfly populations.

Using the Inventory Results to Develop Woodland Conservation in the 1990s and Beyond

The ancient woodland inventory project was initiated at a time when the prime concern was to halt further attrition of this resource from clearance and conversion of semi-natural stands to plantations of introduced conifers. This has largely been achieved, although some losses to roads, railways and other developments continue.

The reduction in the extent of these gross threats has, however, focused attention on other possible impacts on nature conservation arising from woodland management (Mitchell and Kirby, 1989, 1990) for example:

- the scale of felling in ancient semi-natural woods;
- the planting of native trees instead of allowing them to regenerate naturally, and the introduction of foreign provenances of native trees;
- the removal of dead wood;
- the use of herbicides to control weeds;
- the conversion of coppice stands to high forest of native species;
- changing levels of deer and sheep grazing;
- whether we should be managing some woods at all.

Conversion to another land use or replanting with introduced conifers is clearly damaging in any ancient semi-natural wood and hence simply knowing whether or not a site is on the inventory can be used as a basis to limit such activities. However, with these other impacts, the potential effects on nature conservation values depend much more on the individual characteristics of a wood and on what is happening in surrounding ancient woods.

Taking account of individual woodland characteristics

Within the 'ancient semi-natural' category, as defined and developed for the inventory purposes, are woods whose composition and structure have been manipulated to varying degrees over the past 150 years. This can be illustrated by comparing three eastern oak–ash–maple woods on clay soils (all predominantly *Fraxinus excelsior–Acer campestre–Mercurialis perennis* woodland (W8) in the National Vegetation Classification (Rodwell, 1991), Bradfield Woods (Suffolk), Monks Wood (Cambridgeshire) and Salcey Forest (Northamptonshire).

Maintaining the coppice regime in Bradfield Wood is a very high priority because of the continuity of this treatment on the site since medieval times (Rackham, 1976); it is not appropriate in Salcey Forest which has had a high forest structure for most of the last century following planting of oak in *c.* 1843. In Monks Wood, English Nature's management is a mixture of coppice and high forest. The wood was formerly coppiced and retained many species associated with young coppice until recently, hence some areas are kept as coppice; but over large areas the wood has largely been unmanaged since 1926 and it has developed a high forest structure. Planting of trees in Monks Wood and in Bradfield Wood is undesirable because there is no history of significant planting in these woods. The opportunity exists therefore to maintain local populations and genotypes and natural patterns of species distribution. At Salcey Forest, however, the current oak crop was planted and while natural regeneration is still to be preferred, planting of oak, but not other trees and shrubs, could be accepted as a compromise with wood production. Thus, perpetuating historical management in Bradfield is the priority treatment; in much of Monks Wood minimum intervention has been adopted; in Salcey Forest the aim is to develop a mixture of modern high forest management based on locally native species with some areas of minimum intervention.

A similar range of approaches must be adopted towards grazing and browsing in ancient woods (Kirby *et al.*, 1994, 1995). Where the aim is to keep the historically open structure of former wood-pastures, maintaining or reinstating a high level of grazing is desirable. In former coppice-woods, reducing deer levels is often necessary if the rich flora and coppice regeneration is to survive. In western oak woods the interest depends on quite high grazing densities for much of the time, but periodic reductions in the current intensity or duration of grazing are needed to permit regeneration and expansion of the woods.

Thus management recommendations based on the broad types used in the inventory must be refined by more detailed knowledge of recent woodland changes.

Table 26.1. Is there enough ancient woodland to develop a normal forest structure in 25 km² sample squares in contrasting counties?

	West Sussex	Essex	North Yorkshire
Mean area of ancient woodland per 25 km² (10 samples) (ha)	255	153	47
Mean area of ancient semi-natural woodland per 25 km² (ha)	141	137	16
No. of samples capable of sustaining a *c.* 100 year rotation high forest regime with 0.5 ha per annum felling rate, based on ancient woodland area	10	8	4
No. of samples capable of sustaining a *c.* 100 year rotation high forest regime with 0.5 ha per annum felling rate, based on ancient semi-natural woodland area	10	7	1
No. of samples capable of sustaining a 25–30 year coppice rotation with 1.0 ha per annum felling rate, based on ancient semi-natural woodland area	10	10	7

Relating the treatment of a wood to its surroundings

Different treatments will favour different groups of organisms. On some sites there is one type of management that is much better in nature conservation terms than any other (e.g. coppice at Bradfield Woods, see above), but often there may be several more or less acceptable options. The choice as to which to recommend on any particular site must then be influenced by what is happening in the other ancient woods in the region. In the early 1980s most ancient woods were unmanaged, or being managed unsympathetically, so there were major benefits to be gained from promoting management regimes sympathetic to conservation on Sites of Special Scientific Interest (SSSIs) to provide demonstrations of what could be done. As more ancient woods are brought into acceptable management through forestry grants, so the balance may swing to keeping those SSSIs that are not currently managed as minimum intervention areas. By superimposing information on Woodland Grant Schemes (run by the Forestry Authority) on the Inventory data set different conservation options can be explored.

For example, is the rate of felling in ancient woods in an area likely to be sustainable in terms of providing a continuous supply of wood and a continuous supply of temporary open space/young growth such as may be needed by woodland butterflies? An average of about 0.5 ha a year might be considered the minimum area for felling in woods that are to be managed sustainably as high forest. This implies a minimum woodland area of 50 ha if

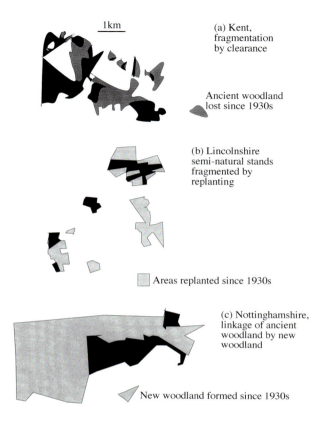

1km

(a) Kent,
fragmentation
by clearance

Ancient woodland
lost since 1930s

(b) Lincolnshire
semi-natural stands
fragmented by
replanting

Areas replanted since 1930s

(c) Nottinghamshire,
linkage of ancient
woodland by new
woodland

New woodland formed since 1930s

Fig. 26.3. (a) Kent, ancient woodland fragmentation; (b) Lincolnshire, fragmentation of ancient semi-natural stands; (c) Nottingham, reduced woodland fragmentation.

the rotation length is 100 years (not unlikely for broad-leaved trees in England). Inventory maps for different parts of the country can be examined to see whether there are such blocks of 50 ha of ancient semi-natural woodland either in one unit or (more likely) as clusters of small woods (Table 26.1). If not then the 'ideal' of maintaining the open-space requirement via the normal silvicultural practice is not possible. Changes to the system or special measures such ride-widening may be needed (Fuller and Warren, 1990). In a densely wooded county such as West Sussex, there are good opportunities for species to move from one site to another, so it is less important to manage each individual ancient wood to maintain continuity of habitat conditions in that wood (Edwards, 1996). Where woods are more isolated as in West Yorkshire it becomes essential to manage for habitat continuity at the site scale.

In Table 26.1 all the sample areas in West Sussex contained sufficient ancient and ancient semi-natural woodland for a regular average felling rate of 0.5 ha per annum to be maintained across a 100 year high forest rotation.

Table 26.2. Internal fragmentation of ancient woods (11–20 ha size class) by clearance or replanting in Kent and Cumbria.

	Total no. of woods in 11–20 ha class	Semi-natural area unchanged	(i) Loss of semi-natural area, but connectivity maintained	(ii) Surviving semi-natural area fragmented	(iii) Semi-natural stands totally lost
Kent (number affected by clearance)	265	93	113 (82)	41 (30)	18 (5)
Cumbria (number affected by clearance)	132	34	55 (17)	18 (5)	25 (1)

Thus both the supply of timber and open space for butterflies could be sustained within a small landscape area. In North Yorkshire this was true for only four of the ten samples if all ancient woodland sites were considered and for only one for ancient semi-natural stands. Coppice regimes, because the rotations are much shorter, could be sustained in more samples even with a higher rate of felling.

Increasing or decreasing isolation of ancient woods

Habitat fragmentation and isolation have increasingly been identified as a cause of species decline and conservation efforts are needed to offset their effects (Kirby, 1995). However the degree to which species associated with ancient woods have been affected by habitat fragmentation or benefit from current woodland expansion is not always easy to assess.

Figure 26.3 illustrates areas where major changes in degree of woodland fragmentation have occurred. These include major losses of woodland habitat to urban development (initially an airfield that was later built upon), and also to agriculture (Fig. 26.3a). Fragmentation of semi-natural stands by coniferous plantation also occurs (Fig. 26.3b). Conversely, some ancient woods are less isolated now than in the past because of new planting and natural regeneration on to heaths and grassland (Fig. 26.3c). This is explored further using data from contrasting upland and lowland counties, Kent and Cumbria, and focusing on woods in the 11–20 ha class. This class was selected because their size makes them vulnerable; even quite small losses may leave the surviving individual fragments too small to contain woodland interior conditions.

The Ancient Woodland Inventory database provides data on total loss or replanting of ancient woodland, but not on where those changes occur in relation to surviving semi-natural stands in the wood. The original data for woods in the 11–20 ha size category in Kent and Cumbria were therefore assessed and the changes that had taken place put into three categories

Table 26.3. Mean area of different woodland types associated within different areas of search centred on randomly-selected woods in the 11–20 ha class for Kent and Cumbria (standard errors in brackets).

	Area of search					
	Within single km squares		Within 3 × 3 km squares		Within 5 × 5 km squares	
	Kent	Cumbria	Kent	Cumbria	Kent	Cumbria
No. of squares sampled	27	19	27	19	27	19
Mean area (ha) present within the search area for:						
Ancient semi-natural woodland	15.6	10.5	94.1	46.3	221.2	109.7
	(1.6)	(2.4)	(8.5)	(8.9)	(16.2)	(19.8)
Ancient woodland	17.7	20.6	115.7	80.3	299.3	185.8
	(1.6)	(2.7)	(10.2)	(12.1)	(25.1)	(31.0)
All woodland	23.8	30.2	192.6	202.7	445.0	431.5
	(2.3)	(4.6)	(14.9)	(45.0)	(32.5)	(97.5)

(Table 26.2): (i) some loss of semi-natural stand area (through clearance or replanting) but connectivity remains between the remaining semi-natural area; (ii) surviving semi-natural area split between at least two sections; and (iii) complete loss of semi-natural stands.

A higher proportion of woods no longer had any semi-natural stands in Cumbria compared to Kent (category (iii), above), but otherwise the proportion of stands where some loss of connectivity had occurred (category (i)) compared to where the semi-natural areas were now fragmented (category (ii)) was similar. Clearance of woodland made more of a contribution than replanting to fragmentation of semi-natural stands in Kent than in Cumbria. The implications therefore at the wood scale are likely to be more serious in Kent because: (i) more species can move through or survive in replanted stands than completely cleared areas; and (ii) the potential for restoring links between the semi-natural stands is less. The usual way in which inventory results have been presented to date (total area of ancient or semi-natural woodland per site) thus underestimates the amount of real fragmentation that has occurred. Digitization of the woodland boundaries on a geographic information system (in progress) will allow more analyses of this type.

Fragmentation of semi-natural stands within a site matters less (in terms of restrictions on species movement) in landscapes with a high cover of ancient woods than where they are sparse. For a sample of ancient woods in the 11–20 ha size class in Kent and Cumbria, therefore, the surrounding landscape was assessed to see how much ancient semi-natural woodland, ancient woodland and woodland of any sort were present. These figures were based on the areas measured off 1:50,000 maps of the three types of woodland in the 1 km^2 in which the site occurred, in the 3 × 3 km square centred on that 1 km^2 in which the wood occurred and on the 5 × 5 km square centred on the 1 km^2 sampling point (Table 26.3).

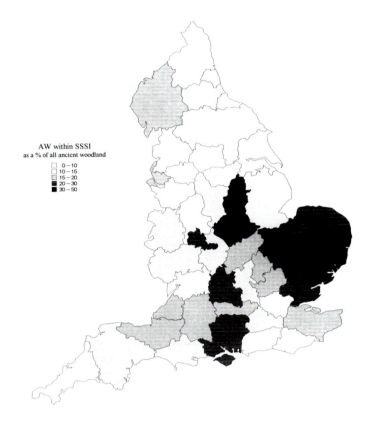

AW within SSSI
as a % of all ancient woodland

☐ 0 – 10
☐ 10 – 15
☐ 15 – 20
■ 20 – 30
■ 30 – 50

Fig. 26.4. Ancient woodland (AW) within SSSI (as a % of all ancient woodland).

There is little difference in the degree of isolation of ancient woods between Kent and Cumbria at the 1 km² level. At the 9 km² scale, ancient semi-natural stands are more isolated (lower mean area in the surrounding squares) in Cumbria than in Kent and this is also shown, but to a lesser degree, for ancient woods. This effect is emphasized further at the 25 km² scale. For 'All woodland' there is no difference between the two counties at any of the three scales. Woodland species would therefore find it equally easy to move between sites in the two counties at very local (1 km²) scales or if they can use all woodland types, but specialist ancient woodland species would be more restricted in their opportunities for movement over larger distances in Cumbria.

Promoting woodland conservation in practice

Woodland conservation can be promoted through designation of special sites, sympathetic land ownership and wider land use policies (Table 26.4) (Thomas *et al.*, 1997). These mechanisms may overlap, in the sense that the same wood may be covered by more than one mechanism, and their significance may vary

Table 26.4. Different ways of promoting woodland conservation.

Designation of special sites
    Sites of Special Scientific Interest
    National Nature Reserves
    Local Nature Reserves
    NGO Reserves

'Sympathetic' land ownership
    Local authorities
    Forest Enterprise
    National Trust

Protected landscapes
    National Parks
    Areas of Outstanding Natural Beauty

General land use policies
    Broadleaves Policy 1985
    Local authority policies

in different areas and for achieving different aims. In some areas of the country many ancient woods are within protected sites (Sites of Special Scientific Interest) (Fig. 26.4). English Nature has therefore a dominant role to play since it is consulted over their management; in other areas wider land use policies and ownership by other bodies become more important because English Nature's direct influence through the special site system is small. Different objectives may also suggest different approaches. If the objective is to promote the restoration of replanted woodland, the Forest Enterprise is a key organization because they have many replanted ancient woods. Restoration of coppice in areas where there are poor markets may be better targeted at voluntary conservation organizations because these may be more willing to accept the costs of such work (because of the wildlife benefits) than private landowners.

Conclusions

George Peterken had the foresight in the early 1980s to realize that the woodland Sites of Special Scientific Interest were not enough to ensure woodland conservation, and that an ancient woodland inventory would provide a mechanism to promote changes in forestry policies. His efforts have been amply rewarded. The inventories are now firmly embedded in all aspects of woodland conservation in England. Their limitations must, however, be understood if they are to be used wisely and further development of them is desirable if they are to remain relevant in the next decade.

Acknowledgements

We thank all those who have worked on the inventory project over the years or have contributed their ideas on how it might be used or developed.

References

Edwards, K.S. (1996) Are current rates of felling in ancient woodland sustainable at a landscape scale? MSc Thesis, Oxford Forestry Institute, Oxford.

Forestry Commission (1985a) *The policy for broadleaved woodlands.* Forestry Commission Policy and Procedure Paper No. 5, Edinburgh.

Forestry Commission (1985b) *Guidelines for the management of broadleaved woodland.* Forestry Commission, Edinburgh.

Forestry Commission (1994) *The management of semi-natural woodlands.* Forestry Practice Guide Nos 1–8. Forestry Commission, Edinburgh.

Fuller, R.J. and Warren, M.S. (1990) *Coppiced woodlands: their management for wildlife.* Nature Conservancy Council, Peterborough.

Goodfellow, S. and Peterken, G.F. (1981) A method for survey and assessment of woodlands for nature conservation using maps and species lists: the example of Norfolk woodlands. *Biological Conservation* 21, 177–195.

HMSO (1994) *Sustainable forestry – the UK programme.* Her Majesty's Stationery Office, London.

Isaac, D. and Reid, C.M. (1996) *Amendments to the Ancient Woodland Inventory for England July 1994–February 1997.* English Nature (Research Report 177), Peterborough.

Kirby, K.J. (1995) *Rebuilding the English countryside.* English Nature (English Nature Science 7), Peterborough.

Kirby, K.J., Peterken, G.F., Spencer, J.W. and Walker, G.J. (1984) *Inventories of ancient semi-natural woodland.* Focus on nature conservation 6. Nature Conservancy Council, Peterborough.

Kirby, K.J., Mitchell, F.J. and Hester, A.J. (1994) A role for large herbivores (deer and domestic stock) in nature conservation management in British semi-natural woods. *Arboricultural Journal* 18, 381–399.

Kirby, K.J., Thomas, R.C., Key, R.J., McLean, I.F.G. and Hodgetts, N. (1995) Pasture woodland and its conservation in Britain. *Biological Journal of the Linnean Society* 56 (suppl.), 135–153.

Mitchell, F.J.G. and Kirby, K.J. (1990) The impact of large herbivores on the conservation of semi-natural woods in the British uplands. *Forestry* 63, 333–354.

Mitchell, P.L. and Kirby, K.J. (1989) *Ecological effects of forestry practices in long-established woodland and their implications for nature conservation.* Oxford Forestry Institute (Occasional Paper 39), Oxford.

Peterken, G.F. (1977) Habitat conservation priorities in British and European woodlands. *Biological Conservation* 11, 223–236.

Peterken, G.F. (1981) *Woodland conservation and management.* Chapman & Hall, London.

Peterken, G.F. and Harding, P.T. (1975) Woodland conservation in eastern England: comparing the effect of changes in three study areas since 1946. *Biological Conservation* 8, 279–298.

Rackham, O. (1976) (1990 2nd edn) *Trees and woodland in the British landscape.* Dent, London

Rackham, O. (1980) *Ancient woodland.* Edward Arnold, London.

Roberts, A.J., Russell, C., Walker, G.J. and Kirby, K.J. (1992) Regional variation in the origin, extent and composition of Scottish woodland. *Botanical Journal of Scotland* 46, 167–189.

Rodwell, J. (1991) *British plant communities: I, woodland and shrubs.* Cambridge University Press, Cambridge.

Spencer, J.W. and Kirby, K.J. (1992) An inventory of ancient woodland for England and Wales. *Biological Conservation* 62, 77–93.

Thomas, R.C. and Phillips, P.M. (1994) *Amendments to the Ancient Woodland Inventory up to June 1994.* Research Report No 72. English Nature, Peterborough.

Thomas, R.C., Kirby, K.J. and Reid, C.M. (1997) The conservation of a fragmented ecosytem within a cultural landscape – the case of ancient woodland in England. *Biological Conservation* 82, 243–252.

Walker, G.J. and Kirby, K.J. (1987) An historical approach to woodland conservation in Scotland. *Scottish Forestry* 41, 87–98.

Walker, G.J. and Kirby, K.J. (1989) *Inventories of ancient, long-established and semi-natural woodland for Scotland.* Nature Conservancy Council (Research and Survey in Nature Conservation No. 22), Peterborough.

Watkins, C. (1990) *Britain's ancient woodland: woodland management and conservation.* David & Charles, Newton Abbot.

Bibliography

Aaby, B. (1992) Sjællands kulturlandskaber i jernalderen. *Arkæologiske skrifter* 6. 209–236.

Adams, J.M. and Woodward, F.I. (1989) Patterns in tree species richness as a test of the glacial extinction hypothesis. *Nature* 339, 699–701.

Agnoletti, M. (1989) La fattoria di Gargonza fra '700 e '800. Elementi per una storia del bosco in Valdichiana. *L'Italia Forestale e Montana* XLIV, 67.

Akademie der Wissenschaften der DDR (ed.) (1981) *Atlas der DDR.* Gotha, Haack, 53 map sheets.

Al, E.J., Koop, H. and Meeuwissen, T. (1995) *Natuur in bossen. Ecosysteemvisie bos.* IKC – Natuurbeheer, Wageningen.

Alexander, K.N.A. (1988) The development of an index of ecological continuity for dead wood associated beetles. *Antenna* 12, 69–70.

Alexander, K.N.A. (1994) The use of freshly downed timber by insects following the 1987 storm. In: Kirby, K.J. and Buckley, G.P. (eds) *Ecological responses to the 1987 Great Storm in the woods of south-east England.* English Nature (Science Report 23), Peterborough, pp. 134–150.

Alexander, K.N.A. (1996) The value of invertebrates as indicators of ancient woodland and especially pasture woodland. *Transactions of the Suffolk Naturalists Society* 32, 129–137.

Alexander, K.N.A., Green, E.E. and Key, R. (1996) The management of over mature tree populations for nature conservation – the basic guidelines. In: Read, H.J. (ed.) *Pollard and veteran tree management II.* Corporation of London, Burnham Beeches, pp. 122–135.

Alfonso XI (*c.* 1350) *Libro de la montería.* Seniff, D.P. (ed.), Hispanic seminary of medieval studies (1983), Madison.

Allen, D.E. (1984) *Flora of the Isle of Man.* The Manx Museum and National Trust, Douglas.

Anderson, S.T. (1970) The relative pollen productivity and pollen representation of North European trees and correction factors for tree pollen spectra. *Danmarks Geologiske Undersogelse, Series II* 96, 1–99.

Andersen, S.T. (1984) Forests at Løenholm, Djursland, Denmark at present and in the past. *Det Kongelige Danske Videnskabevnes Selskab. Biologiske Skrifter* 24, 1–210.

Anonymous (1873–1967) *Overijsselsche Stad-, Dijk- en Markeregren*, Derde deel, *Vereeniging tot beoefening van Overijsselsch regt en geschiedenis*. Zwolle, 1–23.

Anonymous (1974) *Devlet Meteoroloji İsleri Genel Müdürlüğü Ortalama ve Ekstrem Kiymetler Bülteni*. The Ministry of Sources of Energy and Nature Press, Ankara.

Anonymous (1995) *North York Moors National Park: Phase II Woodland Survey: Year 2*. Ecological Advisory Service, Keighley.

Arroyo, J. (1996) Plant diversity in the region of the Strait of Gibraltar: a multilevel approach. *Lagascalia* 19, 393–404.

Ashcroft, M.Y. and Hill, M. (1980) *Bilsdale Survey, 1637–1851*. North Yorkshire County Council, Northallerton.

Austad, I. (1988) Tree pollarding in western Norway. In: Birks, H.H. and H.J.B., Kaland, P.E. and Moe, D. (eds) *The cultural landscape – past, present and future*. Cambridge University Press, Cambridge, pp. 11–30.

Avila, D. (1988) *Explotaciones agropecuarias en Sierra Morena Occidental*. IDR, Universidad de Sevilla, Sevilla.

Bakker, J.A. (1982) TRB settlement patterns on Dutch sandy soils. *Analecta Praehistorica Leidensia* 15, 87–124.

Bakker, M. and Groot, L. (in prep.) *Historische ecologie van de bossen van Twente*. DLO – Staring Centrum, Wageningen.

Balla, G. (1758) *Nagykőrösi Krónika*. (Chronicle of Nagykőrös). Nagykőrös.

Barker, S. (1985) The woodlands and soils of the Coniston Basin, Cumbria. PhD thesis, University of Lancaster, Lancaster.

Barkham, J.P. (1992) The effects of coppicing and neglect on the performance of the perennial ground flora. In: Buckley, G.P. (ed.) *Ecology and management of coppice woodlands*. Chapman & Hall, London, pp. 115–196.

Basaran, S. (1988) *Enez'in Tarihi Önemi, Arkeolojisi ve Turizm Bakimindan Oynayacagi Rol*. Ulusal Enez Kongresi.

Bauer, E. (1991) *Los montes de España en la Historia*, 2nd edn. MAPA, Fundación Conde del Valle de Salazar, Madrid.

Beevor, H.E. (1925) Norfolk woodlands from the evidence of contemporary chronicles. *Quarterly Journal of Forestry* 19, 87–110.

Behre, K.E. (1988) The role of man in European vegetation history. In: Huntley, B. and Webb, T. III (eds) *Vegetation history*. Handbook of vegetation science Volume 7, Kluwer, Dordrecht, pp. 633–672.

Bekker, D.L. and Bakker, J.P. (1989) Het Westerholt IX: veranderingen in vegetatiesamenstelling en -patronen na 15 jaar beweiden. *De Levende Natuur* 90, 114–119.

Bellamy, B. (1986) *Geddington Chase – the history of a wood*. Bellamy, Geddington.

Bergendorff, C. and Emanuelsson, U. (1990) Den skånska stubbskottängen. *Nordisk Bygd* 4, 14–19.

Berglund, B.E. (ed.) (1991) The cultural landscape during 6000 years in southern Sweden – the Ystad Project. *Ecological Bulletin* 41, Copenhagen, 495 pp.

Bernátsky, J. (1901) Növényföldrajzi megfigyelések a Nyírségen. *Természettudományi Közlöny* 53, 203–216.

Bernetti, G. (1987) *I boschi della Toscana*. Edagricole, Bologna.

Bertolotto, S. (1997) Storia e copertura vegetale nell'Appennino: effetti delle coltivazioni temporanee in eta' storica in val d'Aveto. Tesi di laurea in Geografia storica dell'Europa, Universita' di Genova.

Best, J.A. (1983) *King's Wood, Corby – description, history, explanation of habitats and wildlife*. Nene College, Northampton.

Best, J.A. (1989) The changing vegetation. In: Colston, A. and Perring, F. (eds) *The nature of Northamptonshire*. Barracuda, Buckingham, pp. 40–50.

Best, J.A. (1994) *The past, present and future of Rockingham Forest.* The Wildlife Trust for Northamptonshire, Northampton.

Best, J.A. and Logue, W.L. (1991) *The grasslands of Hazel and Thoroughsale Woods, Corby.* The Northamptonshire Wildlife Trust, Corby.

Bieleman, J. (1987) *Boeren op het Drentse zand 1600–1910; Eeen nieuwe visie op de 'oude' landbouw.* Dissertation, Wageningen Agricultural University, Wageningen, 883 pp.

Bilt, E. van der (1986) Begrazing van natuurreservaten in Drenthe. *Huid en Haar* 5, 169–178.

Birch, J.W. (1964) *The Isle of Man: a study in economic geography.* Cambridge University Press, Cambridge.

Biró, M. and Molnár, Zs. (1998) *Landscape types, their distribution, vegetation and land use history in the sand dunes of the Duna-Tisza köze (Kiskunság sensu lato) from the 18th century.* Történeti Földruyzi Füzetek, (in press).

Blanco, R., Clavero, J., Cuello, A., Marañón, T. and Seisdedos, J.A. (1991) *Sierras del Aljibe y del Campo de Gibraltar.* Diputación de Cádiz, Cádiz.

Blink, H. (1929) *Woeste gronden, ontginning en bebossching in Nederland voormaalsch en thans.* Mouton, 's-Gravenhage.

Boddy, L. (1994) Wood decomposition and the role of fungi: implications for woodland conservation and amenity tree management. In: Spencer, J. and Feest, A. (eds) *The rehabilitation of storm damaged woods.* University of Bristol, Bristol, pp. 7–30.

BOE (1989) *Ley 4/89, de 27 de marzo, de conservación de los espacios naturales y de la flora y fauna silvestre.* BOE, Madrid.

Bohnke, S.J.P. (1991) Palaeohydrological changes in the Netherlands during the last 13,000 years. PhD Thesis, Free University of Amsterdam.

BOJA (1989) *Ley 2/89, de 18 de julio, por la que se aprueba el inventario de Espacios Naturales Protegidos de Andalucía y se establecen medidas adicionales para su protección.* BOJA, Sevilla.

Bokdam, J. and Gleichman, J.M. (1989) De invloed van runderbegrazing op de ontwikkeling van Struikheide en Bochtige smele. *De Levende Natuur* 90, 7–14.

Bokdam, J. and Meurs, C.B.H. (1991) Is bijvoedering van heideschapen nodig? *De Levende Natuur* 92, 110–116.

Bokdam, J. and Wallis de Vries, M.F. (1992) Forage quality as a limiting factor for cattle grazing in isolated Dutch nature reserves. *Conservation Biology* 6, 399–408.

Borbás, V. (1886) *A magyar homokpuszták növényvilága, meg a homokkötés.* Pesti könynyomda-rèszvèny-tàrsasàg, Budapest.

Borhidi, A. (1995) Social behaviour types, naturalness and relative ecological indicator values of the Hungarian Flora. *Acta Botanica Hungarica* 39, 97–181.

Boros, Á. (1918–1934) *Utinapló* (Travel diary) – 1918, 1919, 1920, 1922, 1934. History of Science Collection of the Botanical Department of the Hungarian Natural Museum, Budapest.

Boros, Á. (1935) A nagykőrösi homoki erdők növényvilága. *Erdészeti Kísérletek* 37, 1–24.

Boros, Á. (1952) A Duna-Tisza köze növényföldrajza. *Földrajzi Értesítő* 1, 39–53.

Borrow, G. (1843) *The Bible in Spain* (many editions). London.

Bottema, S. (1980) Palynological investigations on Crete. *Review of Palaeobotany and Palynology* 31, 193–217.

Bottema, S. and Clason, A.T. (1979) *Het schaap in Nederland.* Thieme, Zutphen.

Bourlière, F. and Hadley, M. (1983) Present-day savannas: an overview. In: Bourlière, F. (ed.) *Tropical savannas – Ecosystems of the world 13.* Elsevier Scientific Publications, Amsterdam, pp. 1–7.

Bowles, G. [ie w.] (1783) *Introduzione alla storia naturale e alla geografica fisica di spagna* (trans. f. milizia), Parma.

Boycott, A.E. (1934) The habitats of land mollusca in Britain. *Journal of Ecology* 22, 1–138.

Bozkurt, Y. (1992) Odun Anatomisi. *Üniversite yayin no. 3652, Fakülte yayin no. 415.*

Bradshaw, R.H.W. (1981a) Modern pollen representation factors for woods in south-east England. *Journal of Ecology* 69, 45–70.

Bradshaw, R.H.W. (1981b) Quantitative reconstruction of woods in south-east England. *Journal of Ecology* 69, 941–955.

Bradshaw, R.H.W. and Browne, P. (1987) Changing patterns in the Post-glacial distribution of *Pinus sylvestris* in Ireland. *Journal of Biogeography* 14, 237–248.

Braun, M., Sümegi, P., Szűcs, L. and Szöőr, Gy. (1993) The history and development of the Nagy-Mohos fen at Kállósemjén: man induced fen formation and the 'archaic' fen concept. *Jósa Múzeum Évkönyve,* 33–34, 335–366.

Braun-Blanquet, J. and Tüxen, R. (1952) Irische Pflanzengesellschaften. In: Lüdi, W. (ed.) *Die Pflanzenwelt Irlands.* Verlag Hans Huber, Bern, pp. 224–420.

Broad, J. and Hoyle, R. (eds) (1997) *Bernwood – the life and afterlife of a forest.* University of Central Lancashire, Preston.

Brockmann-Jerosch, H. (1936) Futterlaubbaume und Speiselaubbaume. *Berichte der Schweizerischen Botanischen Gesellschaft* 46, 594–613.

Broderick, G. and Stowell, B. (1973) *Chronicle of the Kings of Mann and the Isles,* Recortys Reeaghyn Vannin as ny h Ellanyn. Learmonth Printers, Stirling.

Brongers, J.A. (1976) *Air photography and Celtic field research in the Netherlands.* ROB, Amersfoort.

Brown, A.H.F. and Warr, S.J. (1992) The effects of changing management on seed banks in ancient coppices. In: Buckley, G.P. (ed.) *Ecology and management of coppice woodlands.* Chapman & Hall, London, pp. 147–166.

Brunet, J. (1993) Environmental and historical factors limiting the distribution of rare forest grasses in southern Sweden. *Forest Ecology and Management* 61, 263–275.

Brunet, J. (1994) *Der Einfluß von Waldnutzung und Waldgeschichte auf die Vegetation südschwedischer Laubwälder NNA-Ber. 7. Jg., H. 3,* 96–101.

Buis, J. (1985) *Historia forestis, Nederlandse bosgeschiedenis. Deel 1. Bosgebruik, bosbeheer en boswetgeving tot het midden van de negentiende eeuw.* HES, Utrecht.

Bullock, H.A. (1816) *History of the Isle of Man, with a comparative view of the past and present state of society and manners; containing also biographical anecdotes of eminent persons connected with that island.* Longman, Hurst, Rees, Orme and Brown, London.

Bunce, R.G.H., Ryszkowski, L. and Paoletti, M.G. (1993) *Landscape ecology and agroecosystems.* Lewis Publishers, London.

Bunch, B. and Fell, C. (1949) A stone-axe factory at Pike of Stickle, Great Langdale, Westmorland. *Proceedings of Prehistorical Society* 15, 1–17.

Bürgi, M. (1997a) Waldentwicklung im 19. und 20. Jahrhundert. Veränderungen in der Nutzung und Bewirtschaftung des Waldes und seiner Eigenschaften als Habitat am Beispiel der öffentlichen Waldungen im Zürcher Unter- und Weinland. Diss. ETH Nr. 12'152.

Bürgi, M. (1997b) Benutzung und Bewirtschaftung der Wälder im 19. und 20. Jahrhundert – eine Fallstudie über das Zürcher Unter- und Weinland. *News of Forest History* 25/26, 119–130.

Calcote, R. (1995) Pollen source area and pollen productivity: evidence from forest hollows. *Journal of Ecology* 83, 591–602.

Campodonico, P.G. (1975) Ricerche fitosociologiche ed ecologiche sulla vegetazione ad *Alnus incana* (L.) Moench in val d'Aveto. Tesi di laurea in Scienze naturali, Universita' di Genova.

Casparie, W.A. (1972) Bog development in southeastern Drenthe, the Netherlands. *Vegetatio* 4, 1–272.

Casparie, W.A. and Groenman-Van Waateringe, W. (1980) Palynological analysis of Dutch barrows. *Palaeohistoria* 22, 7–65.

Castel, Y. (1991) Late Holocene eolian drift sands in Drenthe (the Netherlands). PhD Thesis. University of Amsterdam.

Cate, C.L. ten (1972) *Wan god mast gift ... Bilder aus der Geschichte der Schweinezucht im Walde.* Pudoc, Wageningen.

Cavanilles, A.J. (1795) *Observaciones sobre la historia natural, geografia, agricultura, poblacion y frutos del Reyno de Valencia.* Imprenta Real, Madrid.

Çepel, N. (1988) Orman Ekolojisi, İ.Ü. *Orman Fakültesi Yay. İ.Ü. Yay. No. 3518, Or. Fak. Yay. No: 399.*

Cevasco, R. (1995) Storia e Geografia della copertura vegetale nell'Appennino: l'alnocoltura (*Alnus* sp.) nelle alte valli Aveto e Trebbia (secoli XVIII-XX). Tesi di laurea in Geografia storica dell'Europa, Universita' di Genova.

Chatters, C. and Sanderson, N. (1994) Grazing lowland pasture woods. *British Wildlife* 6, 78–88.

Chiappori, A. (1876) *La Silvicultura in Liguria.* Tip. Sorolo-Muti, Genova.

Christensen, K. and Rasmussen, P. (1991) Styning af træer. *Eksperimentel Arkæologi, studier i teknologi og kultur* 1, 23–30.

Clapham, A.R., Tutin, T.G. and Moore, D.M. (1989) *Flora of the British Isles.* Cambridge University Press, Cambridge.

Clerkx, A.P.P.M., Dort, K.W. van, Hommel, P.W.F.M., Stortelder, A.H.F., Vrielink, J.G., de Waal, R.W. and Wolf, R.J.A.M. (1994) *Broekbossen van Nederland.* DLO Instituut voor Bos – en Natuuronderzoek/DLO Staring Centrum, Wageningen.

Clerkx, A.P.P.M., Broekmeyer, M.E.A., Scabo, P.J., Hees, A.F.M. van, Os, L.J. van and Koop, H.G.J.M. (1996) *Bosdynamiek in bosreservaat Galgenberg.* DLO Instituut voor Bos – en Natuuronderzoek, Wageningen.

Collingwood, R.G. (1933) An introduction to the prehistory of Cumberland, Westmorland and Lancashire, North-of-the-Sands. *Transactions of the Cumberland and Westmorland Antiquarian and Archaeological Society (N.S.)* 33, 163–200.

Collingwood, W.G. (1896) Furness, a thousand years ago. *Barrow Field Club Proceedings (N.S.)* 3, 36–44.

Collingwood, W.G. (1898) Reports on excavations at Springs Bloomery near Coniston Hall, Lancashire with notes on probable ages of Furness Bloomeries. *No 11 Transactions Cumberland and Westmorland Archaelogical and Antiquarian Society* 15, 223–228.

Collingwood, W.G. (1925) *Lake District history.* Wislon and Son, Kendal.

Collingwood, W.G. (1927) Ancient industries of the Crake Valley. *Barrow Field Club Proceedings Volume III, New Series,* 36–44.

Collins, E.J.T. (1985) Agriculture and conservation in England: an historical overview, 1880–1939. *Journal of the Royal Agricultural Society* 146, 38–46.

Colón, F. [1517–] *Descripción y cosmografia de España.* Patronato de Huérfanos de Administración Militar, 1910, Madrid.

Cooke, A.S. (1994) Colonisation by muntjac deer and their impact on vegetation. In: Massey, M. and Welch, R.C. (eds) (1994) *Monks Wood National Nature Reserve: the experience of forty years 1953–1993.* English Nature, Peterborough, pp. 45–61.

Cooke, A.S., Farrell, L., Kirby, K.J. and Thomas, R.C. (1995) Changes in abundance and size of dog's mercury, apparently associated with grazing in muntjac. *Deer* 9, 429–433.

Cooney, T. (1996) Vegetation changes associated with Late-Neolithic copper mining in

Killarney. In: Delaney, C. and Coxon, P. (eds) *Central Kerry. Field Guide No. 20.* Irish Association for Quaternary Studies, Dublin, pp. 28–32.

Countryside Commission (1994) *Countryside Character Programme.* John Dower House, Cheltenham.

Countryside Commission and English Nature (1997) *The character of England: landscape, wildlife and natural features.* English Nature, Peterborough.

Cowles, H.C. (1899) Dune floras of Lake Michigan. *Botanical Gazette* 27, 95–117, 167–202, 281–308, 361–391.

Cowper, H.S. (1899) *Hawkshead: its history, archaeology, industries, folklore, dialect etc.* Bernrose and Sons Ltd, London.

Creus, J., Fillat, F. and Gomez, D. (1984) El fresno de hoja ancha como arbol semi-salvaje en el Pirineo de Huesca (Aragon). *Acta Biologica Montana,* 445–454.

Croce, G.F. (1987) Effetti geografici della legislazione forestale in Liguria (XIX secolo). Tesi di laurea in Storia, Universita' di Genova

CSIC (1963) *Estudio agrobiológico de la provincia de Cádiz.* CSIC, Sevilla.

Cubbon, A.M. (1994) *Early maps of the Isle of Man, a guide to the collection in the Manx Museum.* Manx National Heritage, The Manx Museum and National Trust, Douglas.

Dackombe, R. (1990) Solid geology. In: *The Isle of Man, celebrating a sense of place.* Liverpool University Press, Liverpool.

Daniere, C., Capellano, A. and Moiroud, A. (1986) Dynamique de l'azote dans un peuplement naturel d'*Alnus incana* (L.) Moench. *Acta Oecologica* 7, 165–175.

Darby, H.C. (1956) The clearing of the woodland in Europe. In: Thomas, W.L. (ed.) *Man's role in changing the face of the earth.* University of Chicago Press, Chicago, pp. 183–216.

Davies, G. (1990) Agriculture, forestry and fishing in the Isle of Man. In: *The Isle of Man, celebrating a sense of place.* Liverpool University Press, Liverpool.

Dawkins, R.M. (1929) *The sanctuary of Artemis Orthisa at Sparta.* Macmillan, London.

Dawson, D. (1994) *Are habitat conduits for animals and plants in a fragmented landscape? A review of the scientific evidence.* English Nature (Research Report 94), Peterborough.

De Notaris, J. (1844) *Repertorium Florae Ligusticae.* Reg. Typogr. Taurini, Torino.

Didon, J., Durand-Delga, M. and Kornprobst, J. (1973) Homologies geologiques entre les deux rives du détroit de Gibraltar. *Bulletin Societé Geologique Francaise* 15, 77–105.

Dirkx, G.H.P. (1994) *Enschede-Noord; een historish-geografisch onderzoek voor landinrichting.* DLO-Staring Centrum (Rapport 32), Wageningen.

Dirkx, G.H.P. (1997) men sal van een erve ende goedt niet meer dan een trop schaepe holden…" *Historishe begrazing van gemeenschappelijke weidegronden in Gelderland en Overijssel.* DLO-Staring Centrum (Rapport 499),Wageningen.

Dirkx, G.H.P., Hommel, P.W.F.M. and Vervloet, J.A.J. (1992) Historische ecologie. Een overzicht van achtergronden en mogelijke toepassingen in Nederland. *Landschap* 9, 39–51.

Dodson, J.R., Mitchell, F.J.G., Bögeholz, H. and Julian, N. (1998) Dynamics of temperate rainforest from fine resolution pollen analysis, Upper Ringarooma River, north-eastern Tasmania. *Australian Journal of Ecology* (in press).

DOE and MAFF (1993) *Ecological consequences of land use change.* Department of the Environment. HMSO, London.

DOE and MAFF (1995) *Rural England: A nation committed to a living countryside.* Department of the Environment and Ministry of Agriculture, Fisheries and Food. HMSO, London.

Druce, G.C. (1930) *The flora of Northamptonshire.* T. Buncle & Co, Arbroath.

Du Rietz, G.E. (1932) Vegetationsforschung auf soziationsanalytischer Grundlage. *Handbuch biol. Arbeitsmethoden* 11, 293–480.

Duelli, P. (1994) *Rote Listen der gefährdeten Tierarten in der Schweiz.* BUWAL, Bern.

Dupont, L.M. (1985) Temperature and rainfall variation in a raised bog ecosystem. A palaeoecological and isotope-geological study. PhD Thesis. University of Amsterdam.

During, R. and Joosten, J.H.J. (1992) Referentiebeelden en duurzaamheid. Tijd voor beleid. *Landschap* 9, 285–295.

During, R. and Schreurs, W. (1995) *Historische ecologie.* Stichting Uitg. Kon. Ned. Natuurhist. Ver., Utrecht.

Dzwonko, Z. and Loster, S. (1989) Distribution of vascular plant species in small woodlands on the Western Carpathian foothills. *Oikos* 56, 77–86.

Ebdon, D. (1995) *Statistics in Geography,* 2nd edn. Blackwell Scientific Publishers, London.

Edlin, H.L. (1970) *Trees, woods and man,* 3rd edn. Collins, London.

Edwards, K.S. (1996) Are current rates of felling in ancient woodland sustainable at a landscape scale? MSc thesis, Oxford Forestry Institute, Oxford.

Egloff, F.G. (1991) Dauer und Wandel der Lägernflora. *Vierteljahrsschrift der Naturforschenden Gesellschaft in Zürich* 136, 207–270.

Elerie, J.N.H., Jager, S.W. and Spek, T. (1993) *Landschapsgeschiedenis van de Strubben-Kniphorstbos: archeologische en historisch-ecologische studies van een natuurgebied op de Hondsrug.* Stichting Historisch Onderzoek en Beleid/Regio-en landschapsstudies, Van Dijk and Foorthius REGIO-PRoject, Groningen.

Ellenberg, H. (1963) *Vegetation Mitteleuropas mit den Alpen.* Ulmer, Stuttgart.

Ellenberg, H. (1990) *Bauernhaus und Landschaft in ökologischer und historischer Sicht.* Eugen Ulmer, Stuttgart.

Ellenberg, H., Weber, H., Düll, R., Wirth, W., Werner, W. and Paulissen, D. (1991) Zeigerwerte von Pflanzen in Mitteleuropa. *Scripta Geobotanica* 18, 1–248.

Ellis, R.C. (1985) The relationships among eucalyptus forest, grassland and rainforest in a highland area in north-eastern Tasmania. *Australian Journal of Ecology* 10, 291–314.

Emanuelsson, U. and Bergendorff, C. (1990) Löväng, stubbskottäng, skottskog och surskog. *Bebyggelsehistorisk Tidskrift* 19, 109–115.

English Nature (1993) *Strategy for the 1990's – Natural Areas.* English Nature, Peterborough.

Eroskay, O. and Aytuğ, B. (1982) Doğu Ergene Çanağinin Petrifiye Ağaçlari, İ.Ü. *Orman Fakültesi Dergisi Seri A., Cilt 32, Sayi 2.*

Evans, M.N. and Barkham, J.P. (1992) Coppicing and natural disturbance in temperate woodlands – a review. In: Buckley, G.P. (ed.) *Ecology and management of coppice woodlands.* Chapman & Hall, London, pp. 79–98.

Falinski, J.B. (1986) *Vegetation dynamics in temperate lowland primeval forest. Ecological studies in Bialowieza forest.* Junk, Dordrecht.

Falinski, J.B. (1988) Succession, regeneration and fluctuations in the Bialowieza forest (NE Poland). *Vegetatio* 77, 115–128.

Fanta, J. (1982) *Natuurlijke verjonging van het bos op de droge zandgronden.* Rijksinstituut voor onderzoek in bos – en landschapsbouw De Dorschkamp, Wageningen.

Fekete, G. (1992) The holistic view of succession reconsidered. *Coenoses* 7, 21–30.

Fell, A. (1908) *The early iron industry of Furness and district.* Hume Kitchen, Ulverston.

Feltham, J. (1798) *A tour through the island of Man in 1797 and 1798*. Crutwell, Bath.

Fengel, D. (1991) Aging and fossilization of wood and its component. *Wood Science Technology* 25, 153–177.

Finke, L. (1994) *Landschaftsökologie*. Das Geographische Seminar – 2. Aufl., Braunschweig.

Fischer, W. (1963) *Flora der Prignitz*. Heimatmuseum des Kreises Pritzwalk, Pritzwalk, 135 pp.

Flower, N. (1980) The management history and structure of unenclosed woods in the New Forest, Hampshire. *Journal of Biogeography* 7, 311–328.

Forestry Authority (1993) *Woodland Grant Scheme*. April 1993. Forestry Authority, Corstorphine Road, Edinburgh.

Forestry Authority and Countryside Commission (1996) *Woodland creation: needs and opportunities in the English countryside. A discussion paper*. Forestry Authority and Countryside Commission, Northampton.

Forestry Commission (1985a) *The policy for broadleaved woodlands*. Forestry Commission Policy and Procedure Paper No. 5, Edinburgh.

Forestry Commission (1985b) *Guidelines for the management of broadleaved woodland*. Forestry Commission, Edinburgh.

Forestry Commission (1994) *The management of semi-natural woodlands*. Forestry Practice Guide Nos 1–8. Forestry Commission, Edinburgh.

Forman, R.T.T. (1995) *Land mosaics: the ecology of landscapes and regions*. Cambridge University Press, Cambridge.

Foster, D.R. (1992) Land use history (1730–1990) and vegetation dynamics in central New England, USA. *Journal of Ecology* 80, 753–771.

Franks, J.W. and Pennington, W. (1961) The late-glacial and post-glacial deposits of the Esthwaite Basin, North Lancashire. *New Phytologist* 60, 27–42.

Frisnyák, S. (1990) Magyarország történeti földrajza. Tankönyvkiadó, Budapest.

Fritzsche, B., Lemmenmeier, M., König, M., Kurz, D. and Sutter, E. (1994) *Geschichte des Kantons Zürich*. Band 3, 19. und 20. Jahrhundert. Werd Verlag, Zürich, pp. 519.

Fuller, R.J. and Warren, M.S. (1990) *Coppiced woodlands: their management for wildlife*. Nature Conservancy Council, Peterborough.

Fuller, R.J., Gough, S.J. and Marchant, J.H. (1995) Bird populations in new lowland woods: Landscape design and management perspectives. In: Ferris-Kaan, R. (ed.) *The ecology of woodland creation*. Wiley and Sons, London.

Galgóczy, K. (1896) *Nagykőrös város monográfiája*. Nagykőrös.

Gandolfo, G. (1994) Indagine storico-biologica sull'insediamento della vegetazione forestale in consequenza della cessata coltivazione delle pendici terrazzate nel Ponente ligure (Valle Arroscia, Imperia). Tesi di laurea, Università degli studi di Firenze.

Garrad, L.S. (1972a) *The naturalist in the Isle of Man*. David & Charles, Newton Abbot.

Garrad, L.S. (1972b) Oak woodland in the Isle of Man. *Watsonia* 9, 59–60.

Garrad, L.S., Bawden, T.A., Qualtrough, J.K. and Scatchard, W.J. (1972) *The industrial archaeology of the Isle of Man*. David & Charles, Newton Abbot.

Geel, B. van and Groenman-van Waateringe, W. (1987) Palynological investigations. In: Groenman-van Waateringe, W. and van Wijngaarden-Bakker, L.H. (eds) *Farmlife in a Carolingian village*. Van Gorcum, Assen/Maastricht, pp. 6–30.

Gellert, J.H. (1954) *Bemerkungen zur Karte der physisch-geographischen Gliederung der Deutschen Demokratischen Republik im Maßstab 1: 1,000,000*. Petermanns Geographische Mitt., Jg. 1954, H. 1.

Gentile, S. (1982) *Note illustrative della carta della vegetazione dell'alta Val d'Aveto*

(Appennino ligure). CNR Collana del Programma finalizzato 'Promozione della Qualita' dell'ambiente'. Aq/1/123. Roma.

Gilbert, J.M. (1959) Forest succession in the Florentine Valley, Tasmania. *Papers and Proceedings of the Royal Society of Tasmania* 93, 129–205.

Gilgen, R. (1994) Pflanzensoziologisch-ökologische Untersuchungen an Schlagfluren im schweizerischen Mittelland über Würmmoränen. *Veröff. Geobot. Inst. Eidgenöss. Tech. Hochsch.* 116, 127.

Goblet d'Alviella (1927–1930) *Histoire des bois et forêts de Belgique*. Lechevalier P., Paris and Lamertin M., 4 parts, Brussels.

Göçmen, K. (1976) Asagi Meriç Vadisi Taskin Ovasi ve Deltasinin Alüviyal Jeomorfolojisi, İst. Ün. Yay. 1990, Cografya Enstitüsü Yayini 80.

Godwin, H. (1975) *History of the British flora*, 2nd edn. Cambridge University Press, Cambridge.

Gomez, D. and Fillat, F. (1981) La cultura ganadera del fresno. *Revista Pastos* 11, 295–302.

Gomez, D. and Fillat, F. (1984) Utilisation du Frêne comme arbre fourrager dans les Pyrénées de Huesca. *Écologie de Milieux Montagnards et de Haute Altitude. Documents d'Écologie Pryénéenne* 3–4, 481–489.

González, A., Torres, E., Montero, G. and Vázquez, S. (1996) Resultados de cien años de aplicación de la selvicultura y la ordenación en los montes alcornocales de Cortes de la Frontera (Málaga), 1890–1990. *Montes* 43, 12–22.

Goodfellow, S. and Peterken, G.F. (1981) A method for survey and assessment of woodlands for nature conservation using maps and species lists: the example of Norfolk woodlands. *Biological Conservation* 21, 177–195.

Green, E.E. (1995) Creating decaying trees. *British Wildlife* 6, 310–311.

Greenwood, E. (1818) *Map of the county of Lancashire*. Lancashire Record Office, Preston.

Groenman-van Waateringe, W. and van Wijngaarden-Bakker, L.H. (1990) Medieval archaeology and environmental research in the Netherlands. In: Besteman, J.C., Bos, J.M. and Heidinga, H.A. (eds) *Medieval archeology in the Netherlands*. Van Gorcum, Assen/Maastricht, pp. 283–297.

Grøntved, P. (1936) Om floraen i nogle nordvestsjællandske skove. *Bot. Tidsskr* 43, 325–356.

Grove, A.T. and Rackham, O. (1998) *Ecological history of Southern Europe* [provisional title]. Yale University Press (in press).

Guidoni, E. and Marino, A. (1972) *Territorio e città della Valdichiana*. Roma.

Gulliver, R. (1995) Woodland history and plant indicator species in North-East Yorkshire, England. In: Butlin, R.A. and Roberts, N. (eds) *Ecological relations in historic times: human impact and adaptation*. Blackwell, Oxford, pp. 169–189.

Gutiérrez, A., Nebot, M. and Díez, M.V. (1996) Introducción al estudio polínico de sedimentos del Parque Natural de "Los Alcornocales". *Almoraima* 15, 87–92.

Haas, J.N. and Schweingruber, F.H. (1993). Wood anatomical evidence of pollarding in ash stems from the Valais, Switzerland. *Dendrochronologia* 11, 35–43.

Hæggström, C.-A. (1983) Vegetation and soil of the wooded meadows in Nåtö, Åland. *Acta Botanica Fennica* 120, 1–66.

Hæggström, C.-A. (1987) Löväng. In: Emanuelsson, U. and Johansson, C.E. (eds) *Biotopvern i Norden. Biotoper i det nordiska kulturlandskapet*. Nordiska ministerrådet, Miljörapport 6, pp. 69–88.

Hæggström, C.-A. (1988) Protection of wooded meadows in Åland – problems, methods and perspectives. *Oulanka Reports* 8, 88–95.

Hæggström, C.-A. (1992a) Wooded meadows and the use of deciduous trees for fodder, fuel, carpentry and building purposes. *Protoindustries et histoire des forêts. Gdr Isard-Cnrs. Les Cahiers de l'Isard* 3, 151–162.

Hæggström, C.-A. (1992b) Skottskogar och skottskogsbruk. *Nordenskiöld-samfundets tidskrift* 51, 81–112.

Hæggström, C.-A. (1995) Lövängar i Norden och Balticum. *Nordenskiöld-samfundets tidskrift* 54, 21–58.

Hammond, P.M. and Harding, P.T. (1991). Saproxylic invertebrate assemblages in British woodlands: their conservation significance and its evaluation. In: Read, H.J. (ed.) *Pollard and veteran tree management.* Corporation of London, Burnham Beeches, pp. 30–37.

Hanák, P. (ed.) (1991) *The Corvina History of Hungary – From the earliest times until the present day.* Corvina Books, Budapest.

Hansen, K. (ed.) (1981) *Dansk feltflora.* Gyldendal, Copenhagen.

Hardin, G. (1968) The tragedy of the commons. *Science* 162, 1243–1248.

Harding, P.T. and Alexander, K.N.A. (1994) The use of saproxylic invertebrates in the selection and evaluation of areas of relic forest in pasture-woodlands. *British Journal of Entomology and Natural History* 7 (Suppl. 1), 21–26.

Harding, P.T. and Rose, F. (1986) *Pasture-woodlands in lowland Britain: a review of their importance for wildlife conservation.* Institute of Terrestrial Ecology (NERC), Huntingdon.

Hargitai, Z. (1937) Nagykőrös növényvilága. I. A flóra. *Debreceni Református Kollégium Tanárképző Intézet Dolgozatai* 17, 1–55.

Hargitai, Z. (1940) Nagykőrös növényvilága. II. A homoki növényszövetkezetek. (Vegetation of Nagykőrös II. The plant communities on sand). *Botanikai Közlemények* 37, 205–240.

Havinga, A.J. (1984) Pollen analysis op podzols. In: Buurman, P. (ed.) *Podzols.* Van Nostrand Reinhold Soil Science Series 3, New York, pp. 313–323.

Henderson, N.R. and Long, J.N. (1984) A comparison of stand structure and fire history in two black oak woodlands in north-western Indiana. *Botanical Gazette* 145, 222–228.

Heringa, J. (1982) *De buurschap en haar marke.* Drentse Historische Studien 5, Assen.

Heringa, J. (1985) Lijnen en stippelijnen in de geschiedenis van de buurschap. *Nieuwe Drentse Volksalmanak* 102, 69–93.

Hermy, M., van den Bremt, P. and Tack, G. (1993) Effects of site history on woodland vegetation. In: Broekmeyer, M.E.A., Vos, W. and Koop, H. (eds) *European forest reserves.* Pudoc, Wageningen, pp. 219–232.

Hermy, M., Honnay, O., Firbank, L., Grashof-Bokdam, C.J. and Lawesson, J. (1998) An ecological comparison between ancient and other forest plant species of Europe, and the implications for forest conservation. *Conservation Biology* (in press).

Hervouet, J.P. (1980) *Du Faidherbia à la brousse. Modifications culturales et dégradation sanitaire.* Orstom, Ouagadougou.

Hesmer, H. (1938) *Die heutige Bewaldung Deutschlands.* Paul Parey, Berlin.

HMSO (1994) *Sustainable forestry – the UK programme.* Her Majesty's Stationery Office, London.

Hofmann, G. (1994) *Der Wald. Sonderheft Waldökosystem-Katalog.* Deutscher Landwirtschaftsverlag, Berlin.

Hollós, L. (1896) Kecskemét növényzete. In: Bagi, L. (ed.) *Kecskemét múltja és jelene.* Tóth L. Nyomdája, Kecskemét, pp. 77–147.

Hopkins, J.J. (1996) Scrub ecology and conservation. *British Wildlife* 8, 28–36.

Horváth, F., Dobolyi, K., Morschhauser, T., Lőkös, L., Karas, L. and Szerdahelyi, T. (1995) *Flóra adatbázis 1.2, Taxonlista és attributumállomány.* Flora Workgroup,

Institute of Ecology and Botany, Botanical Department of the Hungarian Natural Museum, Vácrátót.

Huntington, E. (1915) *Civilisation and climate*, Yale University Press.

Isaac, D. and Reid, C.M. (1996) *Amendments to the Ancient Woodland Inventory for England July 1994–February 1997*. English Nature (Research Report 177), Peterborough.

Iversen, J. (1949) The influence of prehistoric man on the vegetation. *Danmarks Geologiske Undersogelse, Series (IV)* 3, 1–25.

Iversen, J. (1973) *The development of Denmark's nature since the last Glacial.* Geology of Denmark III. Danmarks Geologiske Undersogelse V. Raekke 7c.

JA (1995) *Plan de Medio Ambiente de Andalucía.* Junta de Andalucía, Sevilla.

Jackson, S.T. and Wong, A. (1994) Using forest patchiness to determine pollen source areas of closed-canopy assemblages. *Journal of Ecology* 82, 89–99.

Jackson, S.T., Futyma, R.P. and Wilcox, D.A. (1988) A palaeoecological test of a classical hydrosere in the Lake Michigan dunes. *Ecology* 69, 928–936.

Jackson, W.D. (1968) Fire, air, water and earth – an elemental ecology of Tasmania. *Proceedings of the Ecological Society of Australia* 3, 9–16.

Jacobson, G.L. and Bradshaw, R.H.W. (1981) The selection of sites for paleovegetational studies. *Quaternary Research* 16, 80–96.

Jagt, J.L. van der (1976) Boswachterij Amerongse berg. *Nederlands Bosbouwtijdschrift* 48, 218–224.

Jaquiot, C., Trenord, Y. and Dýrol, D. (1973) *Atlas d'Anatomie des Bois des Angiospermes.* Centre Technique du Bois.

Járai-Komlódi, M. (1987) Postglacial climate and vegetation in Hungary. In: Pécsi, M. and Kordos, L. (eds) *Holocene Environment in Hungary.* Geographic Research Institute, Budapest, pp. 37–47.

Jenny, H. (1941) *Factors of soil formation.* McGraw-Hill, New York/London.

Jenny, H. (1980) *The soil resource. Origin and behaviour.* Springer, New York.

Jones, M. and Warburton, D.R. (1993) *Sheffield's woodland heritage*, 2nd edn. Green Tree Publications, Rotherham.

Jongerius, A. and Heintzberger, G. (1975) *Methods in soil micromorphology. A technique for the preparation of large thin sections.* Soil Survey Papers, No 10. Netherlands Soil Survey Institute, Wageningen.

Kalb, J. (1984) *Van stuifzand naar woudreus; bosontwikkeling in Drenthe.* Staatsbosbeheer, Assen.

Kalkhoven, J.T.R. (1993) Survival of populations and the scale of the fragmented agricultural landscape. In: Bunce, R.G.H., Ryszkowski, L. and Paoletti, M.G. (eds) *Landscape ecology and agroecosystems.* Lewis Publishers, London, pp. 83–90.

Kay, S. (1993) Factors affecting severity of deer browsing damage within coppiced woodlands in the south of England. *Biological Conservation* 63, 217–222.

Kelly, D.L. (1981) The native forest vegetation of Killarney, south-west Ireland: an ecological account. *Journal of Ecology* 69, 437–472.

Kelly, D.L. (1995) A short guide to Irish oaks. *Moorea* 11, 24–28.

Kennedy, C.E.J. and Southwood, T.R.E. (1984) The number of insects associated with British trees: a re-analysis. *Journal of Animal Ecology* 53, 455–478.

Kerber, D., Reikat, A., Specking, I. and Sturm, H.J. (1996) Les terroirs et la végétation. Paradigmes d'exploitation du sol chez les Mosi et les Bisa dans la province de Bougou. In: *Berichte des Sonderforschungsbereich* 268 Band 7. Frankfurt am Main, pp. 83–91.

Kerney, M. and Stubbs, A. (1980) *The conservation of snails, slugs and freshwater mussels.* Nature Conservancy Council, Shrewsbury.

Kettner, A.J. (1997) Veranderingen in het bosareaal in Drenthe 1970–1993; Een GIS-studie naar veranderingen in het bodemgebruik met behulp van ARC-INFO. MSc thesis. Wageningen Agricultural University, Wageningen.

Kirby, K.J. (1992) *Woodland and wildlife.* Whittet, London.

Kirby, K.J. (1995) *Rebuilding the English countryside: habitat fragmentation and wildlife corridors as issues in practical conservation.* English Nature (English Nature Science 7), Peterborough.

Kirby, K.J. and Drake, C.M. (1993) *Dead wood matters: the ecology and conservation of saproxylic invertebrates in Britain.* English Nature (Science Report 7), Peterborough.

Kirby, K.J., Peterken, G.F., Spencer, J.W. and Walker, G.J. (1984) *Inventories of ancient semi-natural woodland.* Nature Conservancy Council. (Focus on nature conservation 6), Peterborough.

Kirby, K.J., Mitchell, F.J. and Hester, A.J. (1994) A role for large herbivores (deer and domestic stock) in nature conservation management in British semi-natural woods. *Arboricultural Journal* 18, 381–399.

Kirby, K.J., Thomas, R.C., Key, R.S., McLean, I.F.G. and Hodgetts, N.G. (1995) Pasture-woodland and its conservation in Britain. *Biological Journal of the Linnaean Society* 56 (suppl.), 135–153.

Kirby, K.J., Thomas, R.C. and Dawkins, H.C. (1996) Monitoring woodland changes in tree and shrub layers in Wytham Woods (Oxfordshire) 1974–1991. *Forestry* 69 319–334.

Klein, J. (1920) *The Mesta: a study in Spanish economic history 1273–1836.* Harvard University Press, Cambridge, Massachusetts.

Klingen, L.A.S. and Litjens, G.J.J.M. (1985) *Beheersplan Amerongse Bos 1985–1995.* De Bilt: Stichting Het Utrechts Landschap.

Knoppersen, G. (1995) *Algemene informatie van het bosreservaat 3, Amerongen Galgenberg.* IKC-natuurbeheer, Wageningen.

Koeman (1970) Een Franse topografische kaart van Drenthe uit de jaren 1811–1813. *Nieuwe Drentse Volksalmanak* 88, 89–101.

Koop, H. (1989) *Forest dynamics (SILVI-STAR: A comprehensive monitoring system).* Springer-Verlag, Berlin.

Koop, H.G.J.M. and Clerkx, A.P.P.M. (1995) *De vegetatie van bosreservaten in Nederland. Deel 3. Bosreservaat Galgenberg, Amerongen.* DLO Instituut voor Bos- en Natuuronderzoek, Wageningen.

Koster, E.A. (1978) *De stuifzanden van de Veluwe; een fysisch-geografische studie.* Publicaties van het Fysisch geografish en Bodemkundig Laboratorium van de Universiteit van Amsterdam nr. 27.

Kroon, H. de (1986) De vegetaties van Zuidlimburgse hellingbossen in relatie tot het hakhoutbeheer. Een rijke flora met een onzekere toekomst. *Natuurhistorisch Maandblad* 75, 167–192.

Krumbiegel, D. and Schwinge, W. (1991) *Witterung – Klima, Datenzusammenstellung für Mecklenburg-Vorpommern, Brandenburg und Berlin.* Potsdam.

Kuhn, N. (1990) Veränderungen von Waldstandorten. *Ber. Eidgenöss. Forsch.anst. Wald Schnee Landsch.* 319, 47.

Kuile, G.J. ter (1908) *Geschiedenis van den hof Espelo, zijne eigenaren en bewoners.* Tijl, Zwolle.

Kull, K. and Zobel, M. (1991) High species richness in an Estonian wooded meadow. *Journal of Vegetation Science* 2, 711–714.

Künzel, R.E., Blok, D.P. and Verhoeff, J.M. (1988) *Lexicon van Nederlandse toponiemen tot 1200.* Publicaties van het P.J. Meertens-Instituut, Amsterdam.

Laar, J.N. van and De Vries, I.G. (1985) Heidebebossingen na 1900 in Drenthe en Overijssel. MSc thesis, Landbouwhogeschool, Wageningen.

Lachaux, M., de Bonneval, L. and Delabraze, P. (1987). Pratiques anciennes et perspectives d'utilisation fourragere des arbres. *Fourragere* (special issue on *L'animal, les friches et la foret: la foret et l'elevage en region mediterraneenne francaise*), 83–104.

Lake District Special Planning Board (1978) *Lake District National Park Plan.* Kendal.

Landesamt für Datenverarbeitung und Statistik Brandenburg (Hrsg.) (1996) *Statistische Berichte. Bodennutzung im Land Brandenburg 1995.* Potsdam.

Landolt, E. (1991) *Rote Liste. Gefährdung der Farn- und Blütenpflanzen in der Schweiz.* BUWAL, Bern, 185 pp.

Latham, R.E. (1965) *Revised medieval Latin word-list.* Oxford University Press, London.

Leuschner, Chr. and Immenroth, J. (1994) Landschaftsveränderungen in der Lüneburger Heide 1770–1985. Dokumentation und Bilanzierung auf der Grundlage historischer Katen. *Arch. für Natursch. u. Landschaftspfl.* 33, 85–139.

Lier, C.S.J. van and Tonkens, J. (eds) (1975) (Facsimile) *Hedendaagse Historie of Tegenwoordige Staat van het Landschap Drenthe.* (First print 1792–1795), B.V. Foresta, Groningen.

Link, H.F. (1801) *Travels in Portugal, and through France and Spain.* Trans. J. Hinckley. Longman, London.

Little, D.J., Mitchell, F.J.G., von Engelbrechten, S.S. and Farrell, E.P. (1996) Assessment of the impact of past disturbance and prehistoric *Pinus sylvestris* on vegetation dynamics and soil development in Uragh Wood, S.W. Ireland. *The Holocene* 6, 90–99.

López Ontiveros, A., Valle Buenestado, B. and García Verdugo, F.R. (1991) *Caza y paisaje geográfico en las tierras béticas según el Libro de la Montería.* Junta de Andalucía, Córdoba.

Ludwig, J.A. and Reynolds, J.F. (1988) *Statistical ecology.* John Wiley and Sons, New York.

Maccà, G. (1815) *Storia del territorio vicentino.* Anastatic edition. Libreria Alpina Degli Espositi, Bologna.

Majer, A. (1988) *Fenyves a Bakonyalján.* Akadémia; Kiadó, Budapest.

Mantel, K. (1990) *Wald und Forst in der Geschichte: ein Lehr- und Handbuch.* Alfeld-M. u. H. Schaper, Hanover.

Marañón, T. (1988) Agro-sylvo-pastoral systems in the Iberian Peninsula: *Dehesas* and *Montados. Rangelands* 10, 255–258.

Marren, P. (1992) *The wild woods.* David & Charles, Newton Abbot.

Marshall, J.D. (1958) *Furness and the Industrial Revolution: an economic history of Furness (1711–1900) and the town of Barrow (1757–1897) with an epilogue.* James Milner, Barrow-in-Furness.

Marshall, J.D. and Davies-Shiel, M. (1969) *The industrial archaeology of the Lake Counties.* David & Charles, Newton Abbot.

Marshall, W. (1788) *The rural economy of Yorkshire.* Cadgewell, London.

Mattheck, C. and Breloer, H. (1994) *The body language of trees. A handbook for failure analysis.* HMSO, London.

McCarroll, D. (1990) The Quaternary Ice Age in the Isle of Man, an historical perspective. In: *The Isle of Man, celebrating a sense of place.* Liverpool University Press, Liverpool.

McCarroll, D., Garrad, L.S. and Dackombe, R. (1990) Lateglacial and Postglacial environment history. In: *The Isle of Man, celebrating a sense of place.* Liverpool University Press, Liverpool.

McDonnell, J. (1992) Pressures on Yorkshire woodland in the later middle ages. *Northern History* 28, 110–125.

McNeill, J.R. (1992) *The mountains of the Mediterranean World. An environmental history.* Cambridge University Press, New York.

Meeuwissen, T.W.M. (1991) Basisdocumenten *Bosvisie Noord-Nederland; mogelijkheden, ontwikkelingen, kansen.* Ministerie van Landbouw, Natuurbeheer en Visserij.

Mensema, A.J. (1978) *Inventaris van de archieven van de marken in de provincie Overijssel 1300–1942.* Rijksarchief Overijssel, Zwolle.

Metaille, J.-P. (1992) *Protoindustries et histoire des forests.* Les Cahiers de L'Isard, Toulouse.

Meyer, D. and Debrot, S. (1989) Insel-Biogeographie und Artenschutz in Wäldern. *Schweiz. Z. Forstwes.* 140, 977–985.

Miles, J. (1981) *Effect of birch on moorlands.* Institute of Terrestrial Ecology, Cambridge.

Millward, R. and Robinson, A. (1974) *The Lake District,* 2nd edn. Eyre and Methuen, London.

Ministerium für Ernährung, Landwirtschaft und Forsten (Hrsg.) (1994) *Wald und Forstwirtschaft im Land Brandenburg.* UNZE-Verlagsgesellschaft, Potsdam.

Ministerium für Umwelt, Naturschutz und Raumordnung (Hrsg.) (1993) Rote Liste. Gefährdete Farn- und Blütenpflanzen, Algen und Pilze im Land Brandenburg. UNZE-Verlagscgesellschaft, Potsdam.

Ministerium für Umwelt, Naturschutz und Raumordnung (Hrsg.) (1995) *Landschaftsprogramme Brandenburg. Erläuterungen.* Selbstverlag, Potsdam.

Mitchell, G.F. (1958) A Lateglacial deposit near Ballaugh, Isle of Man. *New Phytologist* 57, 256–263.

Mitchell, F.J.G. (1988) The vegetational history of the Killarney Oakwoods, SW Ireland: evidence from fine spatial resolution pollen analysis. *Journal of Ecology* 76, 415–436.

Mitchell, F.J.G. and Cole, E. (1998) Reconstruction of long-term successional dynamics in Bialowieza Forest, Poland. *Journal of Ecology* (in press).

Mitchell, F.J.G. and Kirby, K.J. (1990) The impact of large herbivores on the conservation of semi-natural woods in the British uplands. *Forestry* 63, 333–353.

Mitchell, F.J.G., Bradshaw, R.H.W., Hannon, G.E., O'Connell, M. Pilcher, J.R. and Watts, W.A. (1996) Ireland. In: Berglund, B.E., Birks, H.J.B., Ralska-Jasiewiczowa, M. and Wright, H.E. (eds) *Palaeoecological events during the last 15,000 years.* John Wiley, Chichester, pp. 1–13.

Mitchell, P.L. (1989) Repollarding large neglected pollards: a review of current practice and results. *Arboricultural Journal* 13, 125–142.

Mitchell, P.L. and Kirby, K.J. (1989) *Ecological effects of forestry practices in long-established woodland and their implications for nature conservation.* Oxford Forestry Institute (Occasional Paper 39), Oxford.

Molnár, Zs. (1966) A Pitvarosi-puszták vegetáció- és a tájtörténete az Árpád-kortól napjainkig. *Natura Bekesiensis* 21, 65–97.

Molnár, Zs. (1997) The land use historical approach to study vegetation history at the century scale. In: Tóth, E. and Horváth, R. (eds) *Research, Conservation, Management Conference.* Agglelek, pp. 345–354.

Monk, M.A. (1993) People and environment: in search of the farmers. In: Twohig, E.S. and Ronayne, M. (eds) *Past perceptions. The prehistoric archaeology of south west Ireland.* Cork University Press, Cork, pp. 35–52.

Montanari, C. (1974) Flora e vegetazione del greto del T.Aveto da Farfanosa alla confluenza con il F. Trebbia. Tesi di laurea in Scienze naturali, Universita' di Genova.

Moreno, D. (1984) The agricultural uses of tree-land in the North-Western Apennines since the middle ages. *Supplement du Journal Forestier Suisse* 74, 77–88.

Moreno, D. (1990) *Dal documento al terreno. Storia e archeologia dei sistemi agro-silvo pastorali.* Il Mulino, Bologna.

Moreno, D. (1995) Une source pour l'histoire et l'archeologie des ressources vegetales, les cartes topographiques de la montagne ligure (Italie). In: Bousquet and Bressolier (ed.) *L'oeil du cartographe et la representation geographique du Moyen Age a nos jours.* CTHS, Paris, pp. 175–198.

Moreno, D. and Poggi, G. (1996) Storia delle risorse boschive nelle montagne mediterranee: modelli di interpretazione per le produzioni non legnose in regime consuetudinario. In: Cavaciocchi, S. (ed.) *L'Uomo e la Foresta: secc. XIII–XVIII.* Le Monnier, Prato, pp. 635–654.

Moreno, D. and Raggio, O. (1990) The making and fall of an intensive pastoral land-use-system. Eastern Liguria, 16–19th centuries. *Rivista di Studi Liguri A56*, 193–217.

Morschhauser, T. (1990) A Remete-szurdok flórája, vegetációja és degradáltsági állapotának felmérése. MSc Thesis, Eötvös Lóránd University, Budapest.

Mount, A.B. (1979) Natural regeneration processes in Tasmanian forests. *Search* 10, 180–186.

National Trust (1982) *Biological survey, Lake District woodlands. Coniston (ii) summary and species list.* National Trust, Cirencester.

Nemcsik, J. (1861) *A nagykőrösi erdő, annak kezelése és haszna.* Ballagi Nagykőrösi naptár, Nagykőrös.

Nisbet, J. (1906) Forestry. In: Serjeantson, R.M. and Adkins, W.R.D. (eds) *The Victoria History of the counties of England – Northamptonshire II.* Constable, Westminster, pp. 341–352.

Norušis, M.J. (1993) *SPSS for Windows.* Release 6.0. Chicago, USA.

O'Connell, M. (1990) Origins of Irish lowland blanket bog. In: Doyle, G.J. (ed.) *Ecology and conservation of Irish peatlands.* Royal Irish Academy, Dublin, pp. 49–71.

Ogden, J. and Powell, J.A. (1979) A quantitative description of the forest vegetation on an altitudinal gradient in the Mt Field National Park, Tasmania, and a discussion of its history and dynamics. *Australian Journal of Ecology* 4, 293–325.

Ojeda, J.F. (1989) El bosque andaluz y su gestión a través de la Historia. In: García, G.C. (ed.) *Geografía de Andalucía.* Tartessos, Sevilla, pp. 315–355.

Ojeda, F., Arroyo, J. and Marañón, T. (1994) Biodiversity components and conservation of Mediterranean heathlands in Southern Spain. *Biological Conservation* 72, 61–72.

Ojeda, F., Marañón, T. and Arroyo, J. (1996) Patterns of ecological, chorological and taxonomic diversity at both sides of the Strait of Gibraltar. *Journal of Vegetation Science* 7, 63–72.

Oldfield, F. and Statham, D.C. (1963) Pollen analytical data from Urswick Tarn and Ellerside Moss, North Lancashire. *New Phytologist* 62, 53–66.

Olson, J.S. (1958) Rates of succession and soils changes on southern Lake Michigan sand dunes. *Botanical Gazette* 119, 125–169.

O'Neill, R.V., Krummel, J.R., Gardiner, R.H., Sigihara, G., Jackson, B., De Angelis, D.L., Milne, B.T., Turner, M.G., Zygmunt, B., Christensen, S.W., Dale, V.H. and Graham R.L. (1988) Indices of landscape patterns. *Landscape Ecology* 1, 153–162.

Oosterhuis, J.W. and Rademaker, B.F.M. (1988) *Historisch onderzoek naar het beheer van bossen en natuurterreinen. Deelproject 11.1: 19e-eeuwse droge heidebebossing Boswachterij de Amerongse Berg.* Staatsbosbeheer, Utrecht.

Ordnance Survey (1974) *1:25000 outdoor leisure map. The English Lakes, south-west sheet.* Ordnance Survey, Southampton.

Otto, H.J. (1994) *Waldökologie.* Ulmer, Stuttgart.

Ouden, J.B. den (1997) *A-locatiebossen in Drenthe.* DLO Institute for Forestry and Nature Research, Wageningen.

Ouden, J.B. den and Roosenschoon, O.R. (1994) Van Meerbosch tot Oosterbos; historisch onderzoek in de Boswachterij Emmen. MSc thesis, Wageningen Agricultural University, Wageningen.

Ouedraogo, S.J. (1995) *Les parcs agroforestiers au Burkina Faso.* Afrena Report No. 79, ICRAF, Nairobi.

Owen, J.A. (1989) An emergence trap for insects breeding in dead wood. *British Journal of Entomology and Natural History* 2, 65–67.

Owen, J.A. (1992) Experience with an emergence trap for insects breeding in dead wood. *British Journal of Entomology and Natural History* 5, 17–20.

Pálfai, I. (1994) Összefoglaló tanulmány a Duna-Tisza közi talajvízszint-süllyedés okairól és a vízhiányos helyzet javításának lehetőségeiről. In: Pálfai, I. (ed.) *A Duna-Tisza közi Hátság vízgazdálkodási problémái.* Nagyalföld Alapítvány, Budapest, pp. 111–123.

Paola, G. and Ciciliot, F. (1998) Examples of woodland management related to timber supply for shipbuilding in the 18th century in western Liguria (NW Italy). In: Watkins, C. (ed.) *European Woods and Forests: Studies in Cultural History.* CAB International, Wallingford, pp. 157–163.

Parra, F. (1990) *La dehesa y el olivar.* Debate Ediciones del Prado, Madrid.

Passarge, H. (1966) Waldgesellschaften der Prignitz. *Archiv f. Forstwesen* 15, 475–504.

Pearsall, W.H. (1969) *The Lake District. National Park Guide No. 6.* HMSO, London.

Pearsall, W.H. and Pennington, W. (1947) Ecological history of the English Lake District. *Journal of Ecology* 34, 137–148.

Pearsall, W.H. and Pennington, W. (1973) *The Lake District, a landscape history.* Collins, London.

Pécsi, M. (ed.) (1989) *National atlas of Hungary.* Kartográfiai Vállalat, Budapest.

Pélissier, P. (1980) L'arbre dans les paysages agraires de l' Afrique tropicale. Cah. *Orstom* 17, 131–136.

Pennington, W. (1964) Pollen analyses from six upland tarns in the Lake District. *Philosophical Transactions of the Royal Society* B 248, 204–244.

Pennington, W. (1965) The interpretation of some post-glacial diversities at different Lake District sites. *Proceedings of the Royal Society* B 161, 310–323.

Pennington, W. (1970) Vegetation history in the north-west of England: a regional synthesis. In: Walker, D. and West, R.G. (eds) *Studies in the vegetational history of the British Isles.* Cambridge University Press, Cambridge, pp. 47–49.

Pennington, W. (1983) The vegetation history of the Coniston Basin. Unpublished, University of Leicester, Leicester.

Peterken, G.F. (1969) Development of vegetation in Staverton Park, Suffolk. *Field Studies* 3, 1–39.

Peterken, G.F. (1974) A method for assessing woodland flora for conservation using indicator species. *Biological Conservation* 6, 239–245.

Peterken, G.F. (1976) Long-term changes in the woodlands of Rockingham Forest and other areas. *Journal of Ecology* 64, 123–246.

Peterken, G.F. (1977) Habitat conservation priorities in British and European woodlands. *Biological Conservation* 11, 223–236.

Peterken, G.F. (1981) *Woodland conservation and management.* Chapman & Hall, London.

Peterken, G.F. (1992) Woodland connectivity and design. In: Haines-Young, R. (ed.) *Landscape ecology in Britain.* Department of Geography, University of Nottingham, Nottingham.

Peterken, G.F. (1993) *Woodland conservation and mangement*, 2nd edn. Chapman & Hall, London.

Peterken, G.F. (1996) *Natural woodland – ecology and conservation in northern temperate regions*. Cambridge University Press, Cambridge.

Peterken, G. and Game, M. (1984) Historical factors affecting the number and distribution of vascular plant species in the woodlands of central Lincolnshire. *Journal of Ecology* 72, 155–182.

Peterken, G.F. and Harding, P.T. (1974) Recent changes in the conservation value of woodlands in Rockingham Forest. *Forestry* 47, 109–128.

Peterken, G.F. and Harding, P.T. (1975) Woodland conservation in eastern England: comparing the effect of changes in three study areas since 1946. *Biological Conservation* 8, 279–298.

Peterken, C.P. and Tubbs, C.R. (1965) Woodland regeneration in the New Forest, Hampshire, since 1650. *Journal of Applied Ecology* 2, 159–170.

Peterken, G.F. and Welch, R.C. (eds) (1975) *Bedford Purlieus: its history, ecology and management*. Monks Wood Experimental Station Symposium 7, Huntingdon.

Petersen, P.M. (1988) En botanisk beskrivelse af ni småskove på Røsnæs med præg af tidligere tiders drift. *Flora og Fauna* 94 (1), 15–22.

Petersen, P.M. (1994) Flora, vegetation, and soil in ancient and planted woodland, and scrub on Røsnæs, Denmark. *Nordic Journal of Botany* 14, 693–709.

Pettit, P.A.J. (1968) *The royal forests of Northamptonshire: a study of the economy 1558–1714*. Northamptonshire Record Society, 23, Northampton.

Pfister, Ch. and Messerli, P. (1990) Switzerland. In: Turner, B.L. (ed.) *The earth as transformed by human action. Global and regional changes in biosphere over the past 300 years*. Cambridge University Press, Cambridge, pp. 641–652.

Pignatti, S. (1982) *Flora d'Italia. Vol.* 1, II, III. Edagricole, Bologna.

Pigott, C.D. (1971) Analysis of the response of *Urtica dioica* to phosphate. *New Phytologist* 70, 953–966.

Pigott, C.D. and Huntley, J.P. (1981) Factors controlling the distribution of *Tilia cordata* at the northern limit of its geographical range. III. Nature and causes of seed sterility. *New Phytologist* 87, 817–839.

Pigott, C.D. and Taylor, K. (1964) The distribution of some woodland herbs in relation to the supply of nitrogen and phosphorus in the soil. *Journal of Ecology* 52 (suppl.), 175–185.

Pitt, W. (1809) *General view of the agriculture of the County of Northampton*. Richard Phillips, Bridge Street, London.

Plachter, H. (1991) *Naturschutz*. Gustav Fischer, Stuttgart/Jena, 463 pp.

Ponz, D.A. (1772–1794) *Viaje de España*. Ibarra, Madrid.

Pott, R. (1981) Anthropogene Einflüsse auf Kalkbüchenwalder am Beispiel der Niederholzwirtschaft und anderer extensiver Bewirtschaftungsformen. *Allgemeine Forstzeitschrift* 23, 569–571.

Pott, R. (1988) Impact of human influences by extensive woodland management and former land use in North-Western Europe. In: Salbitano, F. (ed.) *Human influences on forest ecosystems development in Europe*. ESF FERN- CNR, Pitagora Editrice, Bologna, pp. 263–278.

Pott, R. and Hüppe, J. (1991) *Die Hudelandschaften Nordwestdeutschlands*. Westfalisches Museum für Naturkunde, Munster.

Puerto, A. and Rico, M. (1988) Influence of tree canopy (*Quercus rotundiflora* and *Quercus pyrenaica*) in old fields succession in marginal areas of central-western Spain. *Acta Oecologica, Oecologia Plantarum* 9, 337–338.

Pullan, R.A. (1974) Farmed parkland in West Africa. *Savannas* 3, 119–151.

Putnam, R.J. (1986) *Grazing in temperate ecosystems: large herbivores and their effects on the ecology of the New Forest.* Croom Helm/Chapman & Hall.

Putnam, R.J. (1994a) Severity of damage by deer in coppice woodlands: an analysis of factors affecting damage and options for management. *Quarterly Journal of Forestry* 88, 45–54.

Putnam, R.J. (1994b) Effects of grazing and browsing by mammals on woodlands. *British Wildlife* 5, 205–213.

Rackham, O. (1971) Historical studies and woodland conservation. In: Duffey, E. and Watt, A.S. (eds) *The scientific management of animal and plant communities for conservation.* Blackwell, Oxford, pp. 563–580.

Rackham, O. (1972) The vegetation of the Myrtos region. In: Warren, P.M. (ed.) *Myrtos: an Early Bronze Age settlement in Crete.* Thames & Hudson, London, pp. 283–298.

Rackham, O. (1974) The oak tree in historic times. In: Morris, M.G. and Perring, F.H. (eds) *The British oak.* E.W. Classey, Faringdon, pp. 62–79.

Rackham, O. (1975) *Hayley Wood – its history and ecology.* Cambridgeshire and Isle of Ely Naturalists Trust, Cambridge.

Rackham, O. (1976) *Trees and woodland in the British landscape.* Dent, London.

Rackham, O. (1978) Archaeology and land-use history. In: Corke, D. (ed.) *Epping Forest – the natural aspect?* Essex Naturalist N.S.2, pp. 16–57.

Rackham, O. (1980) *Ancient woodland: its history, vegetation and uses in England.* Edward Arnold, London.

Rackham, O. (1982) The Avon Gorge and Leigh Woods. In: Limbrey, S. and Bell, M. (eds) *Archaeological aspects of woodland ecology*, British Archaeological Reports International Series 146, pp. 171–176.

Rackham, O. (1986) *The history of the countryside.* Dent, London

Rackham, O. (1989) *The last forest: the story of Hatfield Forest.* Dent, London.

Rackham, O. (1990). *Trees and woodland in the British landscape*, 2nd edn. Dent, London.

Rackham, O. (1995) Looking for ancient woodland in Ireland. In: Pilcher, J.R. and Mac an tSaoir, S. (eds) *Woods, trees and forests in Ireland.* Royal Irish Academy, Dublin, pp. 1–12.

Rackham, O. (1996) History of woodland and wood-pasture. *Transactions of the Suffolk Naturalists Society* 32, 116–128.

Rackham, O. and Moody, J.A. (1996) *The making of the Cretan landscape.* Manchester University Press.

Rapaics, R. (1918) Az Alföld növényföldrajzi jelleme. *Erdészeti Kísérletek* 21, 1–146.

Rasmussen, P. (1988) Løvfodring af husdyr i Stenalderen. En 40 år gammel teori vurderet gennem nye undersøgelser. In: Madsen, T. (ed.) *Bag Moesgårds maske. Kultur og samfund i fortid og nutid.* Aarhus Universitetesforlag, pp. 187–192.

Rasmussen, P. (1989) Leaf foddering in the earliest neolithic agriculture; evidence from Switzerland and Denmark. *Acta Archeologica* 60, 71–86.

Rasmussen, P. (1990a) Pollarding of trees in the Neolithic: often presumed – difficult to prove. In: Robinson, D.E. (ed.) *Experimentation and Reconstruction in Environmental Archaeology.* Symposia of the Association for Environmental Archaeology No. 9, Roskilde, Denmark. Oxbow Books, Oxford, pp. 77–99.

Rasmussen, P. (1990b) Leaf foddering in the earliest Neolithic agriculture. Evidence from Switzerland and Denmark. *Acta Archaeologica* 60, 71–86.

Rasmussen, P. (1991) Leaf foddering of livestock in the Neolithic: archaeobotanical evidence from Weier, Switzerland. *Journal of Danish Archaeology* 8, 51–71.

Rasmussen, P. (1993) Analysis of goat/sheep faeces from Egolzwil 3, Switzerland:

evidence for branch and twig foddering of livestock in the Neolithic. *Journal of Archaeological Science* 20, 479–502.

Ratcliffe, D.A. (1977) *A nature conservation review, Vol. I.* Cambridge University Press, Cambridge.

Ratcliffe, P.R. (1992) The interaction of deer and vegetation in coppice woods. In: Buckley, G.P. (ed.) *Ecology and management of coppice woodlands.* Chapman & Hall, London, pp. 233–246.

Ráth, I. (1994) Kritikus vízháztartási helyzet a Duna-Tisza közi hátságban. *Ö.K.O.* 5, 29–36.

Rayner, A.D.M. (1993) The fundamental importance of fungi in woodlands. *British Wildlife* 4, 205–215.

Rayner, A.D.M. and Boddy, L. (1988) *Fungal decomposition of wood – its biology and ecology.* Wiley, London.

Read, J. (1985) Photosynthetic and growth responses to different light regimes of the major canopy species of Tasmanian cool temperate rainforest. *Australian Journal of Ecology* 10, 327–334.

Read, H. (ed.) (1991) *Pollard and veteran tree management.* Corporation of London.

Read, J. and Hill, R.S. (1985) Dynamics of *Nothofagus*-dominated rainforest on mainland Australia and lowland Tasmania. *Vegetatio* 63, 67–78.

Rédei, K. (1978) *Adatok Nagykőrös város erdőgazdálkodásának történetéhez.* Kecskemét.

Reid, C.M., Kirby, K.J. and Cooke, R.J. (1996) *A preliminary assessment of woodland conservation in England by natural areas.* English Nature (Research Report 186), Peterborough.

Repetti, E. (1855) *Dizionario geografico, storico della Toscana.* Milano, Cinelli.

Richens, R.H. (1983) *Elm.* Cambridge University Press, Cambridge.

Richter, A. (1957) Zur Entwicklung der Waldverbreitung im Gebiet der DDR während der letzten 150 Jahre. *Arch. Forstwes. Bd. 6, Heft 11/12,* 802–810.

Rijk, J.H. de (1989) De economische geschiedenis van het Edese bos. *Nederlands Bosbouw Tijdschrift,* 61, 106–112.

Rixon, P. (1975) History and former woodland management. In: Peterken, G.F. and Welch, R.C. (eds) *Bedford Purlieus: its history, ecology and management.* Monks Wood Experimental Station Symposium, 7, Huntingdon, pp. 15–38.

Roberts, A.J., Russell, C., Walker, G.J. and Kirby, K.J. (1992) Regional variation in the origin, extent and composition of Scottish woodland. *Botanical Journal of Scotland* 46, 167–189.

Robertson, D. (1794) *A tour through the Isle of Man to which is subjoined a review of the Manks history.* Hodson, London.

Rodwell, J.S. (1991) *British plant communities Vol. 1 Woodlands and scrub.* Cambridge University Press, Cambridge.

Rodwell, J.S. (1992) *British plant communities Vol. 3 Grasslands and montane vegetation.* Cambridge University Press, Cambridge.

Rose, F. (1976) Lichenological indicators of age and environmental continuity in woodlands. In: Brown, D.H., Hawksworth, D.L. and Bailey, R.H. (eds) *Lichenology: progress and problems.* Academic Press, London, pp. 279–307.

Rose, F. (1993) Ancient British woodlands and their epiphytes. *British Wildlife* 5, 83–93.

Rövekamp, C.J.A. and Maes, N.C.M. (1995) *Genetische kwaliteit inheemse bomen en struiken; deelproject: Inventarisatie inheems genenmateriaal in Drenthe.* IKC Natuurbeheer nr W77.

Russell, G. (1988) The structure and vegetation history of the Manx hill peats. In: Davey, P. (ed.) *Man and the environment in the Isle of Man. Bar. Brit. Series* 54, pp. 39–49.

Russell, E. (1997) *People and land through time. Linking ecology and history.* Yale University Press, New Haven and London.

Salbitano, F. (1987) Vegetazione forestale ed insediamento del bosco in campi abbandonati in un settore delle Prealpi Giulie (Taipana-Udine). *Gortania, Atti del Museo Friulano di Storia Naturale* 9, 83–143.

Salbitano, F. (ed.) (1988) *Human influence on forest ecosystems development in Europe.* Pitagora, Bologna.

Salvi, G. (1982) La scalvatura della cerreta nell'alta valle del Trebbia. *Quaderni storici* 49, 148–156.

Sánchez, F.J. (1994) La gestión del Parque Natural de Los Alcornocales (Cádiz-Málaga): influencia en la población rural. In: *Simposio mediterráneo sobre regeneración del monte alcornocal.* IPROCOR, Mérida, pp. 23–26.

Sanderson, N. (1996) The role of grazing in the ecology of lowland pasture woodlands with special reference to the New Forest. In: Read, H.J. (ed.) *Pollard and veteran tree management II.* Corporation of London, Burnham Beeches, pp. 111–117.

Saner, S. (1985) Saros Körfezi dolayinin çökelme istifleri ve tektonik yerlesimi, Kuzeydogu Ege Denizi, Türkiye. Türkiye Jeoloji Kurumu Bülteni C. 28.

SBN (1991) *Tagfalter und ihre Lebensräume.* SBN, Basel.

Scamoni, A. (1964) *Vegetationskarte der Deutschen Demokratischen Republik (1:500,000).* Akademie-Verlag, Berlin.

Schaars, A.H.J. (1974) *De bosbouw van het "Entel" in de tweede helft van de achtiende eeuw.* Gelderse Historische Reeks no. 5, Zutphen.

Schaminée, J.H.J., Stortelder, A.H.F. and Westhoff, V. (1995) *De Vegetatie van Nederland. Deel 1. Inleiding tot de plantensociologie, grondslagen, Methoden, toepassingen.* Opulus, Uppsala/Leiden.

Schauer, W. (1957) Untersuchungen zur Entwicklung der Waldverbreitung in den Bezirken des ehemaligen Landes Brandenburg (1780–1937). Diss. Forstwiss, Fakultät Eberswalde, Humboldt-Univ., Berlin.

Schenk, W. (1996) Waldnutzung, Waldzustand und regionale Entwicklung in vorindustrieller Zeit im mittleren Deutschland. *Erdkundliches Wissen* 117, 325.

Scheper, M. (1989) Drentse bossen: een selectiemethode gericht op natuurwaarden. MSc thesis, Department of Forestry, Wageningen Agricultural University, Wageningen.

Schiess, H. and Schiess-Bühler, C. (1977) Dominauz-minderung als ökologisches Princip: eine Neubewertung der ursprünglichen Waldnutzungen für den Arten- und Biotopschutz am Beispiel der Tagfalter fauna eines Auenwaldes in der Nordschweiz. *Mitt. Eidgenöss. Forsch. aust. Wald Schnee Landsch.* 72, 3–127.

Schmidt, R. and Diemann, R. (eds) (1991) *Erläuterungen zur mittelmaßstäbigen land-wirtschaftlichen Standortkartierung.* Nachdruck im Selbstverlag der Akademie der Landwirtschaftswissenschaften der DDR, Bereich Bodenkunde/Fernerkundung, Eberswalde.

Schweingruber, F.H. (1990) *Anatomie europäischer Hölzer.* Eidgenössische Forchungsanstalt für wald, Schnee und Landschaft, Birmensdorf (Hrsg), Haupt, Bern and Stutgart.

Sereni, E. (1961) *Storia del paesaggio agrario italiano.* Laterza, Bari.

Sernander, R. (1936) The primitive forests of Granskar and Fiby: a study of the part played by storm-gaps and dwarf trees in the regeneration of the Swedish spruce forest. *Acta Phytogeographica Suecica* 8, 1–232.

Shannon, C.E. and Weaver, W. (1962) *The mathematical theory of communication.* University of Illinois Press, Urbana, Illinois, 125 pp.

Sheail, J. (1980) *Historical ecology: the documentary evidence.* ITE, Abbots Ripton.

Sigaut, F. (1982) Gli alberi da foraggio in Europa: significato tecnico ed economico. *Quaderni storici* 49, 49–55.

Simon, T. (1988) A hazai edényes flóra természetvédelmi érték-besorolás. *Abstracta Botanica* 12, 1–23.

Sjöbeck, M. (1932) *Löväng och trädgård. Några förutsättningar för den svenska trädgårdens utveckling.* Fataburen, Nordiska Mus. Skansens Årsbok, pp. 59–74.

Sloet, J.J.S. (1911) *Geldersche Markerechten I*, Werken der vereeniging tot uitgaaf der bronnen van het oud-vaderlandsche recht. Tweede reeks no. 12 's-Gravenhage.

Sloet, J.J.S. (1913) *Geldersche Markerechten II*, Werken der vereeniging tot uitgaaf der bronnen van het oud-vaderlandsche recht. Tweede reeks no. 15 's-Gravenhage.

Slotte, H. (1992) Lövtäkt – en viktig faktor i formandet av Ålands grässvålar. *Svensk Botanisk Tidskrift* 86, 63–75.

Slotte, H. (1993) Hamlingsträd på Åland. *Svensk Botanisk Tidskrift* 87, 283–304.

Smith, A.G. (1970) The influence of mesolithic and neolithic man on British vegetation: discussion. In: Walker, D. and West, R.G. (eds) *Studies in the vegetational history of the British Isles.* Cambridge University Press, Cambridge, pp. 81–86.

Smith, A.J.E. (1978) *The moss flora of Britain and Ireland.* Cambridge University Press, Cambridge.

Somogyi, S. (1965) A szikesek elterjedésének időbeli változásai Magyarországon. *Földrajzi Közlemények* 11, 41–55.

Somogyi, S. (1994) Az Alföld földrajzi képe a honfoglalás és a magyar középkor időszakában. *Észak- és Kelet-Magyarországi Földrajzi Évkönyv* 1, 61–75.

Soó, R. (1929) Die Vegetation und die Entstehung der ungarischen Puszta. *Ecology* 17, 329–350.

Southwood, T.R.E. (1961) The number of species of insects associated with various trees. *Journal of Animal Ecology* 30, 1–8.

Speight, M.C.D. (1989) *Saproxylic invertebrates and their conservation.* Nature & Environment Series No. 42. Council of Europe, Strasbourg.

Spek, Th. (1993) Milieudynamiek en lokatiekeuze op het Drents Plateau (3400 v Chr-1850 na Chr.). In: Elerie, J.N.H. (ed.) *Landschapsgeschiedenis van de Strubben/Kniphorstbos. Archeologische en historisch-ecologische studies van een natuurgebied op de Hondsrug.* Van Dijk and Foorthuis RegioProjekt, Groningen, pp. 167–236.

Spek, Th. (1997) Die bodenkundliche und landschaftliche Lage von Siedlungen, Äckern und Gräberfeldern in Drenthe (nördliche Niederlande). Eine Studie zur Standortwahl in Vor- und Frühgeschichte (3400 v. Chr. – 1000 n. Chr.). *Siedlungsforschung* 14, 200–292.

Spek, Th., Bisdom, E.B.A. and van Smeerdijk, D.G. (1997) *Verdronken dekzandgronden in Zuidelijk Flevoland. Een interdisciplinair onderzoek naar de veranderingen van bodem en landschap in het Mesolithicum en Vroeg-Neolithicum.* Rapport 472 DLO-Staring Centrum, Wageningen.

Spencer, J.W. and Kirby, K.J. (1992) An inventory of ancient woodland for England and Wales. *Biological Conservation* 62, 77–93.

Spurr, S.H. and Barnes, B.V. (1980) *Forest ecology.* Wiley, New York.

Staatsbosbeheer (1985) *Beheersplan voor de boswachterij De Amerongse Berg over de periode 1984–1993. Deel 1. Inventarisatie.* Staatsbasbeheer, Utrecht.

Stace, C. (1991) *New flora of the British Isles.* St Edmundsbury Press, Suffolk.

Statistisches Bundesamt (ed.) (1996) *Statistisches Jahrbuch 1996 für die Bundesrepublik Deutschland.* Metzler & Poeschel, Wiesbaden.

Steane, J.M. (1973) The forests of Northamptonshire in the early Middle Ages. *Northamptonshire Past and Present* 5, 7–17.

Steane, J.M. (1974) *The Northamptonshire landscape*. Hodder & Stoughton, London.

Steele, R.C. (1974) Variation in oakwoods in Britain. In: Morris, M.G. and Perring, F.H. (eds) *The British oak. Its history and natural history*. E.W. Classey, Farringdon, pp. 130–140.

Stubbs, A.E. and Falk, S. (1987) Hoverflies as indicator species. *Sorby Record* (Special Series) 6, 46–49.

Sturm, H.J. (1997) *Nutzbäume in der westafrikanischen Savanne: Der Schibutterbaum* (Vitellaria paradoxa *C.F. Gaertn.*) – *Charakterbaum der Sudanzone*. Der Palmengarten 61, Frankfurt am Main, 41–48.

Sugita, S. (1994) Pollen representation of vegetation in Quaternary sediments: theory and method in patchy vegetation. *Journal of Ecology* 82, 881–897.

Sümengen, M., Terlemez, İ., Sentürk, K. and Karaköse, C. (1983) Gelibolu ve Güneybati Trakya'nin Jeoloji Haritasi.

Sumption, K.J. and Flowerdew, J.R. (1988) The ecological effects of the decline in rabbits *Oryctolagus cuniculus* due to myxomatosis. *Mammal Review* 5, 151–186.

Szabó, A. (1879) Tölgyesek irtása és ákáczosok telepítése a Kecskemét városi erdőkben. *Erdészeti Lapok* 18, 14–26.

Szentpéteri, S. (1990) Ökoszisztéma rekonstrukciós terv a Nagykőrös határában lévő, Strázsadomb" környéki területre. MSc Thesis, Sopron.

Szujkó-Lacza, J. and Kováts, D. (eds) (1993) *The flora of the Kiskunság National Park*. Hungarian Natural Museum, Budapest.

Tack, G.P., van den Bremt, P. and Hermy, M. (1993) *Bossen van Vlaanderen. Een historische ecologie*. Davidsfonds, Leuven.

Tansley, A.G. (1939) *The British Islands and their vegetation*. Cambridge University Press, Cambridge.

Theuws, F.C.W.J. (1988) *De archeologie van de periferie*. Studies naar de ontwikkeling van bewoning en samenlevining in het Maas-Demer-Schelde gebied in de Vroege Middeleeuwen. Universiteit van Amsterdam, Amsterdam.

Thirgood, J.V. (1981) *Man and the Mediterranean Forest*. Academic Press, London.

Thomas, R.C. (1987) The historical ecology of Bernwood Forest. PhD thesis. Oxford Polytechnic, Oxford.

Thomas, R.C. (1997) Traditional woodland management in relation to nature conservation. In: Broad, J. and Hoyle, R. (eds) *Bernwood – the life and afterlife of a forest*. University of Central Lancashire, Preston, pp. 108–125.

Thomas, R.C. and Phillips, P.M. (1994) *Amendments to the Ancient Woodland Inventory up to June 1994*. (Research Report No 72) English Nature, Peterborough.

Thomas, R.C., Kirby, K.J. and Reid, C.M. (1997) The conservation of a fragmented ecosytem within a cultural landscape – the case of ancient woodland in England. *Biological Conservation* 82, 243–252.

Thomasius, H. (1973) Wald. In: *Landeskultur und Gesellschaft*. Theodor Steinkopff, Dresden.

Torres, C. (1974) *El antiguo reino nazarí de Granada (1232–1340)*. Anel, Granada.

Tóth, K. (ed.) (1996) 20 éves a Kiskunsági Nemzeti Park 1975–1995. In: *A tudományos konferencia előadásai és hozzászólásai. Tudományos kutatási eredmények*. Kiskunság National Park, Házinyomda Kft, Kecskemét.

Troels-Smith, J. (1960) Ivy, mistletoe and elm. Climatic indicators – fodder plants. *Danmarks Geologiske Undersogelse Series IV* 4, 1–4.

Tubbs, C.R. (1968) *The New Forest: an ecological history*. David & Charles, Newton Abbot.

Tubbs, C.R. (1986) *The New Forest*. Collins, London.

Tubbs, C. (1996) Wilderness or cultural landscapes – conflicting conservation philosophies? *British Wildlife* 7, 290–296.

Turner, J. (1970) Post-Neolithic disturbance of British vegetation. In: Walker, D. and West, R.G. (eds) *Studies in the vegetational history of the British Isles*. Cambridge University Press, Cambridge, pp. 97–116.

Van Dorp, D. and Opdam, P. (1987) Effects of patch size, isolation and regional abundance on forest bird communities. *Landscape Ecology* 1, 59–73.

Van Gijn, A.L. and Waterbolk, H.T. (1984). The colonization of the salt marshes of Friesland and Groningen: The possibility of a transhumance prelude. *Palaeohistoria* 26, 101–122.

Van Smeerdijk, G.G., Spek, Th. and Kooistra, M. (1995). Anthropogenic soil formation and agricultural history of the open fields of Valthe (Drenthe, the Netherlands) in medieval and early modern times. *Mededelingen Rijks Geologische Dienst* 52, 451–479. Haarlem.

Veen, H.E. van de and van Wieren, S.E. (1980) *Van grote grazers, kieskeurige fijnproevers en opportunistische gelegenheidsvreters*. IVM, Amsterdam.

Verkaar, H.J. (1990) Corridors as tools for plant species conservation. In: Bunce, R.G.H. and Howard, M.C. (eds) *Species dispersal in agricultural habitats*. Belhaven Press, London.

Vervloet, J.A.J. (1977) Cultuurhistorie. In: ten Houten de Lange, S.M. (ed.) *Rapport van het Veluwe-onderzoek*. Pudoc, Wageningen, pp. 76–78.

Vidéki, R. (1993) *A társadalmi beavatkozások hatása a Duna-Tisza köze geomorfológiai, vízrajzi, növénytani viszonyaira*. Kiskunfélegyháza.

Videira, C. (1908) *Memoria historica da muito notável villa de Castello de Vide*. Lisboa.

Visonà, A., Vigolo, G.T. and Cornale, P. (1994) *Valle dell 'Agno: guida alle risorse naturali ed ambientali*. Litovald, Valdagno.

Vos, J.G. (1980) Ontstaan van de staatsbossen in de provincie Drenthe en de rol van de werkverschaffing hierbij. *Nieuwe Drentse Volksalmanak* 97, 50–66.

Vos, W. and Stortelder, A.H.F. (1992) *Vanishing Tuscan Landscapes. Landscape ecology of a submediterranean-montane area*. Pudoc, Wageningen.

Waal, R.W. de (1996) De dynamiek van strooisellagen in bosecosystemen op de overgang van kalkrijk naar kalkarm. In: Kemmers, R.H. (ed.) *De dynamiek van strooisellagen*. DLO Staring Centrum, Wageningen, pp. 67–79.

Wake, J. and Webster, D.C. (eds) (1971) *The letters of Daniel Eaton to the Third Earl of Cardigan, 1725–1732*. Northamptonshire Record Society, Northampton.

Walker, B.H., Ludwig, D., Holling, C.S. and Peterman, R.M. (1981) Stability of semi-arid savanna grazing systems. *Journal of Ecology* 69, 473–498.

Walker, D. (1966) The late Quaternary history of the Cumberland lowland. *Philosophical Transactions of the Royal Society* B 251, 1–210.

Walker, G.J. and Kirby, K.J. (1987) An historical approach to woodland conservation in Scotland. *Scottish Forestry* 41, 87–98.

Walker, G.J. and Kirby, K.J. (1989) *Inventories of ancient, long-established and semi-natural woodland for Scotland*. Nature Conservancy Council (Research and Survey in Nature Conservation No. 22), Peterborough.

Waring, P.M. (1990) Abundance and diversity of moths in woodland habitats. PhD thesis. Oxford Polytechnic, Oxford.

Waring, P.M. (1993) BENHS Field Meeting – Bernwood Forest, 31 July 1993. *British Journal of Entomology and Natural History* 6, 183–188.

Warren, M.S. and Fuller, R.J. (1990) *Woodland rides and glades: their management for wildlife*. Nature Conservancy Council, Peterborough.

Waterbolk, H.T. (1964) Podsolierungserscheinungen bei Grabhügeln. *Palaeohistoria* 10, 87–101.

Waterbolk, H.T. (1984) Bossen. In: Abrahamse *et al.* (eds) *Het Drentse landschap.* De Walburg Pers, Zutphen, pp. 84–87.

Watkins, C. (1988) The idea of ancient woodland in Britian from 1800. In: Salbitano, F. (ed.) *Human influence on forest ecosystems development in Europe.* Pitagora Editrice, Bologna, pp. 237–246.

Watkins, C. (1990) *Britain's ancient woodland: woodland management and conservation.* David & Charles, Newton Abbot.

Watkins, C. (ed.) (1993) *Ecological effects of afforestation: studies in the history and ecology of afforestation in Western Europe.* CAB International, Wallingford.

Watkins, C. (1998a) *European woods and forests: studies in cultural history.* CAB International, Wallingford.

Watkins, C. (1998b) 'A solemn and gloomy umbrage': changing interpretations of the ancient oaks of Sherwood Forest. In: Watkins, C. (ed.) *European woods and forests: studies in cultural history.* CAB International, Wallingford, pp. 93–113.

Watkins, C., Lloyd, T. and Williams, D. (1996) Constraints on farm woodland planting in England: a study of Nottinghamshire farmers. *Forestry* 69 167–176.

Watts, W.A. (1984) Contemporary accounts of the Killarney woods 1580–1870. *Irish Geography* 17, 1–13.

Werf, S. van der (1991) *Natuurbeheer in Nederland. Deel 5. Bosgemeenschappen.* Pudoc, Wageningen.

Westhoff, V. and Held, A.J. den (1969) *Plantengemeenschappen in Nederland.* Thieme, Zutphen.

Whitehead, P.F. (1996) The notable arboreal Coleoptera of Bredon Hill, Worcestershire, England. *The Coleopterist* 5, 45–53.

Wieren, S.E. van *et al.* (1989) *Begrazingsonderzoek in Nederland: samenhang, prioriteiten en samenwerking.* NRLO-rapport 89/31, 's-Gravenhage.

Wieringa, J. (1958) Opmerkingen over het vergand tussen bodemgesteldheid en oudheidkundige verschijnselen naar aanleiding van de NEBO-kartering in Drenthe. *Boor en Spade* 9, 97–113.

Wightman, W.R. (1968) The pattern of vegetation in the Vale of Pickering area *c* 1300 AD. *Transactions of the Institute of British Geographers* 45, 125–142.

Wilbarger, J.W. (1889) *Indian depredations in Texas.* Reprinted 1985. Eakin Press, Austin.

Winter, T. (1993) Deadwood – is it a threat to commercial forestry? In: Kirby, K.J. and Drake, C.M. (eds) *Deadwood matters: the ecology and conservation of saproxylic invertebrates in Britain.* English Nature (Science Report 7), Peterborough, pp. 58–73.

Wolf, R.J.A.M. (1992) *Ontstaansgeschiedenis en beheer van de Nederlandse elzen – en berkenbroekbossen.* DLO Instituut voor Bos- en Natuuronderzoek, Wageningen.

Wolf, R.J.A.M. (1995) *Geschiedenis en beheer van de Nederlandse ooibossen.* DLO Instituut voor Bos- en Natuuronderzoek, Wageningen.

Wolf, R.J.A.M., Dort, K.W. van and Vrielink, J.G. (1996) Groeiplaatsen als basis voor bostypologie. Vaste grond onder de voeten van bosbeheerders. *Nederlands Bosbouwtijdschrift* 68, 177–189.

Wolf, R.J.A.M., Vrielink, J.G. and Waal, R.W. de (1997) Riverine woodlands in the Netherlands. *Global Ecology and Biogeography Letters* 6, 287–295.

Wulf, M. (1995) Sollten Erstaufforstungen an kontinuierlich bewaldete Flächen angrenzen? Ein Beitrag zur Migrationsproblematik typischer Waldpflanzen. *Zalf-Bericht* 52, 52–60.

Wulf, M. and Schmidt, R. (1996) Die Entwicklung der Waldverteilung in Brandenburg in Beziehung zu den naturräumlichen Bedingungen. *Beitr. Forstwirtsch. u. Landschaftsökol* 30, 125–131.

Yates, W. (1786) *A map of the County of Lancashire*. Reprinted by J.B. Harley (1968) Historic Society of Lancashire and Cheshire. Messrs. John Gardner, Liverpool.

Young, C.R. (1979) *The Royal Forests of Medieval England*. Leicester University Press, Leicester.

Zólyomi, B. (1945–1946) Természetes növénytakaró a tiszafüredi öntözőrendszer területén. *Öntözésügyi Közlemények.* 7–8, 62–75.

Zólyomi, B. (1969) Földvárak, sáncok, határmezsgyék és a természetvédelem. *Természet Világa* 100, 550–553.

Zólyomi, B. (1989) Natural vegetation of Hungary. In: Pécsi, M. (ed.) *National atlas of Hungary.* Kartográfiai Vállalat, Budapest, p. 89.

Zólyomi, B. and Fekete, G. (1994) The Pannonian loess steppe: differentiation in space and time. *Abstracta Botanica* 18, 29–41.

Index

Figures in **bold** indicate major references. Figures in *italic* refer to diagrams, photographs and tables.